# METHODS IN MOLECULAR BIOLOGY™

*Series Editor*
John M. Walker
School of Life Sciences
University of Hertfordshire
Hatfield, Hertfordshire, AL10 9AB, UK

For further volumes:
http://www.springer.com/series/7651

# Xenotransplantation

## Methods and Protocols

Edited by

**Cristina Costa**

*Institut d'Investigació Biomèdica de Bellvitge (IDIBELL), Hospital Duran i Reynals, Barcelona, Spain*

**Rafael Máñez**

*Servicio de Medicina Intensiva, Hospital Universitario de Bellvitge, Barcelona, Spain*

*Editors*
Cristina Costa
Institut d'Investigació Biomèdica
de Bellvitge (IDIBELL)
Hospital Duran i Reynals
Barcelona, Spain

Rafael Máñez
Servicio de Medicina Intensiva
Hospital Universitario de Bellvitge
Barcelona, Spain

ISSN 1064-3745 ISSN 1940-6029 (electronic)
ISBN 978-1-61779-844-3 ISBN 978-1-61779-845-0 (eBook)
DOI 10.1007/978-1-61779-845-0
Springer New York Heidelberg Dordrecht London

Library of Congress Control Number: 2012936662



Printed on acid-free paper

Humana Press is a brand of Springer
Springer is part of Springer Science+Business Media (www.springer.com)

# Preface

Despite many technological challenges faced by the xenotransplantation field, many major advances have been made in the last two decades. Moreover, the field continues to progress thanks to the conviction of various scientists of the feasibility of success in multiple clinical applications and the great benefit it can provide to human health. Xenotransplantation research is clearly justified by the long waiting lists for allogeneic organs and the impossibility of the current system to meet the demand. As many of the limitations and difficulties in organ procurement also apply to human cells and tissues, the development of new therapies based on cell and engineered-tissue xenotransplantation are certainly objects of intense study in the field. Xenogeneic cells are simpler than solid organs and seem to pose less hurdles to attain long-term graft survival. In this accordance, recent accomplishments at the level of clinical trials for pancreatic islet xenotransplantation further encourage to continue with these efforts and lead us to believe that xenotransplantation will be for long object of interest and research.

The complexity of the xenotransplantation field requires specialization of scientists and experts in a variety of aspects that need to be simultaneously studied and addressed to develop a successful clinical application. All potential xenogeneic therapies benefit from the basic work done to characterize the xenogeneic interactions at the cellular and molecular levels. The major contribution of innate immunity to xenograft rejection is well known thanks to multiple studies that have opened new ground to reveal the relevance of cells and molecules of the innate immune system in transplant rejection. Thus, not only proteins, but also carbohydrates, play critical roles in triggering the xenogeneic immune responses. In addition, differences in the coagulation systems of donor and recipient may affect xenograft function. Importantly, the technologies developed to engineer the xenogeneic cells and organs, even at the germ-line level, are in place and should allow to modify the xenografts in order to prevent rejection and facilitate function. To test the new developments, the xenotransplantation field requires a great amount of preclinical work that includes transplantation in small animal models as well as major work in non-human primates, such as the baboon. Each cell type, tissue and organ of interest for xenotransplantation has its particularities. Thus, teams usually specialize and acquire a profound understanding of each system and the associated technology to help them progress efficiently. In the way towards the clinical application of xenotransplantation, it is also important to keep high safety standards and develop all the necessary tools to prevent potential zoonosis. A great advance has been done in this regard. Finally, the regulatory, legal, ethical and educational aspects of the xenotransplantation field are key for the success of the xenogeneic applications and its acceptance by society in general.

With this methodological book, we intend to cover many subjects related to xenotransplantation that can specially help the scientific community dedicated to one or another aspect of xenotransplantation. The field will certainly benefit from another tool that helps them integrate the great amount of knowledge needed for progress. Moreover, the methodological focus of this project will make this approach different and of special interest. It can be helpful to all scientists in the field, but specially so to the young researchers and students. This support can facilitate their work and encourage them to continue in this challenging area. Nonetheless, other scientific communities, such as basic researchers and clinicians working in allotransplantation and/or regenerative medicine, can learn and obtain valuable information from this volume.

*Barcelona, Spain*

*Cristina Costa*
*Rafael Máñez*

# Contents

# Contributors

LUIZ ANASTÁCIO ALVES • *Laboratório de Comunicação Celular, Instituto Oswaldo Cruz, Fundação Oswaldo Cruz, Rio de Janeiro, Brazil*
MARCELO ALVES PINTO • *Laboratório de Desenvolvimento Tecnológico em Virologia, Instituto Oswaldo Cruz, Fundação Oswaldo Cruz, Rio de Janeiro, Brazil*
AGNES M. AZIMZADEH • *Department of Surgery, University of Maryland, Baltimore, MD, USA*
SUYAPA BALL • *Revivicor, Inc., Blacksburg, VA, USA*
ANDREAS BAUER • *Department of Anaesthesiology, Ludwig-Maximilians University Munich, Munich, Germany; Walter-Brendel-Centre of Experimental Medicine, Ludwig-Maximilians University Munich, Munich, Germany*
BLANCO FJ • *Osteoarticular and Aging Research Laboratory, Hospital Universitario A Coruna, Spain*
ANDRE GUSTAVO BONAVITA • *Laboratório de Comunicação Celular, Instituto Oswaldo Cruz, Fundação Oswaldo Cruz, Rio de Janeiro, Brazil*
KENNETH R. BONDIOLI • *Louisiana State University Agricultural Center, School of Animal Sciences, Baton Rouge, LA, USA*
PAOLO BRENNER • *Department of Cardiac Surgery, Ludwig-Maximilians University Munich, Munich, Germany*
CRISTIANA BULATO • *Department of Cardiologic, Thoracic and Vascular Sciences, University of Padua, Padua, Italy*
LARS BURDORF • *Department of Surgery, University of Maryland, Baltimore, MD, USA*
DALE CHRISTIANSEN • *Department of Surgery (Austin Health), University of Melbourne, Heidelberg, VIC, Australia*
IRINA CHODNEVSKAJA • *Experimental Transplantation Immunology and Experimental Surgery, Clinic of General, Visceral, Vascular, and Pediatric Surgery, University of Wuerzburg Hospital, Würzburg, Germany*
CRISTINA COSTA • *New Therapies of Genes and Transplants Group [Institut d'Investigació Biomèdica de Bellvitge] IDIBELL, Hospital Duran i Reynals, Barcelona, Spain*
EMANUELE COZZI • *Department of Surgical and Gastroenterological Sciences, University of Padua, Padua, Italy; CORIT (Consortium for Research in Organ Transplantation), Padua, Italy; Direzione Sanitaria, Padua General Hospital, Padua, Italy*
MATHEUS KAFURI CYTRANGULO • *Laboratório de Comunicação Celular, Instituto Oswaldo Cruz, Fundação Oswaldo Cruz, Rio de Janeiro, Brazil*
DÍAZ PRADO SM • *Department of Medicine, INIBIC-University of A Coruña, A Coruña, Spain; CIBER-BBN-Cellular Therapy Area, A Coruña, Spain*
NIEVES DOMÉNECH • *CHU-A Coruña, A Coruña, Spain*

Sibylle Eber • *Experimental Transplantation Immunology and Experimental Surgery, Clinic of General, Visceral, Vascular, and Pediatric Surgery, University of Wuerzburg Hospital, Würzburg, Germany*

Fuentes-Boquete IM • *Department of Medicine, INIBIC-University of A Coruña, A Coruña, Spain; CIBER-BBN-Cellular Therapy Area, A Coruña, Spain*

Sabine Gahn • *Experimental Transplantation Immunology and Experimental Surgery, Clinic of General, Visceral, Vascular, and Pediatric Surgery, University of Wuerzburg Hospital, Würzburg, Germany*

Angelica M. Giraldo • *Revivicor, Inc., Blacksburg, VA, USA; DeSoto Biosciences Inc., Seymour, TN, USA*

Jorge Guerra González • *Leuphana University of Lüneburg, Lüneburg, Germany*

María Jorqui Azofra • *University of Deusto, Bilbao, Spain; University of the Basque Country, Bilbao, Spain*

Xavier Lévêque • *INSERM, U643, Nantes, Cedex, France*

Katia R.F. Lima-Quaresma • *Laboratório de Comunicação Celular, Instituto Oswaldo Cruz, Fundação Oswaldo Cruz, Rio de Janeiro, Brazil*

Giada Mattiuzzo • *Division of Infection and Immunity, Wohl Virion Centre, University College London, London, UK*

Vasiliy Moskalenko • *Clinic of General, Visceral, Vascular and Pediatric (Surgical Clinic I), University of Wuerzburg Hospital, Würzburg, Germany*

Effie Mouhtouris • *Department of Surgery (Austin Health), University of Melbourne, Heidelberg, VIC, Australia*

Philippe Naveilhan • *INSERM, U643, Nantes, Cedex, France; CHU de Nantes, Institut de Transplantation et de Recherche en Transplantation, ITERT, Nantes, France; Faculté de Médecine, Université de Nantes, Nantes, France*

Véronique Nerrière-Daguin • *INSERM, U643, Nantes, Cedex, France; CHU de Nantes, Institut de Transplantation et de Recherche en Transplantation, ITERT, Nantes, France; Faculté de Médecine, Université de Nantes, Nantes, France*

Isabelle Neveu • *INSERM, U643, Nantes, Cedex, France; CHU de Nantes, Institut de Transplantation et de Recherche en Transplantation, ITERT, Nantes, France; Faculté de Médecine, Université de Nantes, Nantes, France*

Magdiel Pérez-Cruz • *New therapies of genes and transplants group [Institut d'Investigació Biomèdica de Bellvitge] IDIBELL, L'Hospitalet de Llobregat, Barcelona, Spain*

Richard N. Pierson III • *Department of Surgery, University of Maryland, Baltimore, MD, USA*

Johannes Postrach • *Department of Cardiac Surgery, Ludwig-Maximilians University Munich, Munich, Germany; Walter-Brendel-Centre of Experimental Medicine, Ludwig-Maximilians University Munich, Munich, Germany*

Claudia Radu • *Department of Cardiologic, Thoracic and Vascular Sciences, University of Padua, Padua, Italy*

Paul A. Ramsland • *Department of Surgery (Austin Health), University of Melbourne, Heidelberg, VIC, Australia; Centre for Immunology, Burnet Institute, Melbourne, VIC, Australia; Department of Immunology, Monash University, Alfred Medical Research and Education Precinct, Melbourne, VIC, Australia*

BRUNO REICHART • *Department of Cardiac Surgery, Ludwig-Maximilians University Munich, Munich, Germany*
CARLOS MARÍA ROMEO CASABONA • *University of Deusto, Bilbao, Spain; University of the Basque Country, Bilbao, Spain*
MAURO S. SANDRIN • *Department of Surgery (Austin Health), University of Melbourne, Heidelberg, VIC, Australia*
JOSEPH SCALEA • *Organ Transplantation Tolerance and Xenotransplantation Laboratory, Transplantation Biology Research Center, Massachusetts General Hospital, Harvard Medical School, Boston, MA, USA*
MICHAEL SCHMOECKEL • *Department of Cardiac Surgery, Asklepios Klinik St. Georg, Hamburg, Germany*
BIANCA SCHNEIKER • *Experimental Transplantation Immunology and Experimental Surgery, Clinic of General, Visceral, Vascular, and Pediatric Surgery, University of Wuerzburg Hospital, Würzburg, Germany*
LINDA SCOBIE • *Department of Biological and Biomedical Sciences, Glasgow Caledonian University, Glasgow, UK*
PAOLO SIMIONI • *Department of Cardiologic, Thoracic and Vascular Sciences, University of Padua, Padua, Italy*
ROBERTA SOMMAGGIO • *New therapies of genes and transplants group [Institut d'Investigació Biomèdica de Bellvitge] IDIBELL, Barcelona, Spain*
ARMIN STRAUß • *Clinic of General, Visceral, Vascular and Pediatric (Surgical Clinic I), University of Wuerzburg Hospital, Würzburg, Germany*
YASUHIRO TAKEUCHI • *Division of Infection and Immunity, Wohl Virion Centre, University College London, London, UK*
VERONICA TISATO • *Department of Surgical and Gastroenterological Sciences, University of Padua, Padua, Italy*
KARIN ULRICHS • *Experimental Transplantation Immunology and Experimental Surgery, Clinic of General, Visceral, Vascular, and Pediatric Surgery, University of Wuerzburg Hospital, Würzburg, Germany*
MIREIA URIBE-HERRANZ • *New Therapies of Genes and Transplants Group [Institut d'Investigació Biomèdica de Bellvitge] IDIBELL, Barcelona, Spain*
HAO WANG • *Department of General Surgery, Tianjin Medical University General Hospital, Tianjin General Surgery Institute, Tianjin, China; Department of Surgery, Western University, London, ON, Canada; Multi-Organ Transplant Program, University Hospital, London Health Sciences Centre, London, ON, Canada*
KAZUHIKO YAMADA • *Organ Transplantation Tolerance and Xenotransplantation Laboratory, Transplantation Biology Research Center, Massachusetts General Hospital, Harvard Medical School, Boston, MA, USA*

BRUNO REICHART • Department of Cardiac Surgery, Ludwig-Maximilians University Munich, Munich, Germany
CARLOS MARIA ROMEO CASABONA • University of Deusto, Bilbao, Spain; University of the Basque Country, Bilbao, Spain
[illegible]

# Chapter 1

# Xenotransplantation: An Overview of the Field

Veronica Tisato and Emanuele Cozzi

## Abstract

Xenotransplantation, the transplantation of cells, tissues, or organs between different species, has the potential to overcome the current shortage of human organs and tissues for transplantation. In the last decade, the progress made in the field is remarkable, suggesting that clinical xenotransplantation procedures, particularly those involving cells, may become a reality in the not-too-distant future. However, several hurdles remain, mainly immunological barriers, physiological discrepancies, and safety issues, making xenotransplantion a complex and multidisciplinary discipline.

**Key words:** Xenotransplantation, Transgenic animals, Hyperacute rejection, Acute humoral xenograft rejection, Cellular rejection, Cellular transplantation, Molecular incompatibility, Tolerance, Zoonoses

## 1. Introduction

The success of transplantation has resulted in a growing shortage in the supply of organs worldwide. Indeed, there is evidence that, notwithstanding the great efforts that have been put in place by national–international agencies to create efficient networks to increase organ donation, the expansion of living donor programmes and the efforts made by experts in social sciences to increase the public response to organ donation, gap between organ availability, and clinical need continue to widen. The United Network for Organ Sharing (UNOS) estimates that, in the USA alone, more than 110,000 people are waiting for a transplant nationwide (1) and a similar picture has been reported in the European Union (EU), where approximately 50,000 people are on the waiting list and, on average, 12 people die daily while waiting for an organ (2). These figures are facts: the supply of organs does not meet the demand and alternative strategies to fill this gap are needed.

Cristina Costa and Rafael Máñez (eds.), *Xenotransplantation: Methods and Protocols,* Methods in Molecular Biology, vol. 885, DOI 10.1007/978-1-61779-845-0_1, © Springer Science+Business Media, LLC 2012

## 2. Possible Strategies to Meet the Clinical Need

When thinking about possible strategies to overcome the current shortage of human organs, at least three approaches may be considered: the use of artificial organs, regenerative medicine, and xenotransplantation.

With regard to artificial organs, dialysis has certainly increased the life expectancy of patients with kidney failure. However, it has very high social costs, often affords only a limited quality of life, and is associated with reduced life expectancy when compared to transplantation. Similarly, high costs and a large number of side effects due to biological incompatibilities preclude the widespread application of artificial hearts.

Emerging concepts on organ regeneration and tissue engineering are opening the road to new, promising therapeutic approaches. In particular, preliminary results suggest that cardiac repair may benefit from the progress made in the field of regenerative medicine, opening a *scenario* that goes from the implant of cardiac patches seeded with cardiac cells (3–5) to the recent preclinical model of a bioartificial heart proposed by Ott and colleagues (6). Tissue engineering has furthermore shown a potential role for kidney repair with the generation of renal scaffolds (7–9) that could, at least in theory, be repopulated with the appropriate cell lines reconstituting the original structure and features of the organ. Recently, a transplantable liver deriving from a decellularized liver matrix (10) and a bioartificial, fully functional lung have been obtained in preclinical models (11, 12). Despite these encouraging results, however, there are still many open issues that need to be addressed before the clinic use of such bioartificial organs can be contemplated.

In this light, at least theoretically, xenotransplantation has the potential to fulfil the need for transplantable organs within a shorter time frame, and this potential is therefore here discussed.

## 3. Barriers to Xenotransplantation

Before making xenotransplantation a clinical reality, several hurdles must be overcome, principally represented by immunological barriers, physiological discrepancies, and safety issues. Each of these represents a crucial obstacle and a key challenge that need to be satisfactorily addressed by scientists in the field. Significant improvements, however, have been unquestionably achieved in recent years in the comprehension of such issues and several approaches to overcome them have been proposed.

### 3.1. The Immunological Barriers to Solid Organ Xenotransplantation

Hyperacute rejection (HAR) which leads to graft rejection within a few minutes of revascularization of the graft represents the first obstacle to solid organ xenotransplantation. HAR has been overcome thanks to the clarification of the key role played in its onset by the binding of preformed natural antibodies primarily directed to the Galalpha1-3Galbeta1-(3) 4GlcNAc-R epitope (αGal-epitope) expressed on pig endothelial cells (13), with subsequent activation of the complement and coagulation cascades and the consequent, irreversible graft damage. In particular, through the implementation of strategies that prevent complement activation or protect the xenografted organ from complement-mediated damage, HAR is very rarely observed in pig-to-primate xenotransplantation (14). Nonetheless, following HAR, an acute humoral xenograft rejection (AHXR) process triggered by anti-αGal and anti-non-αGal antibodies can lead to loss of the graft within days of transplantation or in the weeks that follow (15). AHXR is characterized by immunoglobulin, complement, and fibrin deposition and it is associated with platelet sequestration and cellular infiltration (15). As a consequence of the complex and multifactorial process underlying AHXR, overcoming this type of rejection requires the synergistic effect of multiple, simultaneous, therapeutic strategies. In this context, the use of genetically modified pig donors, combined with novel immunosuppressive strategies, has undoubtedly led to improved survival, particularly of renal and cardiac grafts (16–19). Nevertheless, signs of humoral rejection with thrombotic microangiopathy (20) and/or consumptive coagulopathy (21) are still observed in xenograft recipients, suggesting that the current strategies are not sufficient. Indeed, additional approaches are needed to improve the control of the vigorous humoral immune response directed to the xenograft and prevent massive activation of the coagulation cascade observed when pig organs are transplanted into primates.

With regard to the role of cell-mediated immune responses directed against a solid organ xenograft, the immunosuppressive regimens currently applied appear to be sufficient to control anti-xenograft T-cell immunity. However, other cellular components of the immune system may be involved. In this context, it has been demonstrated that porcine cells are susceptible to human NK cell-mediated damage because of the failure of swine leukocyte antigen (SLA) to engage and provide inhibitory signals through iNKR (22) or as a result of direct human NK cell activation through interaction of NKp44 or NKG2D receptors with their ligands on porcine cells (23). In addition, monocytes and macrophages have also been described as able to accumulate in xenografts and contribute to AHXR and may, therefore, represent additional targets for strategies aimed at controlling xenograft rejection (24).

### 3.2. The Immunological Barriers to Transplantation of Cellular Xenografts

A different scenario characterizes the rejection process of a cellular xenograft. Referring to pig islet xenotransplantation into primates as a model, the first barrier to survival is represented by the onset of the so-called Instant Blood-Mediated Inflammatory Reaction (IBMIR). IBMIR occurs when pig islets are transplanted into the portal vein (25) and is potentially triggered by surface molecules, such as tissue factor and collagen expressed on pig cells (26). The process involves activation of the coagulation and complement cascades leading to macroscopic coagulation, rapid consumption of platelets, leukocyte infiltration, and complement components' deposition (25, 26). This ultimately results in an early, massive loss of islets starting even prior to their homing in the recipient's liver. In the absence of immunosuppression, the surviving cells in the liver undergo rapid cellular rejection with a predominant role of T cells and macrophages that can infiltrate the graft. Different drug combinations have been tested to achieve long-term pig islet survival in non-human primates. However, as for solid organ xenotransplantation, the main challenge is represented by the identification of strategies enabling long-term graft survival using a clinically acceptable immunosuppressive protocol (27). In this regard, islet encapsulation in the absence of immunosuppression has recently been evaluated in a pig-to-primate model, leading to outstanding results that are discussed below (28).

With regard to pig engineering in view of islet xenotransplantation, it is of interest that adult pig islets are essentially free of αGal residues (29) while foetal and neonatal cells apparently express low levels of the epitope (30–32). In this light, the use of islets isolated from transgenic pigs lacking the αGal-epitope (GalT-KO) will be advantageous only if foetal or neonatal islets are transplanted. In all cases, however, preventive strategies for IBMIR together with the use of islets from pigs transgenic for complement regulatory proteins and anti-thrombotic factors are expected to protect the cells and enable long-term islet survival (26).

### 3.3. Molecular Incompatibilities

Several molecular incompatibilities have been detected between pigs and primates and those which are currently perceived as most problematic in the case of pig-to-primate xenotransplantation regard the coagulation cascade. Indeed, the ability of porcine von Willebrand factor to activate spontaneously quiescent human platelets, as a consequence of a functional incompatibility for human platelet GPIb receptors resulting in a pro-coagulant picture, has been demonstrated to be one of these barriers (33). Furthermore, despite the demonstration that porcine thrombomodulin is able to recognize and bind human thrombin, inhibiting its pro-coagulant properties, the resulting complex is a weak activator of both human Protein C and Thrombin-Activatable Fibrinolysis Inhibitor (34). In this regard, although it had been previously reported that porcine Tissue Factor Pathway Inhibitor

(TFPI) is a non-efficient inhibitor of the human-activated factor X (factor Xa) activity impeding the regulation of human tissue factor (TF)-initiated coagulation (35, 36), recent findings have demonstrated that porcine recombinant TFPI and the components of the human TF pathway do not show any apparent incompatibilities, suggesting that other factors may be involved (37). In all cases, cross-species incompatibilities are involved in the generation of the characteristic pro-coagulant status of solid organ xenograft recipients that is primarily attributable to an impaired anticoagulant activity and expression of pro-coagulant molecules on xenograft endothelium. In this regard, it appears that TF may play a key role in initiating thrombotic microangiopathy and consumptive coagulopathy after xenotransplantation. Porcine aortic endothelial cells are able to induce expression of TF on human platelets and monocytes *via* an immunological independent pathway, suggesting that expression of TF by the recipient may play a significant role in the development of consumptive coagulopathy in xenotransplantation (21). In order to control the biological effects of these incompatibilities and overcome inflammation and thrombotic events, the generation of genetically modified pigs expressing modulators of the clotting cascade (preferably on a GalT-KO background) has been suggested as an additional strategy to prolong xenograft survival. In this context, approaches aiming to over-express molecules, such as TF inhibitors (38–40), human thrombomodulin (41), and CD39 (42, 43), or engineered pigs with low/no levels of pro-coagulant molecules on the vascular porcine endothelium are being considered. Coagulation dysregulation in xenotransplantation has recently been reviewed, with an accurate description of the latest findings that appear to play a role in the onset of this barrier (44). A combined approach of appropriately genetically modified pig donors together with systemic treatment of the recipient will probably be necessary to overcome coagulation dysregulation in xenotransplantation (45).

### 3.4. Xenotransplantation and Physiological Compatibility

When considering xenotransplantation, a fundamental issue is to clarify whether pig organs, cells, or tissues that are not rejected can satisfactorily replace the physiological functions of the human counterpart. It has been reported that primates undergoing heart xenotransplantation are active and energetic during the post-operative period, demonstrating that porcine hearts are able to function normally in a xenogeneic recipient (46). Similarly, renal xenografts have enabled long-term survival of bilaterally nephrectomized monkeys (19), suggesting that renal xenotransplantation is achievable from a physiological standpoint, even though proteinuria could represent a limit to its clinical application (47). Finally, with regard to pig islet xenotransplantation, if the immunological barrier is under control through the use of appropriate immunosuppressive regimens or through xenograft encapsulation, it has been demonstrated by different independent groups that xenografted islets can cure diabetes in non-human primates followed for up to 1 year (28, 48). Therefore, notwithstanding the recognized physiological differences, heart, kidney, and islet xenografts can function in primates, meeting the physiological requirements and enabling long-term survival.

On the other hand, different conclusions have been reached to date with regard to porcine lung and liver xenotransplantation. Even though in vivo experiments have demonstrated that pig lungs are able to perform gas exchanges in the primate recipient (49), pulmonary physiology of xenografts can be affected by the different upright posture of the recipient with a direct influence on blood vessel pressure and blood–gas exchanges (50). Moreover, it seems that hyperacute rejection after lung xenotransplantation probably involves mechanisms of coagulation system activation that are complement- and antibody-independent (51). As far as the liver is concerned, this organ presents a complex physiology secondary to its involvement in many metabolic pathways (52). Nevertheless, orthotopic, transgenic pig liver xenotransplantation experiments in immunosuppressed primates have shown sufficient hepatic function (53, 54) to suggest a hypothetical future use in patients as a bridge to allotransplantation (55).

### 3.5. The Safety Issue

Prior to proceeding to the clinical stage, there is the mandatory need to guarantee xenograft recipients the highest level of safety. The measures to put in place include the exclusion of pathogens from the graft and the development of very sensitive assays to identify rapidly the potential onset of new infections in xenograft recipients. Indeed, the risk of zoonoses is enhanced by the immunosuppressive regimen applied to control graft rejection and this could render apparently harmless biological agents or latent viruses potentially dangerous for the recipient and, possibly, for society at large. In this context, significant progress has been made in the safety field principally with regard to animal breeding and the development of tools to identify new pathogens potentially transmitted through xenotransplantation.

The exclusion of most known pathogens from the xenograft is achievable by housing and breeding donor pigs in closed facilities, where animals are obtained by caesarean derivation and raised under a strict health surveillance system (56). These measures, collectively referred to as specific pathogen-free (SPF) conditions, are not sufficient, however, to remove "challenging" pathogens such as porcine endogenous retroviruses (PERVs) or other as yet unknown agents from the herd (57). It has been demonstrated that the three described replication-competent PERVs (PERV-A, PERV-B, PERV-C) and their identified recombinants are able to infect human cells in vitro (58–60), even though there is no evidence of direct PERV transmission in human and non-human primate studies (61–63).

Additional potential risks that require careful consideration include porcine cytomegalovirus, that has been associated with endothelial injury and possibly with consumptive coagulopathy in xenografted primates (64), and porcine lymphotropic herpesvirus (PLHV), that is the agent responsible for a lymphoproliferative disease in pigs (65). Although there is no evidence of virus replication in xenografted primate recipients, it has been demonstrated that reciprocal molecular interaction between human HHV-8 or EBV and porcine PLHV-1 might occur in immunosuppressed recipients (66), suggesting that a vigilant approach to this virus is required.

Overall, these findings suggest that, although encouraging safety results have been obtained so far both in clinical and preclinical studies, specifically engineered pigs lacking potentially harmful genome sequences and the development of new and more sensitive tools to identify potential infectious agents will significantly enhance the safety profile of xenotransplantation (67–71).

## 4. Where Are We Now with Solid Organ Xenotransplantation?

Looking to translate preclinical research into its clinical application, it is important to review critically the results achieved in non-human primates receiving xenografts from genetically modified pigs. With regard to cardiac transplantation, engineered hearts lacking the $\alpha$1,3-galactosyltransferase gene heterotopically transplanted into baboons enabled up to 6 months' survival (17, 18). Likewise, pig hearts orthotopically transplanted in non-human primates have survived for up to 57 days using a clinically acceptable immunosuppressive regimen (16). Concerning kidney xenotransplantation, transgenic expression of human decay accelerating factor (hDAF) in the donor has allowed up to 90 days' survival on xenografted cynomolgus monkeys (19). It is of interest that, in all these studies, a premature loss of xenografted recipients is observed as a consequence of AHXR or a profound coagulopathy.

Liver and lung xenotransplantation into primates has been less extensively explored so far and additional considerations should be taken into account, especially with regard to the more complex physiological differences between pigs and primates for these types of xenograft. Livers from transgenic pigs expressing hDAF transplanted in two non-human primates survived for 4 and 8 days (53). Likewise, more recently, transplantation of GalT-KO livers or CD46-expressing livers led to a survival of 4–7 days under a clinically acceptable immunosuppressive protocol (72). The severe thrombocytopenia showed by recipient animals within 1 h of reperfusion may represent, however, the main obstacle to primate survival.

Similarly, only a short survival time has been reported following lung transplantation in non-human primates, with a disseminated intravascular coagulation associated with pulmonary xenograft dysfunction being the primary cause of recipient loss (73–76). Furthermore, these data suggest that pig lungs are perhaps more susceptible to innate immune-mediated injury.

Taken together, these studies indicate that long-term survival is achievable following solid organ xenotransplantation, especially for hearts. However, current survival cannot be viewed as sufficient to justify initiating experimental clinical trials. In particular, the persisting immunological barriers, primarily AHXR, and the profound coagulopathy frequently observed in xenografted primates need to be adequately controlled in preclinical studies. To this end, the ongoing research activities to refine the immunosuppressive strategies and the generation of novel lines of specifically engineered pigs are expected to lead to significantly improved results in the not-too-distant future.

## 5. Where Do We Stand with Transplantation of Cellular Xenografts?

In the light of the results obtained to date in pig-to-primate models, cellular xenotransplantation appears closer to a clinical application than solid organ xenograft. Indeed, cellular xenograft appears easier to control and the possibility to isolate the graft physically through encapsulation may confer additional advantages.

Several groups have independently reported the ability of pig pancreatic islets to restore a state of insulin independence in diabetic non-human primates. Intraportal transplantation has been applied and proved effective using either adult or neonatal pig islets. Long-term survival (more than 6 months) has been achieved in immunosuppressed recipients by several independent groups (77, 78) and, more recently, Dufrane and colleagues have been able to prolong adult pig islet survival in non-immunosuppressed primates for up to 6 months using adult pig islets encapsulated in a subcutaneous macrodevice (28). In addition, van der Windt and colleagues have recently demonstrated in immunosuppressed diabetic monkeys that transgenic expression of human CD46 can result in normoglycaemia for up to 1 year (48). Foetal and embryonic pig islets have also been considered as a possible alternative to adult and neonatal cells and in this regard, Hetch and colleagues have demonstrated in streptozotocin-treated cynomolgus monkeys that E-42 embryonic pancreatic tissue can correct hyperglycaemia (79).

Parkinson and Huntington disease patients might benefit from a xenotransplantation of neuronal precursor cells. Indeed, transplantation of allogeneic neuronal precursor cells is precluded due to ethical, religious, and social concerns. It has been reported that the use of porcine, foetal, and neural cells unilaterally grafted into Parkinsonian disease- and Huntington's disease-affected patients resulted in a clinical improvement of up to 30%, with graft survival and a positive readout in terms of safety (80). Recently, Badin and colleagues in our laboratory have reported the preliminary results of a study aimed at evaluating hCTLA-4 transgenic pig neurons transplanted into a non-human primate model of Parkinson disease. The authors describe a clinical improvement associated with partial restoration of dopaminergic activity (as documented by 18F-L-DOPA PET scans) and long-term survival of porcine grafts at euthanasia (81). These studies are currently underway to refine the immunosuppressive strategy and to comprehend fully the translational potential of this new approach.

## 6. Xenotransplantation and Graft Acceptance

The application of non-specific immunosuppressive regimens possibly combined with engineered porcine organs or cells has enabled the above-mentioned graft survivals. However, the toxicity usually associated with the immunosuppressive treatments administered makes these strategies difficult to apply in a clinical context. In this light, alternative approaches to control the anti-graft immune response are under evaluation. In this regard, the establishment of accommodation or the induction of immune tolerance represents the two most attractive options to contrast graft failure. In the first case, the approach aims to confer resistance to the immune-mediated injury caused by xenogeneic antibodies and complement activation (82). Some changes in the graft, such as the up-regulation of protective genes or changes in the antigens targeted by humoral response after transplantation, may, at least in part, explain the mechanisms underlying accommodation (83). Nonetheless, at the moment, accommodation emerges as a phenomenon that still requires further investigations better to identify its potential on xenotransplantation. With regard to tolerance, mixed haematopoietic chimerism has shown encouraging results in preclinical transplantation models (84–87) and in the primate allograft setting (recently reviewed by Murakamy and colleagues (88)). It is of interest that, in an attempt to induce T-cell tolerance, when thymus was transplanted in the form of composite thymus and kidney (thymokidney), grafts from hDAF donor pigs transplanted into primate recipients were rejected by day 30 due to the presence of anti-Gal antibodies, even though signs of T-cell unresponsiveness were detected in in vitro experiments (89). Using GalT-KO donors, the transplantation of vascularized thymic tissue led to xenograft survivals of up to 83 days in a life-supporting renal xenotransplantation model (90). In the same model, the use of GalT-KO transgenic pig donors in the presence of a steroid-free immunosuppressive regimen in the absence of whole body irradiation led to an average survival time of more than 50 days (91), demonstrating that T-cell tolerance is achievable and represents a promising tool for xenotransplantation.

Finally, the role of regulatory T cells (Treg) in xenotransplantation and the current status of research on non-human primate regulatory T cells have been reviewed recently (92, 93). In this context, it is noteworthy that human Treg cells can suppress in vitro immune responses to pig xenogeneic stimulation (94), suggesting that expansion of Treg cells and their adoptive transfer may offer the possibility of increasing xenograft acceptance.

## 7. Social and Ethical Stands

The number of issues that need to be satisfactorily addressed when it comes to the clinical application of xenotransplantation include a thorough analysis from social, legal, and bioethical standpoints, with the ultimate goals of defining good practices, guidelines, and policies to be shared and accepted at an international level. The International Xenotransplantation Association (IXA) was indeed created in 1998 with the aim of facilitating the onset of the appropriate framework to promote xenotransplantation as a potential therapeutic opportunity (95). The balance between risk and benefits related to xenotransplantation procedures is still a matter of discussion. However, while the scientific and ethical debate is taking place between experts, there is also the need for an internationally harmonized framework to initiate potential clinical xenotransplantation trials. In this light, in May 2004, the World Health Organization (WHO) Assembly, as a consequence of the adopted resolution WHA57.18, urged Member States "*to allow xenotransplantation only when effective national regulatory control and surveillance mechanisms overseen by National Health Authorities (NHAs) are in place*" (96). In the same resolution, a list of recommendations for Member States is also reported, underlining in particular the value of creating an inventory of clinical trials of xenotransplantation in the various countries. As a result, the "International Human Xenotransplantation Inventory" has been put in place by the WHO in collaboration with IXA and the Hôpitaux Universitaires de Genève (97). In a preliminary report regarding the data collected to date by this registry, Sgroi and colleagues identified 29 human applications of xenotransplantation procedures performed in the past 15 years worldwide that, in some cases, were carried out in the absence of existing national regulation (98). As suggested by the authors, the availability of a constantly updated inventory of the clinical use of xenotransplantation could help the ongoing debate aimed at minimizing the risks and allowing xenotransplantation to become a safe clinical reality.

In conclusion, the progress made in the xenotransplantation field in the last decade is remarkable and suggests that clinical xenotransplantation procedures may become a reality in the not-too-distant future. Still, xenotransplantation is a complex and multidisciplinary science, whose various aspects are here elegantly discussed by internationally recognized experts. The aim is to provide a reference methodology book for scientists involved in xenotransplantation and, at the same time, to offer a practical source of knowledge for those involved in other disciplines but keen to gain insight into the research aspects underlying this fascinating science.

## Acknowledgments

The authors would like to thank the EU FP6 Integrated Project "Xenome" (www.xenome.eu), contract n°LSHB-CT-2006-037377 and the Consortium for Research in Organ Transplantation (CORIT, Padua, Italy) for their support. Special thanks to Dr. Michela Seveso, Dr. Marta Vadori, and Dr Federica Besenzon for their review of the manuscript.

## References

1. http://www.unos.org
2. http://www.edqm.eu
3. Leor J, Aboulafia-Etzion S, Dar A, Shapiro L, Barbash IM, Battler A, Granot Y, Cohen S (2000) Bioengineered cardiac grafts: a new approach to repair the infarcted myocardium? Circulation 102:III56–III61
4. Zimmermann WH, Schneiderbanger K, Schubert P, Didie M, Munzel F, Heubach JF, Kostin S, Neuhuber WL, Eschenhagen T (2002) Tissue engineering of a differentiated cardiac muscle construct. Circ Res 90:223–230
5. Shimizu T, Sekine H, Yamato M, Okano T (2009) Cell sheet-based myocardial tissue engineering: new hope for damaged heart rescue. Curr Pharm Des 15:2807–2814
6. Ott HC, Matthiesen TS, Goh SK, Black LD, Kren SM, Netoff TI, Taylor DA (2008) Perfusion-decellularized matrix: using nature's platform to engineer a bioartificial heart. Nat Med 14:213–221
7. Ross EA, Williams MJ, Hamazaki T, Terada N, Clapp WL, Adin C, Ellison GW, Jorgensen M, Batich CD (2009) Embryonic stem cells proliferate and differentiate when seeded into kidney scaffolds. J Am Soc Nephrol 20: 2338–2347
8. Baptista PM, Orlando G, Mirmalek-Sani SH, Siddiqui M, Atala A, Soker S (2009) Whole organ decellularization—a tool for bioscaffold fabrication and organ bioengineering. Conf Proc IEEE Eng Med Biol Soc 2009: 6526–6529
9. Nakayama KH, Batchelder CA, Lee CI, Tarantal AF (2010) Decellularized rhesus monkey kidney as a three-dimensional scaffold for renal tissue engineering. Tissue Eng Part A 16:2207–2216
10. Uygun BE, Soto-Gutierrez A, Yagi H, Izamis ML, Guzzardi MA, Shulman C, Milwid J, Kobayashi N, Tilles A, Berthiaume F, Hertl M, Nahmias Y, Yarmush ML, Uygun K (2010) Organ reengineering through development of a transplantable recellularized liver graft using decellularized liver matrix. Nat Med 16:814–820
11. Ott HC, Clippinger B, Conrad C, Schuetz C, Pomerantseva I, Ikonomou L, Kotton D,

Vacanti JP (2010) Regeneration and orthotopic transplantation of a bioartificial lung. Nat Med 16:927–933

12. Petersen TH, Calle EA, Zhao L, Lee EJ, Gui L, Raredon MB, Gavrilov K, Yi T, Zhuang ZW, Breuer C, Herzog E, Niklason LE (2010) Tissue-engineered lungs for in vivo implantation. Science 329:538–541
13. Sandrin MS, Vaughan HA, Dabkowski PL, McKenzie IF (1993) Anti-pig IgM antibodies in human serum react predominantly with Gal(alpha 1-3)Gal epitopes. Proc Natl Acad Sci USA 90:11391–11395
14. Schuurman HJ, Cheng J, Lam T (2003) Pathology of xenograft rejection: a commentary. Xenotransplantation 10:293–299
15. Platt JL (2000) Acute vascular rejection. Transplant Proc 32:839–840
16. McGregor CGA, Keiji Oi WRD, Tazelaar HD, Walker RC, Chandrasekaran K, Byrne GW (2008) Recovery of cardiac function after pig-to-primate orthotopic heart transplant. Am J Transplant 8(Suppl s2):205
17. Kuwaki K, Tseng YL, Dor FJ, Shimizu A, Houser SL, Sanderson TM, Lancos CJ, Prabharasuth DD, Cheng J, Moran K, Hisashi Y, Mueller N, Yamada K, Greenstein JL, Hawley RJ, Patience C, Awwad M, Fishman JA, Robson SC, Schuurman HJ, Sachs DH, Cooper DK (2005) Heart transplantation in baboons using alpha1,3-galactosyltransferase gene-knockout pigs as donors: initial experience. Nat Med 11:29–31
18. Tseng YL, Kuwaki K, Dor FJ, Shimizu A, Houser S, Hisashi Y, Yamada K, Robson SC, Awwad M, Schuurman HJ, Sachs DH, Cooper DK (2005) alpha1,3-Galactosyltransferase gene-knockout pig heart transplantation in baboons with survival approaching 6 months. Transplantation 80:1493–1500
19. Baldan N, Rigotti P, Calabrese F, Cadrobbi R, Dedja A, Iacopetti I, Boldrin M, Seveso M, Dall'Olmo L, Frison L, De Benedictis G, Bernardini D, Thiene G, Cozzi E, Ancona E (2004) Ureteral stenosis in HDAF pig-to-primate renal xenotransplantation: a phenomenon related to immunological events? Am J Transplant 4:475–481
20. Shimizu A, Hisashi Y, Kuwaki K, Tseng YL, Dor FJ, Houser SL, Robson SC, Schuurman HJ, Cooper DK, Sachs DH, Yamada K, Colvin RB (2008) Thrombotic microangiopathy associated with humoral rejection of cardiac xenografts from alpha1,3-galactosyltransferase gene-knockout pigs in baboons. Am J Pathol 172:1471–1481
21. Lin CC, Ezzelarab M, Shapiro R, Ekser B, Long C, Hara H, Echeverri G, Torres C, Watanabe H, Ayares D, Dorling A, Cooper DK (2010) Recipient tissue factor expression is associated with consumptive coagulopathy in pig-to-primate kidney xenotransplantation. Am J Transplant 10:1556–1568
22. Sullivan JA, Oettinger HF, Sachs DH, Edge AS (1997) Analysis of polymorphism in porcine MHC class I genes: alterations in signals recognized by human cytotoxic lymphocytes. J Immunol 159:2318–2326
23. Forte P, Lilienfeld BG, Baumann BC, Seebach JD (2005) Human NK cytotoxicity against porcine cells is triggered by NKp44 and NKG2D. J Immunol 175:5463–5470
24. Schneider MK, Seebach JD (2008) Current cellular innate immune hurdles in pig-to-primate xenotransplantation. Curr Opin Organ Transplant 13:171–177
25. Goto M, Tjernberg J, Dufrane D, Elgue G, Brandhorst D, Ekdahl KN, Brandhorst H, Wennberg L, Kurokawa Y, Satomi S, Lambris JD, Gianello P, Korsgren O, Nilsson B (2008) Dissecting the instant blood-mediated inflammatory reaction in islet xenotransplantation. Xenotransplantation 15:225–234
26. van der Windt DJ, Bottino R, Casu A, Campanile N, Cooper DK (2007) Rapid loss of intraportally transplanted islets: an overview of pathophysiology and preventive strategies. Xenotransplantation 14:288–297
27. Hering BJ, Walawalkar N (2009) Pig-to-nonhuman primate islet xenotransplantation. Transpl Immunol 21:81–86
28. Dufrane D, Goebbels RM, Gianello P (2010) Alginate macroencapsulation of pig islets allows correction of streptozotocin-induced diabetes in primates up to 6 months without immunosuppression. Transplantation 90:1054–1062
29. Komoda H, Miyagawa S, Kubo T, Kitano E, Kitamura H, Omori T, Ito T, Matsuda H, Shirakura R (2004) A study of the xenoantigenicity of adult pig islets cells. Xenotransplantation 11:237–246
30. Bennet W, Bjorkland A, Sundberg B, Davies H, Liu J, Holgersson J, Korsgren O (2000) A comparison of fetal and adult porcine islets with regard to Gal alpha (1,3)Gal expression and the role of human immunoglobulins and complement in islet cell cytotoxicity. Transplantation 69:1711–1717
31. Rayat GR, Rajotte RV, Hering BJ, Binette TM, Korbutt GS (2003) In vitro and in vivo expression of Galalpha-(1,3)Gal on porcine islet cells is age dependent. J Endocrinol 177:127–135
32. Rayat GR, Rajotte RV, Elliott JF, Korbutt GS (1998) Expression of Gal alpha(1,3)gal on neonatal porcine islet beta-cells and susceptibility to human antibody/complement lysis. Diabetes 47:1406–1411

33. Schulte Am Esch J 2nd, Robson SC, Knoefel WT, Hosch SB, Rogiers X (2005) O-linked glycosylation and functional incompatibility of porcine von Willebrand factor for human platelet GPIb receptors. Xenotransplantation 12:30–37

34. Roussel JC, Moran CJ, Salvaris EJ, Nandurkar HH, d'Apice AJ, Cowan PJ (2008) Am Pig thrombomodulin binds human thrombin but is a poor cofactor for activation of human protein C and TAFI. Am J Transplant 8: 1101–1112

35. Kopp CW, Siegel JB, Hancock WW, Anrather J, Winkler H, Geczy CL, Kaczmarek E, Bach FH, Robson SC (1997) Effect of porcine endothelial tissue factor pathway inhibitor on human coagulation factors. Transplantation 63:749–758

36. Kopp CW, Robson SC, Siegel JB, Anrather J, Winkler H, Grey S, Kaczmarek E, Bach FH, Geczy CL (1998) Regulation of monocyte tissue factor activity by allogeneic and xenogeneic endothelial cells. Thromb Haemost 79: 529–538

37. Lee KF, Salvaris EJ, Roussel JC, Robson SC, d'Apice AJ, Cowan PJ (2008) Recombinant pig TFPI efficiently regulates human tissue factor pathways. Xenotransplantation 15: 191–197

38. Lin CC, Ezzelarab M, Hara H, Long C, Lin CW, Dorling A, Cooper DK (2010) Atorvastatin or transgenic expression of TFPI inhibits coagulation initiated by anti-nonGal IgG binding to porcine aortic endothelial cells. J Thromb Haemost 8:2001–2010

39. Lee H, Lee B, Kim Y, Paik N, Rho H (2011) Characterization of transgenic pigs that express human decay accelerating factor and cell membrane-tethered human tissue factor pathway inhibitor. Reprod Domest Anim 46:325–332

40. Chen D, Riesbeck K, Kemball-Cook G, McVey JH, Tuddenham EG, Lechler RI, Dorling A (1999) Inhibition of tissue factor-dependent and -independent coagulation by cell surface expression of novel anticoagulant fusion proteins. Transplantation 67:467–474

41. Miwa Y, Yamamoto K, Onishi A, Iwamoto M, Yazaki S, Haneda M, Iwasaki K, Liu D, Ogawa H, Nagasaka T, Uchida K, Nakao A, Kadomatsu K, Kobayashi T (2010) Potential value of human thrombomodulin and DAF expression for coagulation control in pig-to-human xenotransplantation. Xenotransplantation 17:26–37

42. Crikis S, Lu B, Murray-Segal LM, Selan C, Robson SC, D'Apice AJ, Nandurkar HH, Cowan PJ, Dwyer KM (2010) Transgenic overexpression of CD39 protects against renal ischemia-reperfusion and transplant vascular injury. Am J Transplant 10:2586–2595

43. Dwyer KM, Robson SC, Nandurkar HH, Campbell DJ, Gock H, Murray-Segal LJ, Fisicaro N, Mysore TB, Kaczmarek E, Cowan PJ, d'Apice AJ (2004) Thromboregulatory manifestations in human CD39 transgenic mice and the implications for thrombotic disease and transplantation. J Clin Invest 113: 1440–1446

44. Cowan PJ, Robson SC, D'Apice AJ (2011) Controlling coagulation dysregulation in xenotransplantation. Curr Opin Organ Transplant 16:214–221

45. Lin CC, Cooper DK, Dorling A (2009) Coagulation dysregulation as a barrier to xenotransplantation in the primate. Transpl Immunol 21:75–80

46. Vial CM, Ostlie DJ, Bhatti FN, Cozzi E, Goddard M, Chavez GP, Wallwork J, White DJ, Dunning JJ (2000) Life supporting function for over one month of a transgenic porcine heart in a baboon. J Heart Lung Transplant 19:224–229

47. Soin B, Smith KG, Zaidi A, Cozzi E, Bradley JR, Ostlie DJ, Lockhart A, White DJ, Friend PJ (2001) Physiological aspects of pig-to-primate renal xenotransplantation. Kidney Int 60:1592–1597

48. van der Windt DJ, Bottino R, Casu A, Campanile N, Smetanka C, He J, Murase N, Hara H, Ball S, Loveland BE, Ayares D, Lakkis FG, Cooper DK, Trucco M (2009) Long-term controlled normoglycemia in diabetic non-human primates after transplantation with hCD46 transgenic porcine islets. Am J Transplant 9:2716–2726

49. Daggett CW, Yeatman M, Lodge AJ, Chen EP, Linn SS, Gullotto C, Frank MM, Platt JL, Davis RD (1998) Total respiratory support from swine lungs in primate recipients. J Thorac Cardiovasc Surg 115:19–27

50. Cooper DK, Keogh AM, Brink J, Corris PA, Klepetko W, Pierson RN, Schmoeckel M, Shirakura R, Warner Stevenson L (2000) Report of the Xenotransplantation Advisory Committee of the International Society for Heart and Lung Transplantation: the present status of xenotransplantation and its potential role in the treatment of end-stage cardiac and pulmonary diseases. J Heart Lung Transplant 19:1125–1165

51. Pfeiffer S, Zorn GL 3rd, Blair KS, Farley SM, Wu G, Schuurman HJ, White DJ, Azimzadeh AM, Pierson RN 3rd (2005) Hyperacute lung rejection in the pig-to-human model 4: evidence for complement and antibody independent mechanisms. Transplantation 79:662–671

52. Ibrahim Z, Busch J, Awwad M, Wagner R, Wells K, Cooper DK (2006) Selected physiologic compatibilities and incompatibilities between human and porcine organ systems. Xenotransplantation 13:488–499
53. Ramirez P, Chavez R, Majado M, Munitiz V, Munoz A, Hernandez Q, Palenciano CG, Pino-Chavez G, Loba M, Minguela A, Yelamos J, Gago MR, Vizcaino AS, Asensi H, Cayuela MG, Segura B, Marin F, Rubio A, Fuente T, Robles R, Bueno FS, Sansano T, Acosta F, Rodriguez JM, Navarro F, Cabezuelo J, Cozzi E, White DJ, Calne RY, Parrilla P (2000) Life-supporting human complement regulator decay accelerating factor transgenic pig liver xenograft maintains the metabolic function and coagulation in the nonhuman primate for up to 8 days. Transplantation 70:989–998
54. Ramirez P, Montoya MJ, Rios A, Garcia Palenciano C, Majado M, Chavez R, Munoz A, Fernandez OM, Sanchez A, Segura B, Sansano T, Acosta F, Robles R, Sanchez F, Fuente T, Cascales P, Gonzalez F, Ruiz D, Martinez L, Pons JA, Rodriguez JI, Yelamos J, Cowan P, d'Apice A, Parrilla P (2005) Prevention of hyperacute rejection in a model of orthotopic liver xenotransplantation from pig to baboon using polytransgenic pig livers (CD55, CD59, and H-transferase). Transplant Proc 37:4103–4106
55. Ekser B, Gridelli B, Tector AJ, Cooper DK (2009) Pig liver xenotransplantation as a bridge to allotransplantation: which patients might benefit? Transplantation 88:1041–1049
56. Swindle MM (1998) Defining appropriate health status and management programs for specific-pathogen-free swine for xenotransplantation. Ann N Y Acad Sci 862:111–120
57. Clemenceau B, Lalain S, Martignat L, Sai P (1999) Porcine endogenous retroviral mRNAs in pancreas and a panel of tissues from specific pathogen-free pigs. Diabetes Metab 25: 518–525
58. Patience C, Takeuchi Y, Weiss RA (1997) Infection of human cells by an endogenous retrovirus of pigs. Nat Med 3:282–286
59. Martin U, Kiessig V, Blusch JH, Haverich A, von der Helm K, Herden T, Steinhoff G (1998) Expression of pig endogenous retrovirus by primary porcine endothelial cells and infection of human cells. Lancet 352:692–694
60. Bartosch B, Stefanidis D, Myers R, Weiss R, Patience C, Takeuchi Y (2004) Evidence and consequence of porcine endogenous retrovirus recombination. J Virol 78:13880–13890
61. Meije Y, Tonjes RR, Fishman JA (2010) Retroviral restriction factors and infectious risk in xenotransplantation. Am J Transplant 10: 1511–1516
62. Valdes-Gonzalez R, Dorantes LM, Bracho-Blanchet E, Rodriguez-Ventura A, White DJ (2010) No evidence of porcine endogenous retrovirus in patients with type 1 diabetes after long-term porcine islet xenotransplantation. J Med Virol 82:331–334
63. Di Nicuolo G, D'Alessandro A, Andria B, Scuderi V, Scognamiglio M, Tammaro A, Mancini A, Cozzolino S, Di Florio E, Bracco A, Calise F, Chamuleau RA (2010) Long-term absence of porcine endogenous retrovirus infection in chronically immunosuppressed patients after treatment with the porcine cell-based Academic Medical Center bioartificial liver. Xenotransplantation 17:431–439
64. Mueller NJ, Barth RN, Yamamoto S, Kitamura H, Patience C, Yamada K, Cooper DK, Sachs DH, Kaur A, Fishman JA (2002) Activation of cytomegalovirus in pig-to-primate organ xenotransplantation. J Virol 76:4734–4740
65. Ehlers B, Ulrich S, Goltz M (1999) Detection of two novel porcine herpesviruses with high similarity to gammaherpesviruses. J Gen Virol 80(Pt 4):971–978
66. Santoni F, Lindner I, Caselli E, Goltz M, Di Luca D, Ehlers B (2006) Molecular interactions between porcine and human gammaherpesviruses: implications for xenografts? Xenotransplantation 13:308–317
67. Oldmixon BA, Wood JC, Ericsson TA, Wilson CA, White-Scharf ME, Andersson G, Greenstein JL, Schuurman HJ, Patience C (2002) Porcine endogenous retrovirus transmission characteristics of an inbred herd of miniature swine. J Virol 76:3045–3048
68. Mattiuzzo G, Ivol S, Takeuchi Y (2010) Regulation of porcine endogenous retrovirus release by porcine and human tetherins. J Virol 84:2618–2622
69. Ramsoondar J, Vaught T, Ball S, Mendicino M, Monahan J, Jobst P, Vance A, Duncan J, Wells K, Ayares D (2009) Production of transgenic pigs that express porcine endogenous retrovirus small interfering RNAs. Xenotransplantation 16:164–180
70. Dieckhoff B, Petersen B, Kues WA, Kurth R, Niemann H, Denner J (2008) Knockdown of porcine endogenous retrovirus (PERV) expression by PERV-specific shRNA in transgenic pigs. Xenotransplantation 15:36–45
71. Miyagawa S, Nakatsu S, Hazama K, Nakagawa T, Kondo A, Matsunami K, Yamamoto A, Yamada J, Miyazawa T, Shirakura R (2006) A novel strategy for preventing PERV transmission to human cells by remodeling the viral

envelope glycoprotein. Xenotransplantation 13:258–263

72. Ekser B, Long C, Echeverri GJ, Hara H, Ezzelarab M, Lin CC, de Vera ME, Wagner R, Klein E, Wolf RF, Ayares D, Cooper DK, Gridelli B (2010) Impact of thrombocytopenia on survival of baboons with genetically modified pig liver transplants: clinical relevance. Am J Transplant 10:273–285
73. Gaca JG, Lesher A, Aksoy O, Gonzalez-Stawinski GV, Platt JL, Lawson JH, Parker W, Davis RD (2002) Disseminated intravascular coagulation in association with pig-to-primate pulmonary xenotransplantation. Transplantation 73:1717–1723
74. Gonzalez-Stawinski GV, Daggett CW, Lau CL, Karoor S, Love SD, Logan JS, Gaca JG, Parker W, Davis RD Jr (2002) Non-anti-Gal alpha1-3Gal antibody mechanisms are sufficient to cause hyperacute lung dysfunction in pulmonary xenotransplantation. J Am Coll Surg 194:765–773
75. Nguyen BH, Zwets E, Schroeder C, Pierson RN 3rd, Azimzadeh AM (2005) Beyond antibody-mediated rejection: hyperacute lung rejection as a paradigm for dysregulated inflammation. Curr Drug Targets Cardiovasc Haematol Disord 5:255–269
76. Nguyen BN, Azimzadeh AM, Zhang T, Wu G, Schuurman HJ, Sachs DH, Ayares D, Allan JS, Pierson RN 3rd (2007) Life-supporting function of genetically modified swine lungs in baboons. J Thorac Cardiovasc Surg 133: 1354–1363
77. Cardona K, Korbutt GS, Milas Z, Lyon J, Cano J, Jiang W, Bello-Laborn H, Hacquoil B, Strobert E, Gangappa S, Weber CJ, Pearson TC, Rajotte RV, Larsen CP (2006) Long-term survival of neonatal porcine islets in nonhuman primates by targeting costimulation pathways. Nat Med 12:304–306
78. Hering BJ, Wijkstrom M, Graham ML, Hardstedt M, Aasheim TC, Jie T, Ansite JD, Nakano M, Cheng J, Li W, Moran K, Christians U, Finnegan C, Mills CD, Sutherland DE, Bansal-Pakala P, Murtaugh MP, Kirchhof N, Schuurman HJ (2006) Prolonged diabetes reversal after intraportal xenotransplantation of wild-type porcine islets in immunosuppressed nonhuman primates. Nat Med 12: 301–303
79. Hecht G, Eventov-Friedman S, Rosen C, Shezen E, Tchorsh D, Aronovich A, Freud E, Golan H, El-Hasid R, Katchman H, Hering BJ, Zung A, Kra-Oz Z, Shaked-Mishan P, Yusim A, Shtabsky A, Idelevitch P, Tobar A, Harmelin A, Bachar-Lustig E, Reisner Y (2009) Embryonic pig pancreatic tissue for the treatment of diabetes in a nonhuman primate model. Proc Natl Acad Sci USA 106: 8659–8664
80. Fink JS, Schumacher JM, Ellias SL, Palmer EP, Saint-Hilaire M, Shannon K, Penn R, Starr P, VanHorne C, Kott HS, Dempsey PK, Fischman AJ, Raineri R, Manhart C, Dinsmore J, Isacson O (2000) Porcine xenografts in Parkinson's disease and Huntington's disease patients: preliminary results. Cell Transplant 9:273–278
81. Badin RA, Padoan A, Vadori M, Boldrin M, De Benedictis GM, Fante F, Sgarabotto D, Jan C, Daguin V, Naveilhan P, Neveu I, Soullilou JP, Vanhove B, Plat M, Botté F, Venturi F, Denaro L, Seveso M, Manara R, Zampieri P, D'Avella D, Rubello D, Ancona E, Hantraye P, Cozzi E (2010) Porcine embryonic xenografts transgenic for CTLA4-Ig enable longterm Recovery in parkinsonian macaques. Am J Transplant 10(Suppl s4):208
82. Lynch RJ, Platt JL (2009) Escaping from rejection. Transplantation 88:1233–1236
83. Koch CA, Khalpey ZI, Platt JL (2004) Accommodation: preventing injury in transplantation and disease. J Immunol 172: 5143–5148
84. Ildstad ST, Sachs DH (1984) Reconstitution with syngeneic plus allogeneic or xenogeneic bone marrow leads to specific acceptance of allografts or xenografts. Nature 307:168–170
85. Sharabi Y, Aksentijevich I, Sundt TM 3rd, Sachs DH, Sykes M (1990) Specific tolerance induction across a xenogeneic barrier: production of mixed rat/mouse lymphohematopoietic chimeras using a nonlethal preparative regimen. J Exp Med 172:195–202
86. Yang YG, deGoma E, Ohdan H, Bracy JL, Xu Y, Iacomini J, Thall AD, Sykes M (1998) Tolerization of anti-Galalpha1-3Gal natural antibody-forming B cells by induction of mixed chimerism. J Exp Med 187:1335–1342
87. Abe M, Qi J, Sykes M, Yang YG (2002) Mixed chimerism induces donor-specific T-cell tolerance across a highly disparate xenogeneic barrier. Blood 99:3823–3829
88. Murakami T, Cosimi AB, Kawai T (2009) Mixed chimerism to induce tolerance: lessons learned from nonhuman primates. Transplant Rev (Orlando) 23:19–24
89. Barth RN, Yamamoto S, LaMattina JC, Kumagai N, Kitamura H, Vagefi PA, Awwad M, Colvin RB, Cooper DK, Sykes M, Sachs DH, Yamada K (2003) Xenogeneic thymokidney and thymic tissue transplantation in a pig-to-baboon model: I evidence for pig-specific T-cell unresponsiveness. Transplantation 75: 1615–1624

90. Yamada K, Yazawa K, Shimizu A, Iwanaga T, Hisashi Y, Nuhn M, O'Malley P, Nobori S, Vagefi PA, Patience C, Fishman J, Cooper DK, Hawley RJ, Greenstein J, Schuurman HJ, Awwad M, Sykes M, Sachs DH (2005) Marked prolongation of porcine renal xenograft survival in baboons through the use of alpha1,3-galactosyltransferase gene-knockout donors and the cotransplantation of vascularized thymic tissue. Nat Med 11:32–34
91. Griesemer AD, Hirakata A, Shimizu A, Moran S, Tena A, Iwaki H, Ishikawa Y, Schule P, Arn JS, Robson SC, Fishman JA, Sykes M, Sachs DH, Yamada K (2009) Results of gal-knockout porcine thymokidney xenografts. Am J Transplant 9:2669–2678
92. Muller YD, Golshayan D, Ehirchiou D, Wekerle T, Seebach JD, Buhler LH (2009) T regulatory cells in xenotransplantation. Xenotransplantation 16:121–128
93. Dons EM, Raimondi G, Cooper DK, Thomson AW (2010) Non-human primate regulatory T cells: current biology and implications for transplantation. Transplantation 90:811–816
94. Porter CM, Bloom ET (2005) Human CD4+CD25+ regulatory T cells suppress anti-porcine xenogeneic responses. Am J Transplant 5:2052–2057
95. http://www.tts.org/index.php?option=com_content&view=article&id=51&Itemid=292
96. http://www.who.int/transplantation/xeno/en/
97. http://www.humanxenotransplant.org/
98. Sgroi A, Buhler LH, Morel P, Sykes M, Noel L (2010) International human xenotransplantation inventory. Transplantation 90:597–603

# Chapter 2

# Cloning and Expression Analyses of Pig Genes

Mireia Uribe-Herranz and Cristina Costa

## Abstract

Understanding the molecular bases of xenograft rejection is one of the highest priorities in the xenotransplantation field. Furthermore, the identification of physiological incompatibilities in the xenogeneic setting is also necessary for developing the appropriate strategies to have a long-term functioning xenograft. As the pig is the species of choice for the development of xenogeneic applications, the cloning of pig genes or cDNA is a key step to elucidate the interactions of pig and human molecules. It also provides the necessary information for assessing the level of mRNA expression of relevant proteins in tissues and organs of interest for xenotransplantation. In most cases, the cloning of the cDNA is sufficient to attain these goals. Thus, we describe a basic cloning method that comprises total RNA extraction, reverse transcription (RT), and polymerase chain reaction (PCR) amplification. We also include some links for databases and bioinformatic tools available in the Internet for the subsequent analyses and predictions. Finally, we recommend and explain the procedures of northern blotting and quantitative RT-PCR for conducting the mRNA expression studies.

**Key words:** Pig, cDNA, mRNA, RT-PCR, Northern blotting, Bioinformatic tools

## 1. Introduction

The cloning of genes is the key for understanding the molecular bases of any pathological process and allows the development of biomedical therapies. In the field of xenotransplantation, it is critical to understand the molecular bases of xenograft rejection and identify physiological incompatibilities. This kind of work is also necessary for developing the appropriate strategies to overcome rejection and have a long-term functioning xenograft. In this context, the cloning of pig genes is the first step to elucidate the interactions of pig and human molecules. In most cases, the cloning of the cDNA is sufficient to attain these goals and we are going to

Cristina Costa and Rafael Máñez (eds.), *Xenotransplantation: Methods and Protocols,* Methods in Molecular Biology, vol. 885, DOI 10.1007/978-1-61779-845-0_2, © Springer Science+Business Media, LLC 2012

focus this chapter in describing procedures for isolating and analyzing cDNA and mRNA transcripts. However, the isolation of a gene or a portion of it can be of interest for some specific purposes, such as for homologous recombination for knockout or knock-in technologies (1, 2). There may be groups also interested in identifying unknown proteins/receptors involved in processes related to the survival/function of the graft. In this case, the work may start with the partial sequencing of a protein and finding the coding sequences with the appropriate computer program and gene database. Nevertheless, obtaining the cDNA would still be an important part in this type of project.

The sequence of the whole pig genome will be made public soon. Its size has been estimated to be about 3 Gbp (3). Since 2003, the Swine Genome Sequencing Consortium (http://piggenome.org/) has coordinated the public and private efforts of sequencing the pig genome. Its approaches include hierarchical shotgun sequencing of BAC clones and whole genome shotgun sequencing. The current annotated genome assembly, named Sscrofa9, was published in September 2009 as the first release of high-coverage assembly. An improved assembly, named Sscrofa10, has been made available to the pig genomics community and includes the whole genome shotgun sequence data, thus providing >30× genome coverage (3, 4). Even when the pig genome sequence is fully annotated, cDNA cloning will still be useful to detect the native mRNA variants and work with the more manageable cDNA.

Several strategies can be used to clone a specific pig cDNA, such as hybridization of cDNA libraries, standard reverse transcription (RT) followed by polymerase chain reaction (PCR) or rapid amplification of cDNA ends (RACE) (5–7). The cDNA libraries are especially useful when the whole genome is known (i.e., interpretation of results, high specificity of probes). Unfortunately, there are not many libraries available of pig cDNA, it involves a labor-intensive work, and your success still depends on the level of expression of your gene of interest in the tissue and condition used for creating the cDNA library. In our experience, the standard RT-PCR using degenerate primers and various tissues as a source of cDNA is a very useful approach for obtaining an unknown pig cDNA. Thus, we describe a basic cloning method that comprises total RNA extraction, reverse transcription, and PCR amplification with degenerate primers. Nevertheless, it is still of interest to search first in databases for pig cDNA/gene sequences as it may allow the design and use of fully specific oligoprimers when that particular sequence is available. Note that following the RT-PCR there are the necessary steps of sub-cloning and sequencing multiple clones to obtain a consensus sequence. We recommend to use a topoisomerase-based method for sub-cloning. Regarding sequencing, we are not going to describe the procedure as it is now commonly done in specialized facilities at a very reasonable cost. Finally, the

RACE can also be performed to obtain the full cDNA. It is specially useful to conduct it subsequently to the standard RT-PCR as it can be done when specific sequences are available, provide further confirmation of the sequence obtained, and lead to the discovery of new isoforms. Many genes lead to the generation of multiple mRNA variants as a result of alternative splicing (8) and a combination of techniques can be more effective in detecting such variants. In the case of a gene that produces multiple mRNA variants, it is unquestionable that identification of the majority of variants is important to elucidate their function (5, 8).

Nowadays, with the huge amount of information generated by high-throughput techniques such as large-scale expression profiling or genome annotation, bioinformatic tools have become essential for any laboratory. Over the past years, numerous tools have emerged for storing and organizing information, analysis, and prediction of biological processes (9, 10). There are various major databases, such as those from European Molecular Biology Laboratory-European Bioinformatic Institute (EMBL-EBI), National Center for Biotechnology Information (NCBI, USA), and DNA Databank of Japan (DDBJ). The nucleotide databases GenBank (NCBI), EMBL DataLibrary, and DDBJ share and store every nucleotide sequence already published or made public. Protein databases such as Swiss-prot store primary amino acidic sequences, whereas the protein database from NCBI collects information from multiple sources. Moreover, there are plenty of servers which provide the scientist with high-quality resources of protein sequence, functional information, and tools for analysis (e.g., EBI, ExPASy). They provide tools that are especially useful for the analysis of newly cloned genes and cDNA, such as comparison with homologues and prediction of function of the encoded proteins. Here, we provide useful links to some common tools to be used in a molecular laboratory.

Another step in elucidating the role of specific molecules in a process such as xenograft rejection is studying their expression in the pig donor tissue or organ. Once the cDNA or sequence is available, mRNA expression studies can follow. There are a variety of methods to detect mRNA like northern blotting, in situ hybridization, or RT-PCR. Northern blotting with radioactive probes is the most sensitive technique, but it involves handling radioactivity, is time consuming, and requires large amounts of RNA. Less template and labor are needed for the standard RT-PCR, but this is not a technique appropriate for accurate quantification. Thus, the quantitative (RT-qPCR, which involves real-time PCR) is being used extensively because it provides quantitative results utilizing little amounts of RNA (11). There are various strategies to carry out a real-time PCR. In this chapter, we describe northern blotting and one of the approaches for quantitative RT-PCR that is appropriate when focused on studying a specific gene.

## 2. Materials

Prepare and store all reagents at room temperature unless otherwise indicated. Use gloves to prepare and handle the reagents for RNA and DNA analysis for both protection and to prevent sample contamination. Unless stated specifically, use purified water such as Elix deionized water (Millipore Corporation, Billerica, MA, USA) to prepare all solutions.

### 2.1. Cloning of Pig cDNA

#### 2.1.1. Total RNA Extraction

1. Porcine cells or tissue (see Note 1). Use fresh or frozen.
2. Diethyl pyrocarbonate (DEPC)-treated water: 0.1% v/v DEPC in water (see Note 2).
3. Trizol (Invitrogen, Carlsbad, CA, USA) (see Note 3).
4. Chloroform.
5. Isopropyl alcohol.
6. 75% ethanol (in DEPC-treated water).
7. Sterile RNAse-free tips and tubes.
8. Gloves for handling all reagents and materials for RNA isolation.
9. Equipment: Pipettors, fume hood, Polytron homogenizer or tweezers, mortar and pestle baked in oven to prevent ribonuclease contamination, spectrophotometer (a micro-volume spectrophotometer such as NanoDrop 2000 is especially convenient (Thermo scientific, Waltham, MA, USA)).

#### 2.1.2. Standard RT-PCR

1. Kit for RT (see Note 4) containing various reagents: Avian myeloblastosis virus (AMV) reverse transcriptase, amplification buffer, dithiothreitol (DTT), dNTPmix, RNase, oligo(dT)$_{20}$ or random hexamers, and DEPC-treated water.
2. Kit for PCR (see Note 5) containing various reagents: Thermostable DNA polymerase, dNTP mix, amplification buffer 10×, and DEPC-treated water.
3. Specific degenerate primers (see Note 6).
4. cDNA template (see Note 7).
5. Sterile tips (preferably with filter) and PCR tubes.
6. Gloves for handling all reagents and materials.
7. Ethidium bromide stock solution at 10 mg/mL (handle with care as it is very toxic).
8. DNA size-marker, such as 1 Kb Plus DNA Ladder (Invitrogen).
9. Electrophoresis buffer TAE 50×: 121 g of Tris base, 28.6 mL of glacial acetic acid, and 50 mL of 0.5 M EDTA. Bring solution to a final volume of 500 mL.

10. Electrophoresis running buffer: Dilute the 50× TAE 1/50 in water. Prepare in sufficient quantity for the electrophoresis.
11. Erlenmeyer flask made of heat-resistant glass (such as Pyrex) for preparation of agarose gel.
12. Agarose gel (see Note 8): For a 100-mL 1% agarose gel, use 1 g of agarose in 1× electrophoresis buffer in the Erlenmeyer flask, bring to boil, and mix until the solution is clear. Cool down, add 5 μL of the ethidium bromide stock solution, mix well, and pour in gel box. We conduct this last step and leave the gel to solidify under a chemical hood.
13. Loading buffer 6×: 30% (W/V) sucrose and 0.35% (W/V) orange G sodium salt.
14. Equipment: Pipettors, thermal cycler, horizontal electrophoresis unit with power supply, and imaging system with ultraviolet illumination.

### 2.2. Bioinformatic Analyses

1. Databases: GenBank (NIH genetic sequence database), EMBL-EBI, or DDBJ. The three form the International Nucleotide Sequence Database.
2. Pig genome database at http://www.ensembl.org/Sus_scrofa/Info/Index.
3. Equipment: Personal computer with Internet connection.

### 2.3. Determination of mRNA Expression

#### 2.3.1. Analysis of RNA by Northern Blotting

1. 10 μg of total RNA of each tissue and condition to be analyzed. The purified RNA is stored at −80°C.
2. DNA probes labeled with [$\alpha^{32}$P] dCTP, including one for β-actin as control of RNA load. These can be prepared by restriction-enzyme digestion or PCR followed by purification and random primer labeling with the Prime-it II Random Primer labeling Kit (Agilent Technologies Santa Clara, CA, USA). Label the probe immediately before use or within a few days before and store at −20°C. Note that the handling of the radioactive material has to comply with the corresponding regulations set by the local and national authority.
3. 10× MOPS buffer: 0.4 M morpholinopropanesulfonic acid, 0.1 M sodium acetate-3$H_2O$, and 10 mM EDTA-$Na_2$-2$H_2O$ in water. Adjust pH to 7.2 with NaOH. Prepare 1 L and sterilize. Store at room temperature wrapped in aluminum foil.
4. Electrophoresis gel: 1% agarose, 0.66 M formaldehyde, pH > 4 (5.4 mL of 37% formaldehyde per 100 mL), 1× MOPS, and water. A gel of 12–14-cm long is usually sufficient and can be run during the day. To prepare the electrophoresis gel, dissolve first the agarose in water by boiling it. Once the flask has cooled (~70°C), add the appropriate amount of 10× MOPS and formaldehyde. Pour it into a gel box (slots should allow >50 μL of

volume) under a fume hood. Note that all handling of formaldehyde should be done under the fume hood.

5. To stain the gel: 5 μg/mL ethidium bromide in water.
6. Formamide, deionized: Mix 50 mL of formamide with a mixed-bed resin such as Bio-Rad AG 501-X8 (Bio-Rad, Hercules, CA, USA), stir gently for 30 min, and filter twice through Whatman® n° 1 paper. It can be prepared in bulk and frozen down at −20°C for storage.
7. Formaldehyde loading buffer: Formamide 720 μL and 10× MOPS 160 μL.
8. 37% formaldehyde 260 μL, $dH_2O$ 180 μL, 80% glycerol 100 μL, and saturated bromophenol blue solution 80 μL. Aliquot and store at −20°C.
9. 20× SSC: 3 M NaCl and 0.3 M sodium citrate dihydrate in DEPC-treated water. Prepare 1 L. Dilute ½ to use as transfer buffer.
10. 20× SSPE: 3 M NaCl, 0.2 M $NaH_2PO_4$-$H_2O$, and 20 mM EDTA-$Na_2$ (use 0.5 M stock solution) in DEPC-treated water. Adjust pH to 7.4 with NaOH. Prepare 1 L.
11. 100× Denhardt's solution: 1 g polyvinylpyrrolidone (MW 40000), 1 g bovine serum albumin, and 1 g of Ficoll 400 in 50 mL of water. Sterile filter, aliquot, and store frozen.
12. Northern prehybridization and hybridization solutions: 5× SSPE, 50% (w/v) formamide, 5× Denhardt's solution, 1% SDS, and salmon sperm DNA (ssDNA) at 100 μg/mL, all in DEPC-treated water. This solution can be prepared in advance for short-term storage at 4°C and be preheated at 42°C before use with the exception of the ssDNA. In some protocols, the prehybridization and hybridization solutions differ by one reagent such as the ssDNA (that is not included for hybridization), but we use the same solution with good outcome. The ssDNA is added just prior to use after boiling it for 10 min and placing on ice for a fast cool down. For hybridization, the ssDNA and the labeled probe can be boiled together for 5 min and placed on ice before addition to the solution. Use preferably enough probe to get $0.5–1 \times 10^6$ cpm/mL hybridization solution. Make sure that the plastic 15-mL tubes used can resist boiling and keep the screw cap loose.
13. Consumables: Nitrocellulose or nylon membrane such as GeneScreen Plus® (PerkinElmer, Waltham, MA, USA), chromatography paper, sealable bags, tips, tubes, gloves, and autoradiography film.
14. Equipment: Pipettors, horizontal electrophoresis unit and power supply, fume hood, UV cross-linker or oven, 60°C water bath, rotating platform, facility and equipment to work with $^{32}P$, and cassette with intensifying screen.

*2.3.2. Quantitative RT-PCR by TaqMan Chemistry*

1. RNA from different sources.
2. Kit for RT (see Note 4) containing various reagents: AMV reverse transcriptase, amplification buffer, DTT, dNTPmix, RNase, oligo$(dT)_{20}$ or random hexamers, DEPC-treated water.
3. Two unlabeled primers.
4. FAM dye-labeled TaqMan® MGB probe. Primers and probes can be designed using Primer Express 3.0 Software (Applied Biosystems, Foster City, CA, USA).
5. The cDNA of the target gene subcloned in a plasmid (e.g., pcDNA 3.1, pCR2.1, pCR4 (Invitrogen)) to create standards.
6. Reagents for qPCR: Mastermix reagent, such as TaqMan® Gene Expression Master Mix (Applied Biosystems). Store at 4°C. For the working solutions recommended by the manufacturer, prepare and keep aliquots at –20°C to avoid freezing/thawing.
7. Equipment: Pipettors, spectrophotometer, MicroAmp Optical 96-well Reaction Plate (Applied Biosystems), MicroAmp optical caps (Applied Biosystems), Applied Biosystems 7300 Real Time PCR System.

## 3. Methods

### 3.1. Cloning of Pig cDNA

*3.1.1. Total RNA Extraction*

1. Use 1 mL of TRIzol for every 100 mg of porcine tissue or $10^7$ cells and homogenize the sample. For solid tissues, these can be grinded with a mortar and pestle while kept frozen before addition of TRIzol or with a Polytron homogenizer after addition of the solution.
2. Optional step: Freeze the homogenized sample in TRIzol at –80°C to increase tissue or cell lysis and/or for storage.
3. Separation phase: Incubate homogenized sample for 5 min at room temperature and add 0.2 mL of chloroform per mL of TRIzol. Mix it vigorously and incubate for another 2 min. Centrifuge at 12,000 × *g* for 10 min at 4°C.
4. Precipitation phase: Transfer the aqueous (upper) phase to a new tube. Add 0.5 mL of isopropyl alcohol per mL of TRIzol used and let it incubate for 10 min at room temperature. Centrifuge at 12,000 × *g* for 10 min at 4°C.
5. Wash: Remove supernatant and wash RNA pellet with 75% ethanol (1 mL per mL of TRIzol used initially). Vortex to mix and centrifuge at 7,500 × *g* for 5 min at 4°C.
6. Redissolution phase: Dry RNA pellet but not to the point it could cause solubility problems (use air-dry or vacuum-dry,

but avoid centrifugation under vacuum). Dissolve RNA pellet in DEPC-treated water pipetting up and down and leave it at 4°C for several hours or overnight.

7. Calculate concentration of isolated RNA using a spectrophotometer.

*3.1.2. Standard RT-PCR*

1. In a sterile 0.5-mL tube, mix in the following order: 1 μL of primer Oligo$(dT)_{20}$ (50 μM) or random hexamers (50 ng/μL), *X*μL of RNA (5–10 μg), 2 μL of 10 mM dNTP mix, and up to 12 μL of DEPC-treated water.
2. In a different tube, prepare a mix in the following order: 4 μL of amplification buffer 5×, 1 μL of 0.1 M DTT, 20–40 units of RNase Inhibitor (40 units/μL), DEPC-treated water (up to 8 μL final volume), and 15 units of AMV reverse transcriptase (15–20 units/μL). This mixture can be prepared as a master mix by increasing the amounts proportionally to the number of samples to be processed (and a little more due to pipetting error).
3. Mix both solutions in the 0.5-mL tube for a final volume of 20 μL. Place the tube in a preheated thermal cycler. Use the following temperatures and incubation times to get the cDNA: one cycle at 25°C for 10 min (only if you are using random hexamer), one cycle at 50°C for 30–60 min, and one cycle at 85°C for 5 min (see Note 9).
4. Optional step: Before the RT incubation at 50°C, samples can be incubated at 65°C for 5 min and then immediately placed on ice for at least 1 min. This extra step allows the elimination of RNA secondary structures.
5. In order to amplify the target gene once the cDNA is obtained, mix in a sterile 0.5-mL tube the indicated reagents in the following order: 5 μL of amplification buffer 10×, 1.5 μL of 0.3 mM dNTP mix, 1.5 μL of 0.3 μM of each primer, 5 units of thermostable DNA polymerase, *X*μL of DEPC-treated water (up to 50 μL final volume), and 1–5 μL of cDNA (10–1,000 ng).
6. Place the tubes into a preheated thermal cycler and follow the indicated protocol of temperatures and incubation times. First, apply an initial denaturalization cycle at 94°C for 2 min. Then, repeat between 25 and 35 cycles the following program: 1 cycle of denaturalization at 94°C for 15 s, 1 cycle of annealing at *X*°C (see Note 10) for 30 s, and 1 cycle of extension at 72°C for 1 min. The last step is a final elongation cycle at 72°C for 5 min.
7. Mix half of the reaction with loading buffer and load into the agarose gel slots for electrophoresis for approximately 30–45 min at a constant voltage (up to 10 V/cm) (see Note 11). Use a size marker for comparison in one or two of the lanes.

8. Visualize with ultraviolet illumination and take a picture. Determine the size of the fragments amplified. If a PCR product with the expected size is amplified, the procedure should be followed by sub-cloning into a plasmid for sequencing and expression (see Notes 5 and 12).

### 3.2. Bioinformatic Analyses

We provide here the Internet address and a brief explanation of some of the most common tools for analysis in a molecular laboratory. We simply recommend to locate them and follow their instructions and help links.

1. *The Basic Local Alignment Search Tool (BLAST)* (http://blast.ncbi.nlm.nih.gov/Blast.cgi) allows to find regions of local similarity between your sequence of interest and sequences available at databases (homologues, etc.).
2. *T-Coffee* (http://www.ebi.ac.uk/Tools/t-coffee/index.html) is a multiple nucleotide or amino acidic sequence alignment program.
3. *Ex*pert *P*rotein *A*nalysis *Sy*stem (*ExPASy*) (http://au.expasy.org/). This server contains a variety of proteomic tools, such as:
   - *Translate*: Translates a nucleotide sequence into a protein sequence.
   - *Compute pI/Mw*: Calculates the theoretical isoelectric point (*pI*) and molecular weight (*Mw*) of a protein by providing the sequence.
   - *SignalP 3.0*: Predicts the cleavage sites of the signal peptide of a given sequence.
   - *SMART*: Allows the identification and annotation of genetically mobile domains and the analysis of domain architectures in a given sequence.
   - *TargetP 1.1*: Predicts the subcellular location based on the sequence information.
   - *NetNGlyc 1.0 and NetOGlyc 3.1*: Predict, respectively, the N- and O-glycosylation sites in human proteins.

### 3.3. Determination of mRNA Expression

#### 3.3.1. Analysis of RNA by Northern Blotting

1. Prepare master mix to add 38.75 μL for each sample: 5 μL of 10× MOPS running buffer, 8.75 μL of 37% formaldehyde, and 25 μL formamide. Add RNA sample (10–40 μg) and water to bring final volume to 50 μL. Mix thoroughly, spin briefly in a microcentrifuge, and incubate for 15 min at 55°C.
2. Add 10 μL of formaldehyde loading buffer to each sample and mix well. Spin briefly again and load in the gel slots following an established order. The gel has to be completely covered by the electrophoresis running buffer before loading.

3. Conduct the electrophoresis at a constant voltage of 5 V/cm until the bromophenol blue band has migrated halfway to 3/4 down the gel. Conduct this procedure in the fume hood.
4. Put the gel into a tray and rinse it with water several times to remove the formaldehyde.
5. As it is best to conduct the electrophoresis without ethidium bromide, stain the gel afterwards with 5 μg/mL ethidium bromide in water in a designated covered tray. Leave it rotating slowly for 20–30 min (see Note 13). Wash the gel again in water after the staining.
6. Take a picture of the gel on the ultraviolet illumination system with a ruler beside to locate the migration of the 28-s and the 18-s RNA. It can be appropriate to place the gel on a plastic wrap for this step.
7. Optional step: Place the gel in a tray with sufficient 10× SSC to cover well the gel and rotate slowly for 45 min.
8. Cut the membrane the same size of the gel and label one corner in the absorbent side to control the orientation. Soak it shortly in water until fully hydrated and then in 10× SSC for 10–15 min. In the meantime, soak in 10× SSC the chromatography paper in preparation for the next step.
9. Set up an overnight capillary blot using 10× SSC as transfer buffer (see Note 14): To this end, place a support inside the tray, fill the tray with 10× SSC, and place a wick of chromatography paper pre-wet in 10× SSC. Make sure that the wick reaches well the transfer buffer and is larger than the gel placed on top. Remove bubbles of all elements placed on the capillary set (a plastic 10-mL pipette can be rolled to attain this goal). Place the membrane on top of the gel with absorbent face in touch with it (note that this will produce a mirror image). Place on top a stack of chromatography paper (3–5 sheets slightly smaller than the membrane) pre-wet in 10× SSC. Next, put a 3-cm-thick layer of dry paper towels that do not touch the membrane or the gel and a flat weight that gives stability. Keep the transfer from overnight to 24 h and then recover the membrane. It is possible to check the gel, and even the membrane, on the UV illumination system to assess the efficiency of the transfer.
10. If a UV cross-linker is available, cross-link the RNA to the wet membrane (placed on a dry thick filter paper) with ultraviolet light for 1 min following the instructions of the manufacturer. If not, bake the membrane for 2 h at 80°C in an oven.
11. Place the membrane in a sealable plastic bag of good quality and seal the sides close to the membrane to reduce the amount of solution needed. Leave some space on the top to facilitate subsequent manipulation.

12. Add 8–12 mL of the prehybridization solution (sufficient to cover the membrane), remove all bubbles, and seal the bag leaving ~3 cm above the membrane. Prehybridize for at least 1 h (it can be left overnight) in a water bath at 42°C. We usually place the bag in a covered container with pre-warmed water in a slow-rotating water bath.
13. Cut open a corner of the bag and remove the prehybridization solution. Add the hybridization solution with the probe (see Note 15), remove all bubbles very carefully, and hybridize overnight at the same temperature as prehybridization. We recommend to use double bag to avoid possible leaks and contamination of the container or water bath.
14. After hybridization, the solution can be transferred to a tube to be kept frozen at −20°C and even reused shortly after (if needed). If not, it is left frozen to decay before discarding it.
15. Proceed to wash the membrane. Start with a low-stringency wash in 1× SSC 0.1% SDS for 10 min at room temperature and repeat the process. Discard the liquid in designated containers. Follow with two washes in 0.2× SSC 0.1% SDS for 10 min each at 65°C. Monitor for background reactivity on the blot after each wash with the Geiger counter and stop the washing once it is low. The stringency can be modified as needed by changing the amount of salt and temperature (it will increase by reducing the SSC and/or increasing temperature).
16. After the last wash, remove the excess liquid from the membrane and cover in plastic wrap. Expose the membrane to an X-ray film at −70°C in a cassette with an intensifying screen for one to several days. The films can be replaced to find the right signal intensity.

#### *3.3.2. Quantitative RT-PCR by TaqMan Chemistry*

1. Quantify the concentration of your template in a spectrophotometer. The ratio of absorbances $A_{260}/A_{280}$ can give you an assessment of purity and should be above 1.8 for DNA or 2 for RNA.
2. Prepare the cDNA by RT as described in Subheading 3.1.2.
3. Prepare master mix for each target gene (calculate to run triplicates for samples, standards, and a negative control).
4. Set up a serial dilution for standards. Prepare five different amounts of the standard DNA within a range of 10 and $10^5$ copies (see Note 16).
5. Add the reagents in the following order for a final volume of 25 μL: 12.5 μL TaqMan Universal PCR Master Mix No AmpErase UNG (AB), 50 nM of each primer (see Note 17), 200 nM TaqMan® probe (AB), 1 μg cDNA, and up to 25 μL of molecular-grade water.

6. Place it on a 96-well plate, cover it with optical caps, and spin it down to avoid bubbles. Place the plate in the Applied Biosystems 7300 Real Time PCR System. Follow the cycling conditions for TaqMan® Gene expression Master mix: start with one cycle at 50°C for 2 min, followed by one cycle at 95°C for 10 min. Then, repeat 40 cycles that include 1 cycle at 95°C for 15 s and 1 cycle at 60°C for 1 min.
7. Check that the standard curves obtained are reliable (i.e., $R^2 > 0.9$ and slope −3.2) and then use them to calculate the copy number of your target gene in each sample.

## 4. Notes

1. Isolated tissues are wrapped with aluminum foil labeled directly with marker for identification, snap frozen in liquid nitrogen, and stored at −80°C until use.
2. Water is treated with 0.1% DEPC to inactivate the RNase enzymes in the water. For preparation, add 1 mL of DEPC into 1 L of deionized water and stir overnight under the fume hood. Autoclave the solution afterwards to remove any traces of DEPC. Buy DEPC in small aliquots as it can be used only for 1 month after opening.
3. The Trizol reagent is a mono-phasic solution of phenol and guanidine isothiocyanate. Trizol is one of multiple commercially available reagents developed to extract RNA from cells and tissues.
4. There are various options for carrying out RT. The reverse transcriptase can be, for instance, from AMV or from Moloney Murine Leukemia Virus (MMLV). Both can be obtained from various suppliers. There are also many commercially available kits, such as the Cloned AMV First-Strand cDNA Synthesis Kit (Invitrogen), that facilitate the task as they include all the reagents needed for the RT. There are now improved versions that we have not tested, but are probably worth trying if the amplification fails.
5. There are also multiple options for conducting a PCR. The choice of the thermostable DNA polymerase is especially important as it can be a Taq polymerase or a high-fidelity DNA polymerase, which possesses 3′–5′ exonuclease proofreading activity. A high-fidelity polymerase is recommended (such as the Platinum Pfx (Invitrogen)) when the RT-PCR is intended for cDNA cloning. However, sometimes, the Taq polymerase is more effective in producing amplification. We like the GoTaq® Green Master Mix (Promega, Madison, WI, USA),

**Table 1**
**Letter code used for the generation of degenerate primers**

| Letter code | Mixed bases |
|---|---|
| B | T+C+G |
| H | A+T+C |
| W | A+T |
| S | C+G |
| K | T+G |
| M | A+C |
| Y | C+T |
| R | A+G |

Each letter corresponds to a combination of two or three different bases

which comes as a premixed ready-to-use solution with the appropriate reaction buffer, dNTPmix, and $MgCl_2$. Note that the subsequent sub-cloning system will vary depending on the type of polymerase. We use the pcDNA3.1 directional TOPO expression kit (Invitrogen) when using the high-fidelity polymerase and a TOPO TA cloning kit (Invitrogen) when using a Taq polymerase.

6. Degenerate primers for the amplification of a cDNA of interest are designed based on known sequences of other species. Thus, the sequences of mammalian homologues are aligned and the primers are designed to be located at the ends of the coding sequence and contain all the conserved nucleotides as well as the combinations that cover all the variations observed (Table 1). For cloning a porcine cDNA, use for instance bovine, human, and murine sequences. An appropriate primer size would be 20–22 bp. We usually try to have a conserved nucleotide at the primer end and preferably a C or a G to add stability.
7. Note that the source of cDNA is critical for successful amplification of the cDNA of interest. Thus, it is highly recommended that the specific mRNA is well represented in the chosen cell/tissue. This can be estimated from information available/published for human genes. It is a good idea to start with a battery of tissues as some may work better than others.
8. The percentage of agarose that is appropriate varies depending on the size of the expected PCR product/s, but is commonly

used between 1 and 2%. A 1% agarose gel is appropriate for fragments between 1 and 2 kb, whereas a higher agarose percentage such as 2% will separate better the smaller fragments (e.g., 300–500 bp).

9. Note that the 50°C used for the RT reaction is only appropriate for the AMV reverse transcriptase, as the MMLV reverse transcriptase works at 37°C. In addition, we regularly use oligo(dT)$_{20}$, not the random hexamers, and have good success for the amplification of pig cDNA.

10. The annealing temperature depends on the primer sequence. We estimate the annealing temperature with the following formula: $T(°C) = (4 \times (G + C)) + (2 \times (A + T))$. In the case of degenerate primers, we determine the range of potential temperatures. We usually apply an annealing temperature between 50 and 65°C that corresponds to 5–10° below the calculated annealing temperature. If it fails, reduce the annealing temperature as it will increase the chances of amplification. However, it also increases the chances to obtain undesired DNA products.

11. Note that during the electrophoresis the ethidium bromide migrates in the opposite direction to DNA. Thus, it is convenient not to lose the staining in the area where the PCR product is expected to migrate. This is especially important for small fragments or poor amplifications that produce weak signal. Visualize the gel immediately after electrophoresis and excise the desired fragment if present, as the DNA diffuses and the signal can be lost. For this reason, we do not recommend to stain the gels after electrophoresis.

12. We recommend to use a topoisomerase-based system for subcloning the PCR product as mentioned in Note 5. To this end, we usually purify the PCR fragment directly from the electrophoresis gel using a kit, such as GENECLEAN® II (MP Biomedicals, Santa Ana, CA, USA). Minimally, the vector chosen for sub-cloning should allow sequencing of the insert with common primers, such as M13 forward and/or reverse. In addition, some of the vectors will be adequate to be used for recombinant expression of the insert.

13. As the ethidium bromide is highly toxic and the staining of the gel can generate a lot of toxic residual liquid to dispose, it is possible to add 1 μL of the ethidium bromide stock solution to each sample before loading to the gel. This can be sufficient to visualize the RNA after electrophoresis.

14. The capillary blotting system works well, but it is very slow. If available, we recommend to use a DNA blotter that works with either pressure or vacuum.

15. Note that the handling of the radioactive material has to comply with the corresponding regulations set by the local and

national authority. This includes the use of shields and double gloves, conducting of appropriate disposal of solid and liquid waste, etc.

16. Prepare the standard series in a different room than the rest to avoid contamination of samples and controls with the plasmid encoding the target gene. This approach can reduce false-positive results.
17. We recommend to conduct a test to find the most efficient concentration of primers (e.g., 50, 300, and 900 nM) before carrying out the first set of reactions.

## Acknowledgments

This work was mainly supported by Ministerio de Educación y Ciencia (SAF2005-00472, SAF2008-00499), an aid for emerging groups from Generalitat de Catalunya (2005SGR00897), and a Marie Curie action from the European Commission (MIRG-CT-2005-021293), all to C.C. M.U.-H. was supported by an IDIBELL fellowship.

## References

1. Phelps CJ, Koike C, Vaught TD et al (2003) Production of alpha 1,3-galactosyltransferase-deficient pigs. Science 299:411–414
2. Ramsoondar J, Machaty Z, Costa C, Williams BL, Fodor WL, Bondioli KR (2003) Production of α1,3galactosyltransferase-knockout cloned pigs expressing human α1,2fucosyltransferase. Biol Reprod 69:437–445
3. Archibald AL, Bolund L, Churcher C et al (2010) Pig genome sequence—analysis and publication strategy. BMC Genomics 11: 438–442
4. Fan B, Gorbach DM, Rothschild MF (2011) The pig genome project has plenty to squeal about. Cytogenet Genome Res 134:9–18
5. Uribe-Herranz M, Casinghino SR, Bosch-Presegué L, Fodor WL, Costa C (2011) Identification of soluble and membrane-bound isoforms of porcine tumor necrosis factor receptor 2. Xenotransplantation 18:131–146
6. Mankertz J, Buhk HJ, Blaess G, Mankertz A (1998) Transcription analysis of porcine circovirus (PCV). Virus Genes 16:267–276
7. Bruel T, Guibon R, Melo S et al (2010) Epithelial induction of porcine suppressor of cytokine signaling 2 (SOCS2) gene expression in response to Entamoeba histolytica. Dev Comp Immunol 34:562–571
8. Nygard A-B, Jørgensen CB, Cirera S, Fredholm M (2010) Investigation of tissue-specific human orthologous alternative splice events in pig. Anim Biotechnol 21:203–216
9. Hagen JB (2000) The origins of bioinformatics. Nat Rev Genet 1:231–236
10. Doolittle RF (2010) The roots of bioinformatics in protein evolution. PLoS Comput Biol 6:e1000875
11. Jung R, Soondrum K, Neumaier M (2000) Quantitative PCR. Clin Chem Lab Med 38: 833–8366

# Chapter 3

# Basic Analyses of Proteins of Interest for Xenotransplantation

**Mireia Uribe-Herranz and Cristina Costa**

## Abstract

Proteins are the focus of numerous xenotransplantation studies because they provide structure and function to the graft. Their presence, absence, or even a functional incompatibility among species can compromise the long-term functioning of the xenograft. In particular, many cell-surface and soluble proteins, such as cytokines and chemokines, are involved in triggering rejection. For this reason, the identification and characterization of key proteins for xenografting, of either pig or human origin, are very important. Understanding their role in the xenogeneic setting can set the bases for the development of genetic engineering approaches that prolong graft survival and ensure function. There are multiple ways of determining and attaining protein expression, as well as studying protein interactions. In this chapter, we describe some basic techniques that allow us to detect and characterize pig and human proteins in order to better understand the molecular bases of rejection and function of pig xenografts.

**Key words:** Pig, Human, Protein, Antibody, Fusion protein, Immunofluorescence, Flow cytometry, Surface plasmon resonance

## 1. Introduction

The major hurdles in xenotransplantation are caused by the lack or presence of specific proteins or sugars in the xenogeneic donor cell or tissue (1). Proteins play a major role in mediating the interactions of host and donor cells, thus participating in the rejection process of xenografts. They are also responsible for providing structure and function to the graft. Cell-surface proteins are the focus of our research as many are involved in triggering rejection, such as the cellular responses that are relevant in rejection of cell-based xenografts (2). Moreover, soluble proteins, such as cytokines and chemokines of either pig or human origin, also participate in

Cristina Costa and Rafael Máñez (eds.), *Xenotransplantation: Methods and Protocols*, Methods in Molecular Biology, vol. 885, DOI 10.1007/978-1-61779-845-0_3, 

immune rejection (3). Antibodies and proteins of the complement and coagulation systems play a particular prominent role in this process and will receive especial attention in other chapters. Nevertheless, the techniques explained here can be applied to most proteins. In general, understanding their role in the xenograft can set the bases for the development of genetic engineering approaches that prolong graft survival and ensure function. In fact, many genetic modifications are based on the expression of proteins, such as human complement regulatory proteins. Thus, the identification of key proteins as potential targets for intervention is important for the advancement of the xenotransplantation field.

There are multiple techniques to determine protein expression, many of which rely on the use of specific antibodies. However, there are few commercially available antibodies developed to detect porcine proteins. For this reason, we have organized this chapter to describe the basic procedures to be followed when there are no readily available reagents for detection. Thus, the next step after cloning the corresponding cDNA is to confirm its translation into protein. To this end, there are several useful methods to introduce DNA into cells and express proteins of various types (e.g., soluble or membrane bound, with native sequence or engineered as fusion protein). Engineering the protein to include a tag that facilitates its detection can be especially useful, although it has to be carefully designed to prevent loss of function. In particular, we describe some protocols for the expression and purification of proteins that we find easy to perform and set up. Transfection of an established mammalian cell line should be the first choice for expression testing and specific detection, whereas engineering a primary cultured cell (by retroviral transduction, for instance) should be used for more stringent expression testing and functional assays. In our laboratory, we use retroviral transduction to express proteins of interest in porcine aortic endothelial cells (PAEC) and primary cultured pig cells of potential use for xenotransplantation, such as chondrocytes. Successful expression at this stage can set the bases for the development of transgenic approaches.

Once the protein is expressed on any cell, the system can be used to test available reagents for its detection. A mock-modified cell (with the empty expression vector) is used as a negative control. Similarly, a soluble protein purified thanks to the tag can be utilized for finding or producing specific antibodies. When looking for antibodies to detect a certain pig protein, it is worth testing for cross-reactivity with polyclonal antibodies developed against the human homologue. Monoclonal antibodies can also be assessed for this purpose, although the chances of finding cross-reactivity with the pig protein are lower. When specific antibodies are not available, there are many laboratories that provide services for the production of various types of antibodies (anti-peptide, monoclonal, polyclonal). However, one of the best options is to obtain your own reagent. Therefore, we describe the protocol of producing antisera in rabbits using a purified recombinant protein. We consider that this step is justified when the commercial antibodies have failed to detect the protein of interest and especially appropriate if there are plans of producing the protein recombinantly for functional studies. Moreover, the antisera can be used for assessing expression of the native protein in cells and tissues. If needed, the antisera can be further purified following one of the various methods available for immunoglobulin purification (e.g., with protein A-sepharose). Of course, the techniques we describe here can also be applied to human and murine proteins that may be of interest for xenotransplantation studies. In this case, there are multiple commercial sources that sell specific antibodies for most proteins of these species.

Specific antibodies are the tool of choice for protein detection and can be used for multiple techniques, such as immunoblotting, ELISA, flow cytometry, or immunoprecipitation. The Western blot (immunoblotting after sodium dodecyl sulfate-polyacrylamide gel electrophoresis, SDS-PAGE) and the ELISA are the techniques of choice for the specific detection of soluble proteins. The first is recommended to assess size and integrity and the second for quantification. However, in this chapter, we are going to focus on flow cytometry and immunofluorescence for microscopy because these techniques are key for detecting, quantifying, and localizing proteins in cells and tissues. In particular, we use flow cytometry extensively for detecting cell-surface proteins. The FACS analysis will allow determining the proportion of expressing cells and the intensity of the fluorescence will provide a measure of quantity (semiquantification). The immunofluorescence we describe here is particularly destined to determine the subcellular localization of proteins with a confocal microscope. However, the immunofluorescence can also be useful for detecting protein expression when a flow cytometer is not available and studying the protein distribution in an organ of interest for xenotransplantation when applied to slices of frozen tissue. For more elaborate studies of tissue distribution, immunohistochemistry with paraffin-embedded tissues may be a better tool.

Another point of interest in xenotransplantation is the deficient interaction between molecules of different species. An example in the pig-to-human setting is that SLA I fails to provide inhibitory signaling through the human killer inhibitory receptors on NK cells (4). For this reason, techniques such as surface plasmon resonance (SPR) and calorimetry can provide useful information on kinetics, affinity, stoichiometry, and even thermodynamics of the molecular interactions (5, 6). Thus, we describe the bases and a general protocol for SPR that can be used to determine the binding kinetics of pig and human proteins. Nevertheless, we strongly recommend seeking advice from experts when designing the first

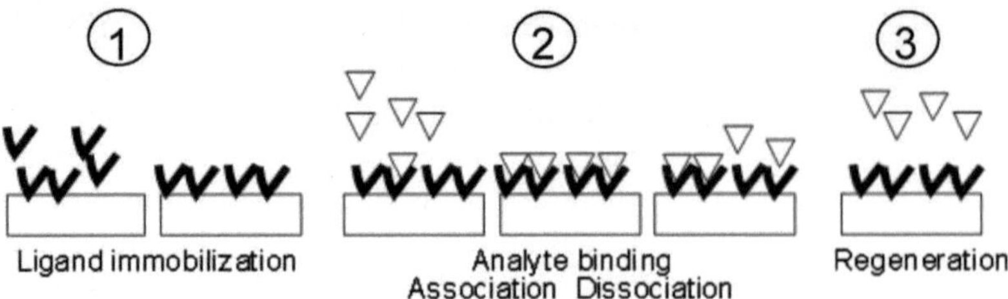


Fig. 1. Scheme of SPR process. (1) Ligand immobilization. (2) Analyte binding which involves association and dissociation to ligand. (3) Regeneration of the chip surface.

SPR assay. The basic procedure involves immobilization of the ligand on the chip, binding of the analyte under flow, and regeneration of the chip surface (Fig. 1). As a readout, the equipment captures changes in refractive index that result from increases in mass at the sensor surface and calculates the affinity and kinetics of the interaction. All this information can be subsequently used for understanding the xenografting process and design new therapeutic strategies.

## 2. Materials

### *2.1. Recombinant Protein Expression and Purification*

#### *2.1.1. Transfection*

1. Cell line, the choice depends on the final objective (see Note 1).
2. Culture media appropriate for the chosen cell line, such as Dulbecco's modified Eagle's medium (DMEM) containing high glucose supplemented with 10% heat-inactivated fetal bovine serum (FBS), 200 U/mL penicillin, and 200 μg/mL streptomycin (see Note 1).
3. Purified DNA of expression vector alone and containing the gene of interest. There are many expression vectors, most available commercially or at molecular biology laboratories. Minimally, they contain a promoter such as the cytomegalovirus (CMV) promoter and a polyadenylation signal.
4. Polyethylenimine (PEI) at 1 mg/mL.
5. Tissue culture plastic: T25 flasks, pipettes, sterile tips.
6. Equipment: Pipettors, tissue culture cabin, 37°C water bath.

#### *2.1.2. Transduction (Retroviral Infection)*

1. Cell line to be infected, such as PAEC.
2. Retroviral plasmid containing the gene of interest and also alone to be used for mock control (e.g., pLXSN) (see Note 2).
3. Packaging cell line (e.g., PG13, PA317) (see Note 2).
4. Polybrene (hexadimethrine bromide) at 8 mg/mL diluted in water and sterilized through a 0.22-μm filter. Keep the solution at 4°C for short-term use.

5. Culture media appropriate for the chosen cell lines. For PAEC, use DMEM/10% FBS with 200 U/mL penicillin and 200 μg/mL streptomycin. We supplement it with a hypothalamus extract, such as Endothelial mitogen (50 mg/L, Biomedical Technologies, Stoughton, MA, USA).
6. Bleach at 50% for cleaning and decontaminating all materials after contact with virus (see Note 2). The liquid waste should be mixed with it to reach 10% bleach for decontamination.
7. Tissue culture plastic: T 25 flask, tips, pipettes.
8. Equipment: Pipettors, tissue culture cabin, 37°C water bath.

#### *2.1.3. Production and Purification of Fusion Proteins in Mammalian Cells*

1. All materials described in Subheading 2.1.1, including an expression vector containing the gene of interest fused to a GST tag (see Note 3).
2. Appropriate matrix to purify the fusion proteins by affinity (see Note 3). For capturing the GST tag, the Bulk glutathione sepharose 4B (GE Healthcare, Waukesha, WI, USA) can be used directly in tubes or packed in columns.
3. Elution buffer for GST: 10 mM glutathione. This mild buffer should suffice for successful elution.
4. Phosphate-buffered saline (PBS).
5. Disposable sterile tubes to be used for bulk purification. As mentioned, there is the alternative to use columns if preferred.
6. Buffer exchange devices, such as protein desalting spin columns (Pierce, Rockford, IL, USA). This is needed for all procedures that require that the protein of interest is resuspended in a specific buffer, such as injection into animals or SPR.

### 2.2. Polyclonal Antibody Production

1. Two rabbits (New Zealand, Male).
2. Complete Freund's adjuvant (CFA).
3. Incomplete Freund's adjuvant (IFA).
4. 300 μg of recombinant purified protein for each administration (see Note 4).
5. Insulin syringes with 21-G needles.
6. Animal facility and equipment approved for housing rabbits.
7. This technique requires specific written approval from the local animal care committee.

### 2.3. Determining Protein Expression

#### *2.3.1. Flow Cytometry*

1. Cells.
2. Solution of trypsin (0.05%) and ethylenediaminetetraacetic acid (EDTA) (1 mM) for cell harvest if cells are adherent. Cell dissociation buffer may be preferable to trypsin when detecting proteins that are particularly sensitive to the trypsin treatment.

3. PBS.
4. Heat-inactivated FBS or bovine serum albumin (BSA) to be used as blocking agent at 1% in PBS for antibody dilutions and incubations. In general, FBS works better to block unspecific reactivity, but BSA can also be utilized for this purpose.
5. Primary antibody.
6. Secondary antibody conjugated to a fluorochrome if the primary antibody is not conjugated.
7. Plastic: 96-well plate (V-bottom), 15-mL sterile tubes, tips, FACS tubes.
8. Equipment: Tissue culture basic equipment and FACS/flow cytometer.

#### 2.3.2. Immunofluorescence

1. PBS.
2. Paraformaldehyde, 4% in cold PBS.
3. PBS/20% FBS with or without Triton X-100 0.2% (see Note 5).
4. PBS/1% FBS with or without Triton X-100 0.2% (see Note 5).
5. Primary antibody.
6. Secondary antibody conjugated to a fluorochrome if the primary antibody is not conjugated.
7. Optional: Tracker/s for subcellular localization studies (e.g., DAPI, phalloidin, ER trackers).
8. Mounting solution (e.g., Mowiol 4-88 (Merk4Biosciences, Gibbstown, NJ, USA)).
9. Tissue culture plastic: 6-well plates and pipettes.
10. Sterile 12-mm coverslips, microscope glass slides, and tweezers.
11. Equipment: Tissue culture equipment, horizontal shaker, and epifluorescence microscope for basic immunofluorescence analyses. For subcellular studies, a confocal microscope is needed.

### 2.4. Surface Plasmon Resonance

1. Ligand (protein) (see Note 6).
2. Analyte (protein) (see Note 6).
3. HBS-N buffer: 10 mM HEPES, pH 7.4, 150 mM NaCl (see Note 7).
4. HBS-EP buffer: 10 mM HEPES, pH 7.4, 150 mM NaCl, 3 mM EDTA, and 0.005% vol/vol polysorbate 20 (Surfactant P20, GE Healthcare) (see Note 7).
5. *N*-ethyl-*N'*-(dimethyl-aminopropyl)-carbodiimide-hydrochloride (EDC) at 100 mM.
6. *N*-hydroxysuccinimide (NHS) at 400 mM. Store at –20°C.

7. Ethanolamine 1 M. Store at 4°C.
8. Filters of 0.22 μm.
9. Sterile Erlenmeyer flasks or glass bottles with connection for degassing with vacuum system.
10. Sensor surface Biacore chip (e.g., CM5) and Biacore tubes (GE Healthcare).
11. Equipment: Biacore system (e.g., T100) (GE Healthcare).

## 3. Methods

### *3.1. Recombinant Protein Expression and Purification*

#### *3.1.1. Transfection*

1. For transfection of adherent cells, these should be between 50 and 90% confluence (depending on the cell line) and maintained at 37°C in a 5% $CO_2$ atmosphere.
2. In a sterile 0.5-mL tube, put 48 μL of PEI (1 mg/mL) followed by 100 μL of media without antibiotic and mix (see Note 8). Incubate the solution for 1 min at room temperature.
3. Add 8 μg of DNA and mix gently. Incubate for another 5 min at room temperature.
4. Add this mixture to the flask containing the cells and 3 mL of medium without antibiotic. Incubate it at 37°C in 5% $CO_2$ during 6 h.
5. Remove transfection media and add fresh media (5 mL) for subsequent culture.
6. Cells should be ready for analyses and studies 48 h after the transfection. To obtain stably transfected cells, add selection media at this time point (48 h) (see Note 9).

#### *3.1.2. Transduction (Retrovirus Infection)*

1. Transfect the packaging cell line (as described in Subheading 3.1.1) with a retroviral plasmid (e.g., pLXSN) containing the gene of interest. In parallel, follow the same procedure with the empty vector to be used as mock control.
2. After 48 h, harvest the culture supernatant (containing the viral particles) and filter it through a 0.45-μM filter (see Note 2). Note that the viral supernatant can be stored at −80°C for future use. To obtain a stably transfected virus-producing cell line, add selection media and culture for the necessary amount of time (see Note 9).
3. For transduction, adherent cells should be at 40–50% confluence as the retrovirus will infect dividing cells.
4. In a sterile 15-mL tube, mix 2 mL of culture medium, 1 mL of viral supernatant, and 48 μL polybrene (1 mg/mL) in the indicated order. The polybrene increases the infection by murine and avian subgroup E retroviruses.

5. Replace the culture medium with this mixture in the T25 flask containing the cells to be infected and incubate for 6–8 h at 37°C in 5% $CO_2$.
6. Remove transduction media and add fresh media (5 mL).
7. To generate stable transductants, add selection media after culturing for 48 h (see Note 9). If not considered necessary, proceed to assess expression at this time point.

*3.1.3. Production and Purification of Fusion Proteins in Mammalian Cells*

1. Transfect the chosen cell line with the appropriate expression vector containing the cDNA of the fusion protein following the method described in Subheading 3.1.1.
2. Harvest the culture supernatant if the recombinant fusion protein is secreted (contains signal peptide) or lyse the cells following standard methods if the protein is intracellular.
3. Prepare the affinity matrix for bulk method following the instructions of the manufacturer. The procedure depends on the matrix chosen.
4. Add the culture supernatant or cell lysate to the tube containing the prepared resin and incubate with agitation for 30 min at room temperature.
5. Centrifuge at 500 × *g* for 5 min and wash the matrix three times with PBS by aspirating the supernatant and resuspending in PBS after centrifugation (see Note 10).
6. Incubate the matrix with the appropriate elution buffer (glutathione 10 mM for GST fusion proteins) at room temperature with agitation during 10 min.
7. Centrifuge and keep the supernatant.
8. Analyze the protein yield following standard chromogenic methods of quantification, such as Bradford, or by measuring the absorbance at 280 nm. Purity can be assessed by SDS-PAGE, gel staining, and western blotting.
9. Use an exchange buffer device, such as protein desalting spin columns, following the instructions of the manufacturer when it is necessary to change the buffer (e.g., into PBS for injection in animals or specific buffer for SPR).

### 3.2. Polyclonal Antibody Production

1. First week: Blood extraction from the marginal ear vein (3–5 mL) to get pre-immune serum. Do not use any anticlotting agent as blood should be allowed to clot at room temperature for at least 30 min and serum collected after centrifugation (2,200 × *g* for 10 min at 4°C). Store serum at −80°C, preferably in aliquots.
2. Second week: Subcutaneous injection of 300 μg of purified recombinant protein mixed with CFA as an emulsion (volume 1:1). Between 1 and 3 mL of final volume is appropriate for injection. We prepare the emulsion by simply passing the

mixture of protein and adjuvant through a 21-G needle multiple times until it is homogeneous.

3. Fourth week: Subcutaneous injection of 300 μg of purified recombinant protein mixed with IFA as an emulsion (volume 1:1).
4. Fifth week: Blood extraction from the marginal ear vein (3–5 mL) to get serum as described. Testing of the antisera can be initiated at this time point or be done with all serum samples at the end. We recommend to test each sample separately and with multiple dilutions (titration) by flow cytometry using a target cell known to express the protein of interest (native or recombinant expression).
5. Sixth week: Subcutaneous injection of 300 μg of purified recombinant protein mixed with IFA as described (volume 1:1).
6. Seventh week: Blood extraction from the marginal ear vein (3–5 mL) to get serum as described.
7. Eighth week: Subcutaneous injection of 300 μg of purified recombinant protein mixed with IFA as described (volume 1:1).
8. Ninth week: Blood extraction by terminal cardiocentesis on deeply anesthetized animals. Get as much blood as possible and do not allow the animal to recover from anesthesia. Collect serum as described.

### 3.3. Determining Protein Expression

#### 3.3.1. Flow Cytometry

1. Harvest cells using trypsin or cell dissociation buffer (if they are adherent) and centrifuge at 260 × *g* for 5 min.
2. Resuspend with PBS/1% FBS and centrifuge at 260 × *g* for 5 min.
3. Aspirate the supernatant and repeat washing step.
4. Load the cells into a 96-V-bottom-well plate (0.5–1 × $10^6$ cells/well) and centrifuge at 260 × *g* for 5 min.
5. Aspirate the supernatant, always being careful not to touch the cellular pellet.
6. Resuspend the cells with the specific primary antibody in a final volume of 100 μL/well and incubate during 30 min at 4°C. An appropriate antibody concentration is 10 μg/mL diluted in PBS/1% FBS.
7. Wash the cells by adding 100 μL of PBS/1% FBS and centrifuge at 260 × *g* for 5 min. Aspirate the supernatant.
8. Resuspend and incubate with the secondary conjugated antibody (final volume 100 μL) during 30 min at 4°C (see Note 11).
9. Wash the cells twice by adding 100 μL of PBS (without FBS or BSA) and centrifuge at 260 × *g* for 5 min. Aspirate the supernatant.

10. Resuspend the cells with 100 μL of PBS and transfer to a FACS tube. Keep the cells in the dark at 4°C until use.
11. Acquire and analyze the data generated with a flow cytometer.

#### 3.3.2. Immunofluorescence

1. Lay three sterile coverslips at the bottom of each well (triplicate) in a 6-well plate and seed cells in an appropriate number to reach 70–80% confluence for the procedure.
2. Optional: The cells can be transfected, once they are seeded and growing, following the procedure described (Subheading 3.1.1). In this case, perform the immunofluorescence staining 24–72 h after transfection.
3. Wash the cells with PBS for 5–10 min placing the plate on a horizontal shaker at a slow rotation rate. Aspirate the PBS at the end.
4. Add 1 mL of 4% paraformaldehyde in cold PBS and keep rotating for 20–30 min.
5. Wash the cells three times with PBS for 5–10 min using the shaker at low rpm. Aspirate the PBS each time.
6. Add 1 mL of PBS/20% FBS/0.2% Triton X-100 (see Note 5) and keep rotating at low rpm for 1 h at room temperature.
7. Lay down a piece of parafilm flat on a smooth surface and put on the parafilm sheet one drop (20 μL) of primary antibody diluted in PBS/1% FBS/0.2% Triton X-100 (see Note 5) for each staining to be conducted.
8. Then, with the help of thin tweezers, place each coverslip on the desired drop with the cells facing the solution and incubate for 2 h at room temperature. Figure 2 shows a scheme of how it is set up.
9. Place the coverslips again in the plate wells (without mixing conditions) and wash three times with PBS for 10 min using the shaker at low rpm.
10. Place one drop (20 μL) of secondary conjugated antibody (diluted in PBS/1% FBS/0.2% Triton X-100 (see Note 5)) on the parafilm sheet and put on top the coverslips as described

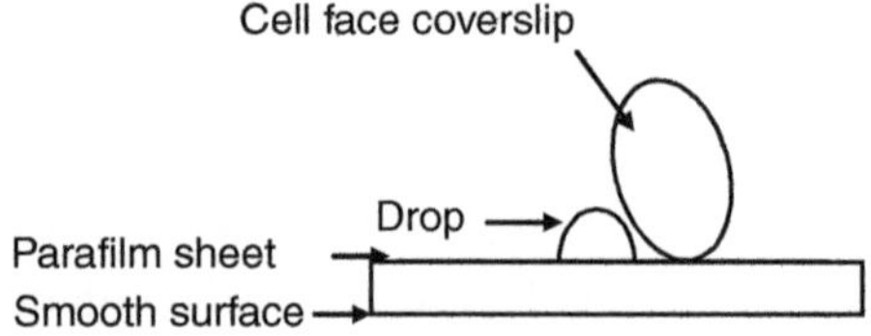


Fig. 2. Scheme of methodology. A piece of parafilm is placed on a smooth surface with one drop of primary antibody or tracker on the parafilm sheet. Then, with the help of thin tweezers, the coverslip is put on top with the side containing the cells facing the drop.

before. From this step to the end, avoid direct light exposure to keep the fluorescence. Remember to include cells only stained with secondary antibody as control of background.

11. Incubate for 30–45 min at room temperature with the conjugated secondary antibody.
12. Place the coverslips again in the wells and wash three times with PBS for 10 min in a shaker at low rpm.
13. Optional: At this point, it is possible to stain with a tracker to localize the protein followed by washes. The exact protocol depends on the tracker chosen.
14. Put 5–7 μL of mounting solution on a microscope glass slide and place on top the coverslip with the stained cells inside. Leave it dry overnight at room temperature in the dark.
15. Observe the cells under an epifluorescence microscope. For subcellular localizations, use a confocal microscope.

### 3.4. Surface Plasmon Resonance

1. First, it is important to conduct a pH scouting to find experimentally the best pH to bind the ligand to the sensor surface. To this end, use one channel of the sensor chip to test three to four conditions for binding (the surface can be regenerated between tests) and follow the resonance units (RU) as a guide. As the pH appropriate for binding is commonly close to the ligand isoelectric point (pI), it is recommended to assess at least one unit above and one below the pI (e.g., use 6, 7, and 8 for a pI of 7).
2. Activate the sensor chip with NHS and EDC following the instructions of the manufacturer.
3. Immobilize the ligand (protein) on the sensor chip using HBS-EP (see Note 7) as the running buffer. A ligand concentration 0.5 μM should be sufficient for a successful immobilization. Remember to dilute the ligand with the appropriate buffer (chosen at the pH scouting step).
4. Leave one channel of the chip empty or loaded with the tag protein alone (e.g., GST) in order to subtract the background.
5. Once you have immobilized the ligand, pass the ethanolamine solution to block the surface of the sensor chip.
6. Dilute the analyte sample with the running buffer (usually HBS-N, but HBS-EP may work better for some proteins) (see Note 7). Set up minimally five different concentrations and include a control with only buffer. Pass them randomly to ensure that there is no additive effect.
7. Data analysis is performed using the Biacore software provided by the manufacturer. To obtain conclusive results, repeat the experiment two more times.

## 4. Notes

1. Multiple established cell lines can be used for expression of pig and human proteins. It may be interesting to test a couple of them simultaneously as it is common to find major differences in expression levels between cell lines. We use extensively Chinese hamster ovary (CHO) and human embryonic kidney 293-EBNA cells. The CHO cells are good for recombinant expression and have normal culture requirements (DMEM/10% FBS). The 293-EBNA cells are cultured as well with DMEM/10% FBS, but it is necessary to add 250 μg/mL G418 to maintain the EBNA-1 system that increases protein expression. This system will only be advantageous when using plasmids with OriP, such as pCEP4 (Invitrogen, Carlsbad, CA, USA).
2. The commonly used retroviral vectors, such as pLXSN, need a packaging cell line containing *gag*, *pol*, and *env* genes to produce infectious particles. The generated viruses are non-replicative and their handling only requires the use of level II safety protocols and equipment.
3. There are several systems available for purification of tagged proteins by affinity chromatography that involve, for instance, pre-packed ready-to-use columns. We explain the procedure for the GST tag because we find it especially efficient compared to others. Nevertheless, other tags such as the His tail combined with nickel columns or other purification systems can be used. The position of the tag in the fusion protein can be important depending on the objectives and the type of protein studied. Most plasmids available place the GST at the aminoterminus of the protein, but this can be a problem for type I proteins (extracellular aminoterminus) if the tag is kept for detection. Thus, the fusion protein has to be designed carefully and always keep the sequence in frame for proper translation.
4. The purified recombinant protein should be endotoxin free to avoid artifacts and undesired uncontrolled effects in cell-based and animal experiments. There are commercially available kits for its detection, such as the QLC-1000 endpoint chromogenic limulus amebocyte lysate (LAL) assay (Lonza, Basel, Switzerland). The levels should be <1 EU/mL.
5. If the protein is extracellular, it is not necessary to permeabilize the cell with Triton X-100.
6. It is convenient to become familiar with the terms used in SPR technology. In particular, the protein (or other type of molecule) immobilized on the sensor surface is called the ligand (independently of its native biological function). The ligand

can be immobilized directly or indirectly via an immobilized capturing molecule such as an anti-GST antibody for GST-fusion proteins. The molecule that is used free in solution to determine binding to the ligand is called analyte. It is recommended that the analyte has a high molecular weight to better detect changes in the refractive index.

7. All buffers for the Biacore should be filtered through a 0.22-μm filter and degassed just before the assay to prevent the formation of aggregates and bubbles. Ready-to-use buffers can be also obtained from the manufacturer.
8. It is important to perform the transfection without antibiotic; otherwise, PEI will become toxic to the cells.
9. Note that neither transfection nor transduction ensures that 100% of the cells express the recombinant protein unless the cells are subsequently selected. If stable and/or long-term recombinant expression is needed, use an expression vector that contains a gene that confers resistance to a selection antibiotic, such as G418 or puromycin. Take into account that different cell lines display different levels of resistance to such antibiotics and that each antibiotic kills in a particular time frame. Thus, it is advisable to check the literature, choose the appropriate antibiotic and conditions, and, if necessary, do some testing. Always treat simultaneously unmodified cells in a small flask or dish to confirm they all die.
10. Keep aliquots from each step to analyze the outcome of the procedure and to ensure that you do not lose your protein of interest in any intermediate step.
11. Once the antibody labeled with the fluorochrome is added, it is better to keep the samples in the dark to avoid the loss of fluorescence.

## References

1. Costa C (2007) Immunobiology of xenograft rejection. In: Ulricker ON (ed) Transplantation immunology research trends. Nova, New York
2. Sommaggio R, Máñez R, Costa C (2009) TNF, pig CD86 and VCAM-1 identified as potential targets for intervention in xenotransplantation of pig chondrocytes. Cell Transplant 18:1381–1393
3. Costa C, Bell NK, Stabel TJ, Fodor WL (2004) Use of porcine tumor necrosis factor receptor 1-Ig fusion protein to prolong xenograft survival. Xenotransplantation 11:491–502
4. Seebach JD, Comrack C, Germana S et al (1997) HLA-Cw3 expression on porcine endothelial cells protects against xenogeneic cytotoxicity mediated by a subset of human NK cells. J Immunol 159:3655–3661
5. Otamiri M, Nilsson KG (1999) Analysis of human serum antibody-carbohydrate interaction using biosensor based on surface plasmon resonance. Int J Biol Macromol 26:263–268
6. Link C (1999) Selection of phage-displayed anti-guinea pig C5 or C5a antibodies and their application in xenotransplantation. Mol Immunol 36:1235–1247

# Chapter 4

# Studies on Carbohydrate Xenoantigens


## Dale Christiansen, Effie Mouhtouris, Paul A. Ramsland, and Mauro S. Sandrin



## Abstract

Naturally occurring and elicited anti-carbohydrate antibodies play a major role in immune responses to xenografts. The original obstacles associated with the Gal antigen have been largely resolved by the generation of knockout pigs. In contrast, much less is known about the nature and role of non-Gal carbohydrate antigens and the antibodies recognizing these. These antibodies can be identified and characterized by enzyme-linked immunosorbent assay. Furthermore, the biological significance of the non-Gal antigen(s) can be determined by expression of the relevant glycosyltransferase(s) by transfection and analyzed by antibody and/or lectin binding.



**Key words:** Non-Gal antigens, Xenotransplantation, Carbohydrate antigens, Transfection, Antibodies


## 1. Introduction

Clinical transplantation of pig organs has the potential to resolve the acute shortage of human organs for transplantation. However, transplantation of vascularized xenografts into humans results in hyperacute rejection (HAR) due to the presence of high levels of natural anti-pig IgM and IgG antibodies (1). Several lines of evidence, including carbohydrate inhibition studies (2–6) and removal of anti-pig antibodies by absorption (7–9), strongly suggested that the major target of these natural antibodies was the terminal carbohydrate epitope Galα(1,3)Gal. This was confirmed by direct molecular evidence, whereby conversion of Galα(1,3)Gal-negative cells to Galα(1,3)Gal-positive cells by transfection with α1, 3galactosyltransferase Galα(1,3)Gal, the enzyme that synthesizes this epitope (1, 10), resulted in lysis by normal human serum (4).

The production of homozygous pigs with a disruption in the gene that encodes the enzyme α1,3GT (11–13) represented a critical


Cristina Costa and Rafael Máñez (eds.), *Xenotransplantation: Methods and Protocols,* Methods in Molecular Biology, vol. 885, DOI 10.1007/978-1-61779-845-0_4, © Springer Science+Business Media, LLC 2012

step towards the clinical reality of xenotransplantation by largely overcoming HAR. However, continuing damage to transplanted tissue in the form of delayed rejection and/or microangiopathy suggests that antigens other than Galα(1,3)Gal are targeted by the immune response. These antigens on α1,3GT(−/−) cells that are recognized by antibodies have been defined as non-Gal antigens (14–16). Many naturally occurring antibodies are directed against carbohydrate epitopes; therefore, identification of novel non-Gal pig antigens that are targets for human natural antibodies is important.

Identification of non-Gal antigens can be determined by several methods, including analysis of antibodies in the serum of individuals before and/or after rejection of xenografts by glycan arrays (17). Indeed, Blixt et al. used a glycan array, composed of 200 natural and synthetic glycan structures, representative of major glycoproteins and glycolipids, to characterize anti-carbohydrate antibody repertoires from patients previously transplanted with fetal pig islets (17). Alternatively, our laboratory has examined the glycan profile of α1,3 GT(−/−) mouse cells by flow cytometry using specific lectins (18). A similar, but somewhat more sophisticated, lectin microarray approach was used by Miyagawa et al. to determine differences in the glycan profile between fibroblasts and endothelial cells from α1,3 GT(−/−) and wild-type pigs (19). However, Kim et al. used positive MALDI-TOF MS, negative ion nano-ESI MS/MS, and normal-phase HPLC to identify and semi-quantify total *N*-glycans derived from specific pathogen-free mini pig kidney membrane proteins (20). Irrespective of the identification method used, molecules identified by glycome profiling become relevant to xenotransplantation if they are recognized by human antibodies, which enzyme-linked immunosorbent assay (ELISA) can demonstrate. Moreover, ELISA can also determine the specificity and affinity/avidity of the antibodies. Biological significance of the candidate non-Gal carbohydrate antigens can then be assessed by expression of the specific glycosyltransferase involved in the generation of the non-Gal epitope by transfection and analysis by antibody or lectin binding using fluorescence microscopy or flow cytometry. There are several Web-accessible databases (e.g., www.functionalglycomics.org) that can be used to aid in the identification of the glycosyltransferase responsible for generation of the identified carbohydrate, including the DNA sequence. Experimental three-dimensional structures of glycosyltransferases are available from the Protein Data Bank (PDB: www.pdb.org) and can be used to guide the design of site-directed mutations to confirm and further understand the biochemical basis for the production of carbohydrate xenoantigens (both Gal and non-Gal). For example, we previously used homology models of porcine α1,3GT, based on the bovine crystal structure of the enzyme (PDB code 1G93, (21), Fig. 1), to identify through mutagenesis

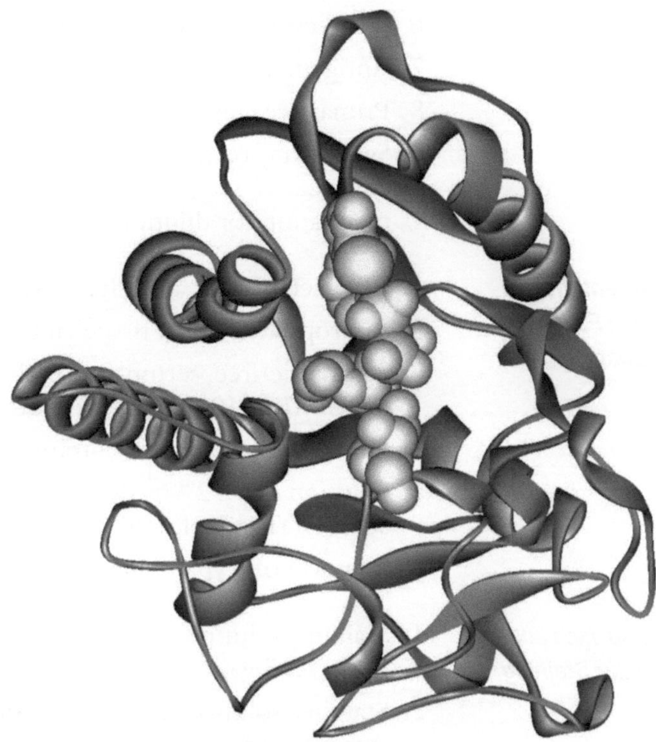

Fig. 1. Crystal structure of the catalytic domain of bovine $\alpha$1,3GT. The polypeptide chain is displayed as a ribbons-style secondary structure representation and the bound UDP-Gal substrate or donor molecule is displayed as a spacefill (*CPK spheres*) representation. The coordinates from PDB code 1G93 (21) were used to generate this view of the glycosyltransferase that is involved in production of the major carbohydrate xenoantigen, i.e., Gal$\alpha$(1,3)Gal terminating carbohydrates.

the functional role of histidine (position 271) in the generation of Gal$\alpha$(1, 3)Gal carbohydrate xenoantigens (22). After transfection, cells can not only be characterized serologically, as outlined below, but can also be used in other assays (e.g., cytotoxicity (4), adhesion (23)).

## 2. Materials

### 2.1. ELISA

1. Plates: Flat-bottom 96-well plates.
2. Coating buffer: 1× phosphate-buffered saline (PBS).
3. Blocking buffer: 1× PBS 2% (w/v) bovine serum albumin (BSA), 0.01% (w/v) Thiomersal, pH 7.2–7.4.
4. Washing buffer: 1× PBS 0.05% (v/v) Tween 20 (0.05%).
5. Substrate: 3,3′,5,5′-TetraMethylBenzidine (TMB).

6. Stop solution: 0.18 M $H_2SO_4$.
7. Antigen: Titration of antigen diluted in 1× PBS.
8. Primary reagent: Biotinylated lectin, antibody, or serum.
9. Secondary reagent: Streptavidin–HRP-conjugated or HRP-conjugated secondary antibody (see manufacturer's recommendation for dilution).

### *2.2. Transient Protein Expression Following Transfection*

1. Lipofectamine™ and Plus™ Reagent.
2. Appropriately sized tissue culture plate, flask, or dish.
3. Antibiotic-free, serum-free Dulbecco's modified Eagle's media or RPMI 1640 media (see Note 1).
4. DNA at a known concentration (μg/μl).
5. Solution of 0.12% trypsin or 1 mM ethylenediamine tetraacetic acid (EDTA) in 1× PBS.
6. Solution of trypan blue (0.4% in 1× PBS).

### *2.3. Cell Surface and Intracellular Staining*

1. Staining solution 1: Prepare 250 mL of 1× PBS/0.5% (w/v) BSA (1.25 g).
2. Staining solution 2: Prepare 100 mL of 1× PBS/0.5% (w/v) saponin (0.5 g).
3. Staining solution 3: Prepare 100 mL of 1× PBS/2% (w/v) paraformaldehyde (2.0 g) (see Note 2).
4. FACS fix: 2% (w/v) glucose (10 g), 2% (v/v) formaldehyde, 0.02% sodium azide. Make up to 500 mL with 1× PBS. Filter solution and store at 4°C.

## 3. Methods

Below, we describe how carbohydrate non-Gal antigens may be identified using several straightforward protocols that our laboratory routinely use.

### *3.1. ELISA*

1. Coat plate with 50 μL of antigen. Double dilute across 12 wells in coating buffer. Leave plate overnight at 4°C in a humidified box (see Note 3).
2. Block plate. Add 100 μL/well of blocking buffer and incubate plate at 37°C for 2 h in a humidified box.
3. Discard washing buffer and remove any excess liquid by vigorously blotting plate on paper toweling.
4. Dilute primary reagent in blocking buffer and add 50 μL/well (see Note 4). Then, incubate at room temperature for 1 h in a humidified box.

5. Wash plate by immersing in wash buffer and flick to remove excess liquid. Repeat this process six times. Remove any remaining liquid by vigorously blotting plate on paper toweling.
6. Add 50 μL/well of an appropriate dilution of streptavidin–HRP- or HRP-conjugated secondary antibody (in blocking buffer). Incubate at room temperature for 45 min in a humidified box.
7. Repeat washing step (step 5 above).
8. Add 50 μL/well of TMB and allow color to develop (typically 5 min).
9. Stop the reaction by adding 50 μL/well of stop solution.
10. Read absorbance values on a plate reader at 450 nm.

### 3.2. Transient Protein Expression Following Transfection

1. The day before performing the transfection, trypsinize or EDTA treat cells, pellet by centrifugation (450×*g*/5 min), then resuspend, and count (see Note 5). Seed cells so they will be ~50–80% confluent on the day of transfection (see Note 6).
2. Pre-complex the DNA with the Plus reagent (*Tube A*). To this end, dilute the appropriate volume of DNA into serum-free medium (see Table 1, i.e., add 1 and 3). To this, add the appropriate volume of PLUS reagent (number 2), mix gently, and allow to incubate at room temperature for 15 min.
3. Dilute appropriate volume of Lipofectamine™ reagent into serum-free medium in a second tube (*Tube B*) (i.e., add 3 and 4), and then mix.

**Table 1**
**Transfection reference table**

| | DNA (μg) | PLUS reagent (μL) | Serum-free dilution medium (SFM, μL) | Lipofectamine reagent (μL) | Transfection medium (serum free) | Transfection volume |
|---|---|---|---|---|---|---|
| Number | 1 | 2 | 3 | 4 | 5 | 6 |
| Culture vessel | | | | | | |
| 96 well | 0.1 | 1 | 10 | 0.5 | 50 μL | 70 μL |
| 24 well | 0.4 | 4 | 25 | 1 | 200w μL | 250 μL |
| 6 well | 1 | 6 | 100 | 4 | 800 μL | 1 mL |
| 60 mm | 2 | 8 | 250 | 12 | 2 mL | 2.5 mL |
| 100 mm | 4 | 20 | 750 | 30 | 5 mL | 6.5 mL |

4. Combine contents of tubes A and B and mix gently (transfection mix). Then, incubate at room temperature for 15 min.
5. While complexes are forming, remove growth medium from pre-seeded cells and wash twice with serum-free medium. Add appropriate volume of serum-free transfection medium (see Table 1, column 5).
6. After 15 min, add transfection mix to cells; this should result in the appropriate transfection volume (see Table 1, column 6). Incubate at 37°C in 5–10% $CO_2$ for 2–24 h (see Note 7).
7. Remove transfection mix and replace with complete growth medium. Cells are incubated for 72 h before determining expression.

### *3.3. Cell Surface and Intracellular Staining*

1. For cell surface staining, wash cells twice with PBS/0.5%BSA, and then go to step 3. For intracellular staining, fix cells in PBS/2% paraformaldehyde. Incubate on ice for 10 min.
2. Remove paraformaldehyde (into a hazardous waste container) and wash cells in PBS/0.5%BSA. Then, permeabilize by incubating with PBS/0.5% saponin for 10 min.
3. Add an appropriate dilution of primary antibody (diluted in PBS/0.5%BSA) (see Note 8) or fluorochrome-conjugated lectin (i.e., FITC or TRICT) to cells and incubate on ice for up to 2 h (typically 30–45 min).
4. Remove antibody or lectin and wash cells twice with PBS/0.5%BSA. For lectin staining, go to step 6. For antibody staining, add appropriate dilution of FITC-conjugated secondary antibody (if required) in PBS/0.5%BSA and incubate covered on ice for 30–45 min.
5. Remove antibody and wash cells twice with PBS/0.5%BSA. Add 1 mL PBS/0.5%BSA (for 1 well of a 6-well plate) and keep the plate covered and on ice prior to analysis by fluorescence microscopy (see Note 9). Example results of both cell surface (Fig. 2a, b) and intracellular staining (Fig. 2c) are shown. If required, cells can be removed from the well, resuspended by pipeting, and analyzed by flow cytometry. An example result is shown in Fig. 2d.
6. Lectin binding can be analyzed by flow cytometry. An example of staining wild-type pig cells and transgenic pig cells expressing the α1,2fucosyltransferase (which produces O blood group substance) is shown in Fig. 3. The level of Galα(1,3)Gal in wild-type pig cells (Fig. 3a) is decreased by expression of the α1,2fucosyltransferase (Fig. 3b). This is expected as the level of cell-surface O blood group is increased (Fig. 3c, d). In contrast, the level of α 2,3sialic acid is unaltered (Fig. 3e, f).

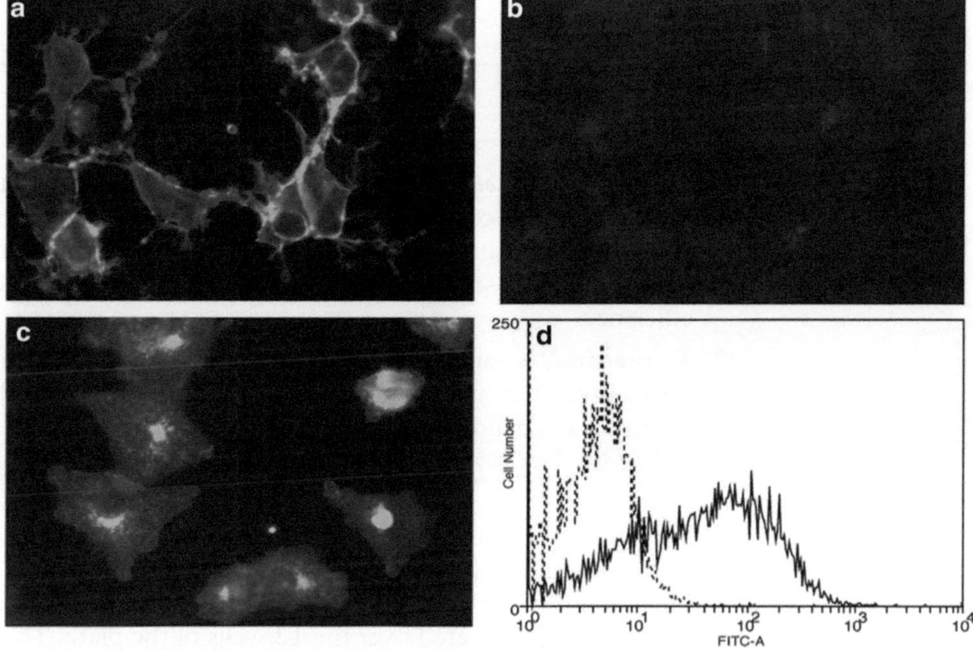


Fig. 2. Immunostaining of transfected cells. CHOP cells transfected with cDNA encoding α1,3galactosyltransferase and examined for expression. (**a**) Cell surface staining using anti-Galα(1,3)Gal antibody; (**b**) mock-transfected CHOP cells show no antibody binding; (**c**) Golgi localization of α1,3galactosyltransferase after permeabilization of cells; (**d**) flow cytometric analysis of cell surface expression of Galα(1,3)Gal epitope (*solid line*); background staining with secondary antibody is shown by a *broken line*.

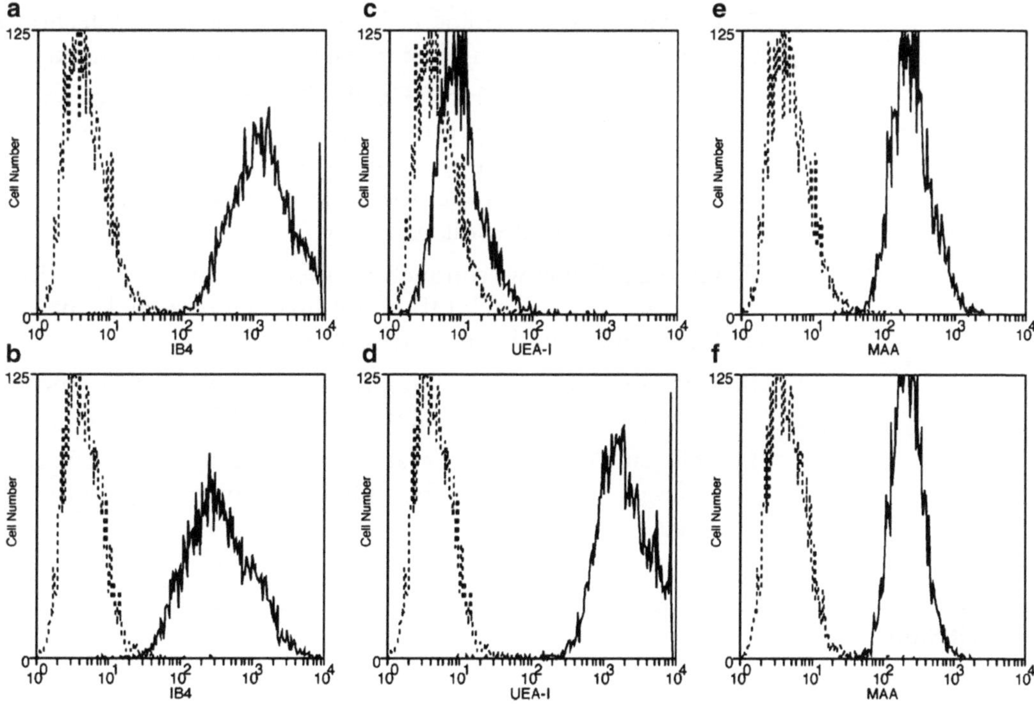


Fig. 3. Flow cytometric analysis of cell surface lectin binding. The parental wild-type pig endothelial cell line (**a**, **c**, and **e**) and the transfected pig endothelial cell line expressing the α1,2fucosyltransferase (**b**, **d**, and **f**) were stained with (**a** and **b**) the IB4 lectin (from *Griffonia simplicifolia*, detects Galα(1,3)Gal); (**c** and **d**) the UEA-I lectin (from *Ulex europaeus*, detects O blood group); (**e** and **f**) the MAA lectin (from *Maackia amurensis*, detects α2,3sialic acid). Flow cytometric analysis of cell surface expression of lectin binding (*solid line*); background staining with irrelevant protein is shown by a *broken line*.

## 4. Notes

1. It is important *not* to have antibiotics in media during transfection as this causes cell death. However, although the manufacturers recommend the use of Opti-Mem reduced serum medium, we have found that serum-free medium gives a similar transfection efficiency.
2. To prepare 2% paraformaldehyde, heat 100 mL 1× PBS in the microwave (1–2 min), and allow to cool slightly before adding 2 g of paraformaldehyde in a fume hood. Swirl constantly until all has dissolved. Once completely cooled, prepare 5-mL aliquots and freeze.
3. We typically use 50 μg/mL of antigen as a starting concentration. For doubling dilutions, the first sample is in a volume of 100 μL, 50 μL of which are then mixed with 50 μL of coating buffer. This is repeated over the 12 wells of the plate. The final 50 μL are discarded, resulting in a final volume of 50 μL in each well.
4. Biotinylated lectins are typically used at a dilution of 1:100. Antibody or serum concentrations have to be determined empirically.
5. We routinely determine cell viability prior to transfection by adding an equal volume of trypan blue and resuspended cells.
6. We routinely transfect CHOP (Chinese hamster ovary cells transformed with polyoma large T antigen (24)); seeding cells at $2 \times 10^5$/well (of a 6-well plate) or at $1 \times 10^6$ (in a 100-mm dish) typically gives the desired level of confluence (50–80%) for transfection the following day.
7. For the cell lines we routinely transfect (pig iliac artery endothelial cells (PIECs), CHOP, and human embryonic kidney cells E293), we typically incubate cells for ~3–5 h.
8. For intracellular staining, primary and secondary antibody (if required) should be diluted in PBS/0.5% saponin.
9. Cells can be analyzed at a later date (typically within a week) by adding 1 mL of FACS fix/well and keeping them in the dark at 4°C.

## 5. Conclusion

The identification of non-Gal carbohydrates is essential. Understanding the role these play in antibody-mediated damage, coagulation, or recognition by cells is fundamental to the development of appropriate strategies to modify/neutralize these antigens. Once new non-Gal carbohydrate xenoantigens have been identified and detailed structure–function studies have been performed similar to those previously undertaken for the Gal antigens, the information should ultimately facilitate clinical xenotransplantation.

## Acknowledgment

This work was supported by funds from the National Health and Medical Research Council of Australia (NHMRC).

### References

1. Sandrin MS, McKenzie IFC (1994) Galα(1,3) Gal, the major xenoantigen(s) recognised in pigs by human natural antibodies. Immunol Rev 141:169–190
2. Sandrin MS, Vaughan HA, Dabkowski PL, McKenzie IFC (1993) Anti-pig IgM antibodies in human serum reacts predominantly with Galα(1,3)Gal epitopes. Proc Natl Acad Sci USA 90:11391–11395
3. Good AH, Cooper DKC, Malcolm AJ, Ippolito RM, Koren E, Neethling FA, Ye Y, Zuhdi N, Lamontagne LR (1992) Identification of carbohydrate structures that bind human antiporcine antibodies: implications for discordant xenografting in humans. Transplant Proc 24:559–562
4. Vaughan HA, Loveland BE, Sandrin MS (1994) Gal α(1,3)Gal is the major xenoepitope expressed on pig endothelial cells recognized by naturally occurring cytotoxic human antibodies. Transplantation 58:879–882
5. Neethling FA, Koren E, Ye Y, Richards SV, Kujundzic M, Oriol R, Cooper DKC (1994) Protection of pig kidney (PK15) cells from the cytotoxic effect of anti-pig antibodies by α-galactosyl oligosaccharides. Transplantation 57:959–963
6. Rieben R, von Allmen E, Korchagina EY, Nydegger UE, Neethling FA, Kujundzic M, Koren E, Bovin NV, Cooper DKC (1995) Detection, immunoabsoption, and inhibition of cytotoxic activity of anti-aGal antibodies using newly developed substances with synthetic Gala1-3Gal disaccharide epitopes. Xenotransplantation 2:98–106
7. Taniguchi S, Neethling FA, Korchagina EY, Bovin N, Ye Y, Kobayashi T, Niekrasz M, Li S, Koren E, Oriol R, Cooper DKC (1996) In vivo immunoadsorption of antipig antibodies in baboons using a specific Galα1-3Gal column. Transplantation 62:1379–1384
8. Kozlowski T, Ierino FL, Lambrigts D, Foley A, Andrews D, Awwad M, Monroy R, Cosimi AB, Cooper DKC, Sachs DH (1998) Depletion of anti-Galα1-3Gal antibody in baboons by specific αGal immunoaffinity columns. Xenotransplantation 5:122–131
9. Xu Y, Lorf T, Sablinski T, Gianello P, Bailin M, Monroy R, Kozlowski T, Awwad M, Cooper DKC, Sachs DH (1998) Removal of anti-porcine natural antibodies from human and nonhuman primate plasma in vitro and in vivo by a Galα1-3Galβ1-4βGlc-X immunoaffinity column. Transplantation 65:172–179
10. Sandrin MS, Dabkowski PL, Henning MM, Mouhtouris E, McKenzie IFC (1994) Characterization of cDNA clones for porcine α1,3 galactosyltransferase: the enzyme generating the Galα(1,3)Gal epitope. Xenotransplantation 1:81–88
11. Dai Y, Vaught TD, Boone J, Chen SH, Phelps CJ, Ball S, Monahan JA, Jobst PM, McCreath KJ, Lamborn AE, Cowell-Lucero JL, Wells KD, Colman A, Polejaeva IA, Ayares DL (2002) Targeted disruption of the alphal,3-galactosyltransferase gene in cloned pigs. Nat Biotechnol 20:251–255
12. Lai L, Kolber-Simonds D, Park KW, Cheong HT, Greenstein JL, Im GS, Samuel M, Bonk A, Rieke A, Day BN, Murphy CN, Carter DB, Hawley RJ, Prather RS (2002) Production of alpha-1,3-galactosyltransferase knockout pigs by nuclear transfer cloning. Science 295:1089–1092
13. Phelps CJ, Koike C, Vaught TD, Boone J, Wells KD, Chen SH, Ball S, Specht SM, Polejaeva IA, Monahan JA, Jobst PM, Sharma SB, Lamborn AE, Garst AS, Moore M,

Demetris AJ, Rudert WA, Bottino R, Bertera S, Trucco M, Starzl TE, Dai Y, Ayares DL (2003) Production of alpha 1,3-galactosyltransferase-deficient pigs. Science 299:411–414

14. Harnden I, Kiernan K, Kearns-Jonker M (2010) The anti-nonGal xenoantibody response to alpha1,3-galactosyltransferase gene knockout pig xenografts. Curr Opin Organ Transplant 15:207–211
15. Puga Yung G, Schneider MK, Seebach JD (2009) Immune responses to alpha1,3 galactosyltransferase knockout pigs. Curr Opin Organ Transplant 14:154–160
16. Yeh P, Ezzelarab M, Bovin N, Hara H, Long C, Tomiyama K, Sun F, Ayares D, Awwad M, Cooper DK (2010) Investigation of potential carbohydrate antigen targets for human and baboon antibodies. Xenotransplantation 17:197–206
17. Blixt O, Kumagai-Braesch M, Tibell A, Groth CG, Holgersson J (2009) Anticarbohydrate antibody repertoires in patients transplanted with fetal pig islets revealed by glycan arrays. Am J Transplant 9:83–90
18. Christiansen D, Mouhtouris E, Milland J, Zingoni A, Santoni A, Sandrin MS (2006) Recognition of a carbohydrate xenoepitope by human NKRP1A (CD161). Xenotransplantation 13:440–446
19. Miyagawa S, Takeishi S, Yamamoto A, Ikeda K, Matsunari H, Yamada M, Okabe M, Miyoshi E, Fukuzawa M, Nagashima H (2010) Survey of glycoantigens in cells from alpha1-3galactosyltransferase knockout pig using a lectin microarray. Xenotransplantation 17:61–70
20. Kim YG, Gil GC, Harvey DJ, Kim BG (2008) Structural analysis of alpha-Gal and new non-Gal carbohydrate epitopes from specific pathogen-free miniature pig kidney. Proteomics 8:2596–2610
21. Gastinel LN, Bignon C, Misra AK, Hindsgaul O, Shaper JH, Joziasse DH (2001) Bovine alpha1,3-galactosyltransferase catalytic domain structure and its relationship with ABO histo-blood group and glycosphingolipid glycosyltransferases. EMBO J 20:638–649
22. Lazarus BD, Milland J, Ramsland PA, Mouhtouris E, Sandrin MS (2002) Histidine 271 has a functional role in pig alpha-1,3galactosyltransferase enzyme activity. Glycobiology 12:793–802
23. Inverardi L, Clissi B, Stolzer AL, Bender JR, Sandrin MS, Pardi R (1997) Human natural killer lymphocytes directly recognize evolutionarily conserved oligosaccharide ligands expressed by xenogeneic tissues. Transplantation 63:1318–1330
24. Heffernan M, Dennis JW (1991) Polyoma and hamster papovavirus large T antigen-mediated replication of expression shuttle vectors in Chinese hamster ovary cells. Nucleic Acids Res 19:85–92

# Chapter 5

# Xenoantibodies and Complement Activity Determinations in Pig-to-Primate Xenotransplantation

Nieves Doménech

## Abstract

Based on the results from different studies, it is recommended to use multiple assays to fully monitor a peripheral antibody response in recipients of pig xenografts. In particular, the complement-dependent hemolytic assay (CH50) determines total endogenous complement activity; the cytolytic activity of anti-pig antibodies (APA) is also measured by hemolytic assay; an ELISA is used for quantification of anti-galactose α1,3galactose (Gal) IgG and IgM antibodies and flow cytometry for determining the antiendothelial cell (anti-EC) antibodies. The APA is recommended because this assay measures cytolytic antibodies unlike the other assays which are binding assays. The anti-Gal ELISA is especially useful as it specifically quantifies antibodies against the major xenoantigen in the pig-to-primate transplant combination. Thus, it provides different information than the APA assay and is particularly useful to follow up treatment modalities aimed at blocking/inhibiting the anti-Gal antibodies. Finally, the anti-EC assay measures IgG and/or IgM antibodies reactive to Gal and non-Gal antigens expressed on endothelial cells. This information is key for understanding the rejection process of vascularized xenografts and finding strategies to overcome rejection. Here, we describe in detail the procedures of these techniques.

**Key words:** Primates, Pig, Serum, Xenoantibodies, Complement, Cytotoxicity, Flow cytometry, Endothelial cell

## 1. Introduction

Clinical application of xenogeneic vascularised organ transplantation (1) depends on strategies that simultaneously eliminate both antibody reactivity and complement activation on the xenogeneic cell surface (2). Initially, antibodies directed against the oligosaccharide Gal were found to be responsible for the hyperacute rejection of pig organs transplanted into nonhuman primates and play a major role in acute humoral xenografts rejection (3, 4). However,

Cristina Costa and Rafael Máñez (eds.), *Xenotransplantation: Methods and Protocols,* Methods in Molecular Biology, vol. 885, DOI 10.1007/978-1-61779-845-0_5, © Springer Science+Business Media, LLC 2012

the continuous depletion of these antibodies by polymers containing multiple Gal epitopes (5) and the use of grafts from pigs lacking this epitope (α1,3 galactosyltransferase knockout; Gal-KO) evidenced that antibodies to non-Gal antigens also contribute to acute humoral xenograft rejection of pig organs in nonhuman primate (6, 7). During pig-to-primate xenotransplantation, strategies are aimed at limiting the reactivity of pig cells and tissues and removing the primate xenoantibodies for prolonging graft survival (8, 9). To evaluate the efficacy of these strategies, several methods have been developed to assay cytotoxicity mediated by antibody and complement and assess the levels of functional xenoantibodies.

Assessing the functional integrity of the complement system (classical pathway) has been accomplished in xenotransplantation by the traditional 50% hemolytic complement (CH50) assays. The titer at which 50% hemolysis occurs (CH50 unit) is proportional to the functional activity of the classical pathway in the primate serum, and it is sensitive to the reduction, absence, and/or inactivity of any component of the pathway (10). Also, different antibody assays are used to measure the antibody levels in serum of primate recipients of a porcine xenograft (11). To date, the hemolytic assay used to measure anti-pig antibodies (APA) has been the most utilized to assess the functional lytic capacity of xenoantibodies with the addition of exogenous complement (12). In a typical APA assay, a curve of points is generated by plotting the percentage of haemoglobin released by lysed red cells against serial dilutions of primate serum (antibody) in the presence of rabbit complement. In addition, antibodies against the Gal antigen have been widely studied, as they account for more than 85% of circulating natural APA, consisting mainly of IgM and IgG subclasses. A specific ELISA is used to measure the anti-Gal antibodies (13). Finally, in pig-to-primate xenotransplantation, flow cytometry assays that measure serum levels of IgM or IgG xenoantibodies using pig endothelial cells (EC) as target cells have been shown to be at least as sensitive as other methods (hemolysis, hemagglutination, and anti-Gal enzyme-immunoassays) in detecting increases or decreases of antibodies in posttransplant periods (14). The relative accuracy of this flow cytometry assay may be due to the existence of xenoantibodies reactive to non-Gal antigens and, perhaps, xenoantibodies that do not activate complement or that are directed against antigens absent from red cells.

All of the assays have the potential to be of "predictive value," and more than one assay should be employed to ensure any increase in serum antibodies is detected (15, 16). So, it is recommended to use multiple assays (Fig. 1) for monitoring a peripheral antibody response in xenograft recipients and this information should be used in conjunction with clinical results and biopsies (17) to fully elucidate if a peripheral antibody response is a good indicator of acute humoral xenograft rejection (18, 19). Moreover, it is important to

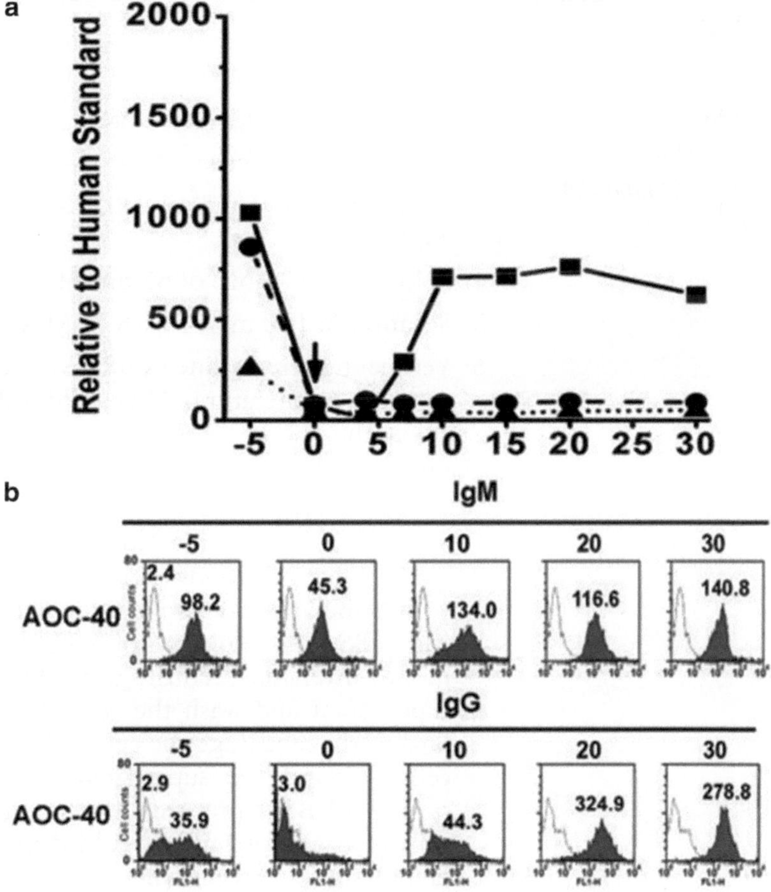


Fig. 1. (**a**) Levels of anti-Gal IgM (*circles*), IgG (*triangles*) and APA (*squares*) in baboons immunized with pig blood. The *arrows* indicate the days of pig blood injection. (**b**) Sequential measurement by flow cytometry of baboon IgM and IgG antibody binding to endothelial cell line AOC-40. Values represent median fluorescence intensity on different days before and after xenotransplantation (Dr. R. Mañez, unpublished data).

take into account that in the case of solid organ xenotransplantation, xenoantibodies are captured by the graft and it is often difficult to detect the increase in titers unless the xenograft is surgically excised (not possible with life-supporting grafts).

Based on the foregoing, we describe in detail the following assay protocols for measurement of complement and antibodies in serum of primates with a pig xenograft: (a) the CH50 assay to measure the functional lytic capacity of serum complement, (b) the APA assay to measure serum antibodies which are capable of lysing pig red blood cells (PRBC) in the presence of heterologous baby rabbit complement (BRC), (c) the anti-Gal assay to measure by titration the serum levels of IgG and IgM xenoantibodies reactive to the Gal antigen, and (d) a method to measure in serum by flow cytometry the binding levels of IgG and IgM xenoantibodies reactive to xenogeneic EC.

## 2. Materials

### 2.1. The CH50 Complement-Dependent Hemolytic Assay

1. Serum test samples (see Note 1).
2. Normal human serum (NHS) as standard (see Note 2).
3. Solution 1: 350 mL of 1.02 M NaCl and 13 mM sodium barbitone.
4. Solution 2: 125 mL of 62.5 mM barbitone.
5. Solution 3: 100 mL of 2.18 M $MgCl_2$ and 440 mM $CaCl_2$.
6. Veronal buffered saline (VBS): Mix solutions 1 and 2 and cool to room temperature. Then, add 1.25 mL of solution 3 and adjust the pH to 7.3–7.5 using 1 M HCl. Adjust the final volume to 500 mL with distilled water to prepare a 5× stock solution.
7. Sheep red blood cells (SRBC) precoated with rabbit anti-sheep red blood cell antibody (hemolysin). For sensitization of SRBC with hemolysin, prepare first the hemolysin by diluting it 1:50 with VBS. Add also 6 mL of VBS to 4 mL of SRBC and gently mix by inversion. Centrifuge at 600 × *g* for 5 min, discard the supernatant and wash the cells another two times. After the final wash, centrifuge the cells at 900 × *g* for 5 min to pack the cells. Discard the supernatant and resuspend the cells in sufficient VBS to prepare a 10% solution (i.e. 0.5 mL of packed cells in 5 mL of buffer). Add dropwise an equal volume of hemolysin to the cells while swirling continuously. Incubate at 30°C for 30 min in a water bath. Gently mix the cells every 15 min. The sensitised SRBC can be stored overnight at 4°C.
8. Variable pipettes (e.g. Gilson) for 200–1,000 μL and 20–200 μL.
9. Appropriate tips.
10. Multichannel pipette.
11. 96-well rigid V-bottomed microtiter plates.
12. 96-well flexible PVC flat-bottom plates.
13. Water bath at 37°C.
14. Standard laboratory centrifuge with microtiter plate holders.
15. Automatic microtiter-plate reader for use at 420 nm. Preferably, containing software capable of exporting data to PC for plotting.

### 2.2. The APA Hemolytic Assay

1. Serum test samples (see Notes 1 and 3).
2. NHS as standard (see Notes 2 and 3).
3. Complement fixation test diluent (CFD) (Oxoid Ltd, Basingstoke, UK). Use one tablet/100 mL of distilled water.

4. PRBC 1% suspension. For preparation of 1% PRBC, the heparinized pig blood is allowed to age in the refrigerator for 1 day before use. Then, 2–3 mL of this blood is placed in a 15-mL tube, which is then filled with CFD buffer. Wash three times with CFD following centrifugation at 750×*g* for 6 min at 4°C. The supernatant is removed and 1 mL of the final PRBC pellet is transferred into a 1.5-mL tube and centrifuged at 17,000×*g* for 2 min (room temperature or cold). Remove the supernatant and take up 100 μL of the cell pellet and dilute into 9.9 mL of CFD buffer. The resulting suspension is taken to be 1% PRBC.
5. Selected BRC (see Note 4). For the BRC preparation, follow the supplier's instructions. It is presented by most suppliers as a lyophilized preparation and usually requires the sample to be dissolved in a small volume of distilled water.
6. Microcentrifuge tubes of 1.5 or 2.0 mL.
7. Variable pipettes for 200–1,000 μL and 20–200 μL.
8. Appropriate tips.
9. Multichannel pipette.
10. 96-well rigid V-bottom microtiter plates.
11. 96-well flexible PVC flat-bottom plates.
12. Water bath at 56°C.
13. Standard laboratory centrifuge with microtiter plate holders.
14. High-speed bench centrifuge (e.g. microfuge).
15. Automatic microtitre-plate reader for use at 420 nm. Preferably, containing software capable of exporting data to PC for plotting.

### 2.3. ELISA for Detection of Anti-Gal IgG and IgM Antibodies

1. Serum test samples (see Notes 1 and 3).
2. NHS as standard (see Notes 2 and 3).
3. Human serum albumin (HSA).
4. Phosphate buffered saline (PBS) such as tablets without Ca++ and Mg++.
5. Gal-HSA: Gal α1, 3Galβ1, 4GlcNAc linked to HSA (an average of 14 Gal residues per HSA molecule with a three-atom spacer) (Dextra Labs, Reading, UK).
6. Gal-HSA stock solution: dissolve the Gal-HSA (white powder) in PBS to a concentration of 1 mg/mL. Prepare small aliquots and store at −20°C.
7. Optional: HSA stock solution: do as described for Gal-HSA.
8. Sodium carbonate–bicarbonate buffer: Mix $NaHCO_3$ (0.1 M) with $Na_2CO_3$ (0.1 M) in proportion 2:1. Adjust to pH 9.6. It can be stored in refrigerator for 1 week.

9. Horseradish peroxidase-conjugated AfiniPure F(ab′) 2 goat anti-human IgG (γ-specific) and horseradish peroxidase-conjugated AfiniPure F(ab′) 2 goat anti-human IgM (μ-specific) (Jackson ImmunoResearch Laboratories, West Grove, PA, USA). Usually, these reagents are lyophilized and need to be dissolved according to the manufacturer's instructions. This stock solution can then be distributed into small aliquots and stored at –20°C.
10. Polyoxyethylenesorbital monolaurate (Tween 20).
11. Plate washing solution (WS): Add 1 mL of Tween 20 per L of PBS (0.1% Tween 20/PBS). Make up fresh for each day.
12. Plate blocking solution: Add 0.5 mL of Tween 20–100 mL of PBS. Make up fresh for each day.
13. O-Phenylenediamine dihydrochloride (OPD) tablets (Sigma, St Louis, MO, USA).
14. Citric acid anhydrous.
15. Sodium phosphate dibasic.
16. Hydrogen peroxide 100 vol/30%.
17. Sulfuric acid (1 M).
18. Substrate solution: OPD tablets are dissolved in citrate (0.15 M)/phosphate (0.2 M) buffer according to supplier's instructions. 5 μL of 30% hydrogen peroxide solution is added per 12.5 mL of substrate solution immediately before use. Make up sufficient substrate solution freshly for each test. Do not store.
19. Quenching solution: 1 M sulfuric acid. This is added at the last step of the ELISA to stop the chromogenic reaction between the peroxidase enzyme and the OPD.
20. Microtiter plates for ELISA Polysorb 96-well flat-bottom plates and lids.
21. Microtiter plates for making dilutions such as 96-well rigid. U-bottom plates.
22. Variable pipettes for 1–10 μL, 20–200 μL and 100–1,000 μL or equivalent.
23. Multichannel pipettes.
24. Appropriate tips.
25. Microcentrifuge tubes of 1.5 or 2 mL.
26. Microfuge.
27. Automatic microtiter-plate reader for use at 490 nm. Preferably, containing software capable of exporting data to PC for plotting.

#### 2.4. Determination of Anti-Porcine EC Antibodies by Flow Cytometry

1. Serum test samples (see Notes 1 and 3).
2. NHS as standard (see Notes 2 and 3).
3. Immortalized porcine aortic endothelial cells (PAEC) or similar (see Note 5).
4. Dulbecco's modified Eagle's medium (DMEM).
5. Fetal bovine serum (FBS).
6. Penicillin–streptomycin (P/S): 5,000 U/mL penicillin and 5,000 μg/mL streptomycin.
7. Trypsin–ethylenediamine tetraacetic acid (EDTA) solution: 0.25% trypsin–0.04% EDTA.
8. Sterile physiologic saline solution.
9. Fluorescein isothiocyanate (FITC)-labeled rabbit anti-human IgG (γ-chain specific) or rabbit anti-human IgM (μ-chain specific). Those stock solutions should be kept at 4°C in the dark.
10. PBS tablets without Ca++ and Mg++ (follow manufacturer's instructions).
11. Bovine serum albumin (BSA).
12. Sodium azide.
13. Fluorescence-activated cell sorter (FACS) buffer: PBS + 1% BSA +0.01% sodium azide.
14. Propidium iodide (PI) at 500 μg/mL.
15. Sterile centrifuge tubes.
16. Sterile Pasteur pipettes.
17. Graduated pipettes and micropipettes.
18. Standard laboratory centrifuge with tube holders.
19. Incubator at 37°C and humidified 5%, $CO_2$ atmosphere.
20. Laminar-air-flow cabinet.
21. Inverted optical microscope.
22. FACS tubes.
23. Becton Dickinson FACScan flow cytometer (or equivalent) with software for analysis.

## 3. Methods

#### 3.1. The CH50 Complement-Dependent Hemolytic Assay

1. Serially dilute test serum samples and NHS standard (twofold, starting at 1:5) with VBS in a 96-well V-bottom plate at 100 μL/well. Test samples in duplicate. A negative control (spontaneous lysis of SRBC in VBS) and a positive control (maximum lysis obtained by lysis of SRBC in water) should be included.

2. Add 100 μL of suspended sensitized SRBC to all wells and mix immediately by repeated pipetting with a multichannel pipette.
3. Incubate at 37°C for 60 min in a waterbath.
4. Centrifuge the samples at 500 × *g* for 10 min to sediment the SRBC.
5. Transfer 100 μL of supernatant from each well to a new well in a 96-well flat-bottom plate.
6. Read the absorbance at 420 nm with an automatic plate reader within 15 min of centrifugation. Then, plot the data as average values with serum dilution on *x* axis and absorbance on *y* axis.
7. For data processing, calculate the mean absorbance for each sample and subtract the negative control absorbance (spontaneous lysis). Calculate the %lysis for each dilution using the following formula: %lysis = OD420 test – OD420 negative control lysis/(OD420 positive control lysis – OD420 negative control lysis) × 100. Plot the percentage lysis (vertical axis) versus the serum dilution on the horizontal axis. Calculate the dilution required for 50% hemolysis for the test sera and NHS.

### 3.2. The APA Hemolytic Assay

1. Serially dilute test serum samples and NHS standard (twofold, starting at 1:5) with CFD in a 96-well V-bottom plate at 50 μL/well. Test samples in duplicate. A negative control (only complement, no antibody) and a positive control (maximum lysis obtained by lysis of erythrocytes in water) should be included.
2. Add 50 μL of 1% PRBC to all wells, including the negative control wells, except to the positive control wells. Cover the plates and place in an orbital incubator at 37°C for 1 h.
3. Remove plates from incubator and add 100 μL CFD buffer to all wells. Centrifuge at 500 × *g* for 10 min at 4°C.
4. Remove the supernatant fluid. It can be done by inverting the plate over a sink and applying one sharp shaking movement, making sure that the PRBC remains firmly attached.
5. Add 200 μL CFD buffer to all wells. To resuspend the PRBC pellet, use a multichannel pipette and pipette up and down several times for each column.
6. Centrifuge the plates at 500 × *g* for 10 min at room temperature and discard the supernatant as described.
7. Add 150 μL of BRC at the preselected dilution in CFD buffer to all wells, including the negative control wells. Add 150 μL of water to the positive control wells instead of CFD buffer.

8. Cover and place the plates in the orbital incubator at 37°C for 1 h.
9. Remove plates from incubator and centrifuge at 500×*g* for 10 min at room temperature.
10. Transfer 100 μL of the supernatant fluid from the assay plate to a suitably-marked flat-bottom plate.
11. Read the absorbance at 420 nm with an automatic plate reader within 15 min of centrifugation. Plot data as average values with serum dilution on *x* axis and absorbance on *y* axis.
12. For data processing, plot the mean absorption of each sample as a function of dilution. The asymptote for each curve is determined and subtracted from each curve, then the area under the curve is calculated and standardized to the human control equal to 1,000 and the value can be expressed as arbitrary units (AU). For samples with low levels of hemolysis, the 50% titer point method cannot be used. In cases where prozones occur at high sample concentration, adjustments must be made to avoid errors in measuring the area under the curve.

### 3.3. ELISA for Detection of Anti-Gal IgG and IgM Antibodies

1. Coat the 96-well microtiter plates overnight at 4°C with 50 μL/well of Gal-HSA in 0.1 M bicarbonate buffer, pH 9.6 at 5 μg/mL.
2. Optional: coat identical number of wells with HSA alone in the same plate to be use as background reactivity following identical conditions.
3. Remove plates from refrigerator and empty the wells by inverting and shaking briskly over a sink. Wash the plates once with plate washing solution (WS) by filling and emptying all the wells. Blot the plates on paper tissue with gentle tapping to make sure the wells are empty. All subsequent procedures are performed at room temperature.
4. The plates are then blocked for 1 h with 0.5% Tween 20 in PBS. Then, empty the plates by shaking and blotting as described.
5. Prepare seven serial twofold dilutions of the test sera and of NHS for each assay (all starting at 1:5 and continued to a final concentration of 1 in 320) in a separate plate at 50 μL/well in duplicate. Use 0.5% Tween 20 in PBS for all sera dilutions. Prepare separate plates for IgG and IgM antibody assays.
6. Transfer 50 μL of sera from the dilution plate into the corresponding wells of the ELISA plates, tap the plates gently, cover with lids and incubate for 1 h at room temperature.
7. Then, wash the plates with WS (three times quickly and then four times leaving the plate for 1 min between washes). Empty and blot the plates.

8. Add 50 μL/well of the peroxidase-labeled anti-IgG and anti-IgM antibodies at 1:1,000 dilution to the corresponding separate test plates to measure the anti-Gal IgG and IgM antibodies. Use 5% Tween20/PBS for the dilution of the secondary antibodies.
9. Tap the plate gently, cover with lid and allow standing for 1 h at room temperature.
10. Remove the secondary antibodies by inverting, shaking, and blotting the plates. Wash the plates with WS six times.
11. Dispense 100 μL of substrate solution to all wells. Allow the color reaction to proceed for a fixed length of time. This is usually about 5 min and must be initially determined empirically. Color development can be allowed to proceed as long as the control wells remain colorless. The time period for color development should subsequently be standardized for all assays.
12. Stop color development by adding 50 μL of 1 M sulfuric acid to all wells in the same order as the substrate was added.
13. Read the plates immediately using an automated plate reader set at 490 nm.
14. Plot the results: $x$ axis = dilution of sample; $y$ axis = absorption at 490 nm.
15. The data processing can be done by integrating the areas under each titration curve using a computer program. The standard sample is given an arbitrary value of 1,000. Alternatively, the dilution of each sample giving an absorbance level of 50% of the standard absorbance titration can be compared. This second method is not feasible for samples with low levels of antibody binding or for samples which show nonparallel titration curves compared to the standard. Relative potencies of samples giving reasonable parallelism of the central sections of their titration curves may be compared by working out the horizontal displacement between titers (point of 50% binding) in terms of the log2 titration scale. Results can also be expressed as a "standardized value" by dividing the absorbance of the 1:5 dilution of the test serum by that for the same dilution of the human serum.

### 3.4. Determination of Anti-Porcine EC Antibodies by Flow Cytometry

1. Culture PAEC as monolayer in DMEM supplemented with 10% FBS and 1% P/S (complete DMEM) in appropriate culture flask. The cells are maintained in an incubator at 37°C in 5% $CO_2$.
2. The day of the assay, detach cells from the culture flask. For this purpose, remove culture medium, wash the cell monolayer with physiologic saline solution and dissociate the cells by treatment at 37°C for 2–3 min with a trypsin–EDTA solution (1×) (add enough volume to cover the cell monolayer). Block enzyme activity by adding complete DMEM.

3. Transfer cell suspension to a 50-mL centrifuge tube and wash carefully the culture flask in order to remove all the remaining cells. Centrifuge PAEC at 300 × *g* for 10 min.
4. Remove supernatant, resuspend cells in complete DMEM and count. Harvest sufficient cells for $1 \times 10^5$ cells/test.
5. Bring final volume of cell suspension up to 20 mL with complete DMEM and spin at 300 × *g* for 10 min.
6. Discard the supernatant and resuspend the pellet in the appropriate volume of DMEM (no FBS from now on) to have $5 \times 10^6$ cell/mL.
7. Pipette 20 μL of cells ($1 \times 10^5$) into each FACS tube and 20 μL of the appropriate solution for staining.
8. Set up the tubes as follows:
   - Control of cells only: 20 μL of DMEM.
   - Controls of secondary antibody (two tubes, one for FITC-rabbit anti-human IgG and one for FITC-rabbit anti-human IgM): 20 μL of DMEM.
   - Control of NHS (two tubes, one for detecting IgG and one for IgM): 20 μL of 1:2 NHS in FACS buffer so that the final serum dilution is 1:4.
   - Test sera samples (two tubes for each, one for detecting IgG and one for IgM): received 20 μL of 1:2 test sera in FACS buffer so that the final serum dilution is 1:4.
9. Incubate for 30 min at room temperature.
10. After incubation, wash the cells twice with 2 mL of FACS buffer and spin at 175 × *g* for 5 min.
11. Resuspend the appropriate pellets (all tubes except control of cells only) in 20 μL of 1/20 dilution of FITC-labeled rabbit anti-human IgG or rabbit anti-human IgM diluted in FACS buffer.
12. Incubate for 30 min at 4°C in the dark.
13. Wash the cells twice with 2 mL of FACS buffer and spin at 175 × *g* for 5 min.
14. Resuspend the final pellets in 400 μL of FACS buffer immediately before acquisition.
15. Optional: add 4 μL of PI (500 μg/mL) to each tube in order to include a dead/live gating.
16. Acquire and analyze results with a Becton Dickinson FACScan flow cytometer and CellQuest software. Collect 5,000 events in the live cell gate and plot as specific median fluorescence intensity (MFI) which is generated by "test MFI- control (secondary antibody only) MFI."

## 4. Notes

1. Serum primate samples are obtained from coagulated blood (allowed to clot for 30 min at room temperature) after centrifugation (800 × *g* for 10 min). Samples should be immediately distributed in aliquots and frozen at −80°C. For complement determinations, serum samples should be thawed immediately before the experiments in order to avoid the loss of activity of endogenous complement.
2. NHS is used as a control. The NHS sample is obtained from a stock of 5–30 pooled screened human sera obtained from healthy adults.
3. For all assays, except the CH50 assay, all serum samples (test and NHS) are heat-treated at 56°C for 30 min in microfuge tubes, then cooled and spun in the microfuge at 17,000 × *g* to remove fine particulates (precipitate) and lipids (surface) before use in the assay. Heat inactivation is performed on the same day of the assay.
4. It is recommended testing a number of small samples of BRC before selecting one or more samples which show maximum complement activity with minimum intrinsic cytotoxicity (in the absence of serum that is the antibody source) towards PRBC targets in the APA hemolytic assay.
5. Primary (PAEC) could be isolated from donor pigs and cultured in the same conditions as the immortalized PAEC line. However, there are some inherent problems with using primary cells to measure primate antibodies on a regular bases such as having sufficient cells, spontaneous dedifferentiation of the cultures, mycoplasma contaminations, etc. For this reason, a pig endothelial cell line (20) is a good substitute for primary EC in this assay system.

### References


1. Pierson RN, Dorling A, Ayares D et al (2009) Current status of xenotransplantation and prospects for clinical application. Xenotransplantation 16:263–280
2. Pierson RN (2009) Antibody-mediated xenograft injury: mechanisms and protective strategies. Transpl Immunol 21:65–69
3. Sandrin MS, McKenzie IFC (1994) Gal α (1,3) Gal, the major xenoantigen(s) recognised in pigs by human natural antibodies. Immunol Rev 141:169–190
4. Ramsland PA, Farrugia W, Yuriev E et al (2003) Evidence for structurally conserved recognition of the major carbohydrate xenoantigen by natural antibodies. Cell Mol Biol 49: 307–317
5. Mañez R, Lopez-Pelaez E, Centeno A et al (2004) Failure to deplete anti-Galα1-3Gal antibodies after pig-to-baboon organ xenotransplantation by immunoaffinity columns containing multiple Galα1-3Gal oligosaccharides. Xenotransplantation 11:408–415

6. Cooper DK, Dorling A, Pierson RN 3rd et al (2007) Alpha1,3-galactosyltransferase gene-knockout pigs for xenotransplantation: where do we go from here? Transplantation 84:1–7
7. Harnden I, Kiernan K, Kearns-Jonker M (2010) The anti-nonGal xenoantibody response to alpha1,3-galactosyltransferase gene knock-out pig xenografts. Curr Opin Organ Transplant 15:207–211
8. Katopodis AG, Warner RG, Duthaler RO et al (2002) Removal of anti-Galα1,3Gal xenoantibodies with an injectable polymer. J Clin Invest 110:1869–1877
9. Klymiuk N, Aigner B, Brem G et al (2010) Genetic modification of pigs as organ donors for xenotransplantation. Mol Reprod Dev 77:209–221
10. Turgeon ML (1996) Soluble mediators of the immune system. In: Immunology and serology in laboratory medicine. 2nd edn. Shanahan, J., Ed, Mosby Year Book Inc., St. Louis, MO, p. 89–107
11. Ekser B, Rigotti P, Gridelli B et al (2009) Xenotransplantation of solid organs in the pig-to-primate model. Transpl Immunol 21:87–92
12. Porcel JM, Peakman M, Senaldi G et al (1993) Methods for assessing complement activation in the clinical immunology laboratory. J Immunol Methods 157:1–9
13. Teranishi K, Manez R, Awwad M et al (2002) Anti-Gal alpha 1-3Gal IgM and IgG antibody levels in sera of humans and old world non-human primates. Xenotransplantation 9: 148–154
14. Richards AC, Davies HF, McLaughlin ML et al (2002) Serum anti-pig antibodies as potential indicators of acute humoral xenograft rejection in pig-to-cynomolgus monkey kidney transplantation. Transplantation 73:881–889
15. Mañez R, Lopez-Pelaez E, Centeno A et al (2004) Transgenic expression in pig hearts of both human decay-accelerating factor and human membrane cofactor protein does not provide an additional benefit to that of human decay-accelerating factor alone in pig-to-baboon xenotransplantation. Transplantation 78:930–933
16. Chen G, Sun H, Yang H et al (2006) The role of anti-non-Gal antibodies in the development of acute humoral xenograft rejection of hDAF transgenic porcine kidneys in baboons receiving anti-Gal antibody neutralization therapy. Transplantation 81:273–283
17. Shimizu A, Yamada K (2010) Histopathology of xenografts in pig to non-human primate discordant xenotransplantation. Clin Transplant 22:11–15
18. Hisahi Y, Yamada K, Kuwaki K et al (2008) Rejection of cardiac xenografts transplanted from alpha1,3-galactosyltransferase gene-knockout (GalT-KO) pigs to baboons. Am J Transplant 8:2516–2526
19. Sprangers B, Waer M, Billiau AD (2008) Xenotransplantation: where are we in 2008? Kidney Int 74:14–21
20. Carrillo A, Chamorro S, Rodriguez-Gago M et al (2002) Isolation and characterization of immortalized porcine aortic endothelial cell lines. Vet Immunol Immunopathol 89:91–98

# Chapter 6

# Studies on Coagulation Incompatibilities for Xenotransplantation

Cristiana Bulato, Claudia Radu, and Paolo Simioni

## Abstract

Microvascular thrombosis, following the activation of clotting cascade, is a hallmark of porcine solid organ xenograft rejection. The analysis of differences between human, monkey, and pig coagulation systems is crucial when monkey is used as animal model and pig as organ donor in xenotransplantation. Thrombosis, according to many authors, may be due to the molecular incompatibilities between natural anticoagulants present on pig endothelium and primate activated coagulation factors. The generation of activated protein C (PC) is critical for the physiological anticoagulation. One of the major incompatibilities may be related to the inability of pig thrombomodulin (TM) and endothelial protein C receptor to activate the recipient (primate) circulating PC in the presence of thrombin. Tissue factor pathway inhibitor (TFPI), is the primary inhibitor of tissue factor (TF)-induced coagulation. TFPI directly inhibits the activated factor X (FXa) and blocks the procoagulant activity of the TF/factor VIIa (FVIIa) complex by forming a quaternary TF/FVIIa/FXa/TFPI complex. Microvascular thrombosis, observed in the organ transplant, may also be due to the failure of pig TFPI to bind human FXa efficiently and inhibit human FVIIa/TF activity. The methods described in this chapter can be useful for the identification and characterization of primate and pig coagulation factors (isolated from a small volume of blood) by using SDS-PAGE and immunoblotting. Differences in molecular weight can help in the identification of the origin (pig or primate) of coagulation proteins in plasma from the recipient of xenografts. On the other hand, in vitro models of PC pathway and TFPI on human umbilical vein endothelial cells (HUVEC) and porcine aortic endothelial cells (PAEC) are described which can be used for studying incompatibilities between primate and pig.

**Key words:** Xenotransplantation, Coagulation incompatibilities, Coagulation factors, Western blotting, Protein C pathway, Tissue factor pathway inhibitor

## 1. Introduction

The cynomolgus monkey (*Macaca fascicularis*) is used as an animal model in several experiments of pig to primate xenotransplantation (1–3). It has long been known that coagulation dysregulation plays

Cristina Costa and Rafael Máñez (eds.), *Xenotransplantation: Methods and Protocols,* Methods in Molecular Biology, vol. 885, DOI 10.1007/978-1-61779-845-0_6, 

a critical role on the failure of pig grafts in primates (4). Coagulation disorders, which occur after xenotransplantation, may be due to molecular incompatibilities between pig and primate coagulation factors (5–7). Since coagulation factors are synthesized by the liver, compatibility of pig coagulation proteins with those of primates is one of the determinants for the success of pig to primate transplantation. Furthermore, several important molecular incompatibilities may arise when primate coagulation proteins interact with pig receptors [TM and endothelial PC receptor (EPCR)] and contribute to coagulation disorders after transplant (8). Investigations of differences between human, monkey and pig coagulation systems are critical to interpret the experimental results and to promote xenotransplantation research. In the first part of this chapter, methods for the isolation of platelet and plasma coagulation proteins from small volumes of blood are described. In addition, immunoblotting methods for the identification and characterization of the coagulation proteins of the three species considered, i.e., human, cynomolgus monkey, and pig, are reported in detail. Attention has been paid particularly on some coagulation factors including factor II (FII), factor V (FV), factor VII (FVII), factor IX (FIX), factor X (FX), PC, and protein S (PS).

The coagulation balance between anticoagulant and coagulant factors is lost by transplantation of pig organs into primate recipients. Activation of the clotting cascade, fibrin deposition and thrombosis are key features of the rejection process that takes place when pig organs are transplanted into primates (1). This hypercoagulable state is a real barrier to the long term survival of pig organs transplanted into primates (8). Several studies demonstrate that the key event in the failure of solid organ xenografts appears to be the activation of graft endothelial cells (9, 10). This prothrombotic environment may predispose to vascular thrombosis because of putative cross-species functional incompatibilities between natural anticoagulants present on the donor endothelium and host-activated coagulation factors (11). In humans, the PC pathway and Tissue factor pathway inhibitor (TFPI) are the principal regulatory mechanisms of blood coagulation since they play a fundamental role in anticoagulation. PC activation is catalyzed on endothelium by a complex composed by thrombin, TM and EPCR. Activated protein C (APC) inhibits factor VIIIa (FVIIIa) and factor Va (FVa), which are critical components of tenase and prothrombinase complexes. TFPI regulates the initiation of coagulation by inhibiting both FXa and the phospholipid-bound complex of tissue factor and FVIIa (TF-FVIIa). Impairment of these two systems results in severe thrombotic manifestations in xenotransplantation (2, 3). However, the exact mechanism by which coagulation activation is induced after xenotransplantation remains unclear (12). The two major molecular incompatibilities described in literature are between human PC and pig TM and between human FXa and pig TFPI. Pig endothelial cells are a poor activator of the human PC pathway. Experiments in which pig TM was cloned and expressed show that the incompatibility is probably due to inefficient presentation of human PC by pig TM to the catalytic site of human thrombin (13, 14). A second molecular incompatibility is believed to exist between porcine TFPI and human FXa (15, 16), although this is based on indirect evidence and awaits confirmation. In the second part of this chapter, in vitro models of PC pathway and TFPI on human umbilical vein endothelial cells (HUVEC) and porcine aortic endothelial cells (PAEC) are described which can be used for studying incompatibilities between these species.

## 2. Materials

All solutions should be prepared using distilled water. The described buffers are stable for 2 months when stored at 4°C.

### 2.1. Preparation of a Platelet Lysate for Detection of Platelet Coagulation FV and PS by Western Blotting

1. Phosphate-buffered saline (PBS): 137 mM NaCl, 2.6 mM KCl, 8 mM $Na_2HPO_4$, 1.4 mM $KH_2PO_4$, pH 7.4.
2. Ethylenediamine tetraacetic acid (EDTA).
3. Triton® X-100.
4. 50% (v/v) glycerol.
5. Tyrod's buffer: 140 mM NaCl, 2.7 mM KCl, 12 mM $NaHCO_3$, 0.36 mM $NaH_2PO_4$ $H_2O$, 1 mM $MgCl_2$ $6H_2O$, 2 mM $CaCl_2$ $2H_2O$, 5 mM dextrose, pH 7.4.

### 2.2. Purification of Vitamin K-Dependent Proteins from Plasma by Precipitation with Barium Chloride

1. 1 M $BaCl_2$ $2H_2O$.
2. 1 M Benzamidine hydrochloride hydrate, adjust the pH to 7.4.
3. 100 mM Tris–HCl, pH 7.4.
4. 0.25 M EDTA.
5. 40% trichloroacetic acid (TCA).
6. Ethanol absolute anhydrous.

### 2.3. Purification of Plasma FV by Immunoadsorption

1. 0.25 M EDTA.
2. 1 M benzamidine hydrochloride hydrate, adjust the pH to 7.4.
3. 10 mM pefabloc.
4. Anti-human FV monoclonal antibody coupled to sepharose (Haematologic Technologies, Essex Junction, Vermont, USA).
5. 1 M Tris–HCl, pH 7.4.

6. 2.5 M NaCl.
7. 100 mM $CaCl_2$.
8. 0.1 M glycine, pH 2.45.
9. 50% (v/v) glycerol.

***2.4. SDS-Polyacrylamide Gel Electrophoresis for Plasma and Platelet Coagulation Proteins***

1. Loading buffer (4×): 250 mM Tris–HCl, pH 6.8, 8% (w/v) sodium dodecyl sulfate (SDS), 40% (v/v) glycerol, 100 mM dithiothreitol (DTT), 0.08% (w/v) bromophenol blue. Store in aliquots at –20°C.
2. Molecular weight markers: prestained SDS-polyacrylamide gel electrophoresis (SDS-PAGE) standards (broad range).
3. 30% acrylamide/bis solution, 29:1 (3.3% C) (see Note 1).
4. Separating buffer: 1.5 M Tris–HCl, pH 8.8. Store at 4°C.
5. Stacking buffer: 0.5 M Tris–HCl, pH 6.8. Store at 4°C.
6. 10% (w/v) SDS solution. Stable at room temperature.
7. 10% (w/v) ammonium persulfate solution (APS). Prepare fresh as required.
8. *N,N,N′,N′*-tetramethylethylenediamine (TEMED).
9. Running buffer: 250 mM Tris base, 192 mM glycine, 0.1% (w/v) SDS. The pH of this buffer is approximately 8.3. Do not adjust pH.
10. A gradient former for preparation of four 5–15% gradient gels of 1.5 mm thickness in a casting chamber.
11. Peristaltic pump.
12. Electrophoresis apparatus attached to a cooling system (see Note 2).

***2.5. Western Blotting for Plasma and Platelet Coagulation Proteins***

1. Protein samples to be separated by SDS-PAGE and transferred to an immobilizing membrane by the semi-dry method.
2. Transfer membrane: Immobilon-P membrane (PVDF), pore size: 0.45 μm.
3. Filter paper: 1 mm thick chromatography paper.
4. Transfer buffer: 25 mM Tris Base (do not adjust pH), 200 mM glycine, 0.1% (w/v) SDS, 20% (v/v) methanol.
5. Tris-buffered saline with Tween (TBS-T): 20 mM Tris–HCl, pH 7.4, 150 mM NaCl, 0.1% Tween20.
6. Blocking buffer: 0.1% (w/v) albumin from chicken egg white in TBS-T.
7. Primary antibodies directly tagged with the horseradish peroxidase (HRP):
   - Sheep anti-human FII IgG (Affinity Biologicals, Ancaster, Ontario, Canada).

- Mouse anti-human FV IgG (Haematologic Technologies, Essex Junction, Vermont, USA).
- Sheep anti-human FVII IgG (Affinity Biologicals, Ancaster, Ontario, Canada).
- Goat anti-human FIX IgG (Affinity Biologicals, Ancaster, Ontario, Canada).
- Goat anti-human FX IgG (Affinity Biologicals, Ancaster, Ontario, Canada).
- Sheep anti-human PC IgG (Affinity Biologicals, Ancaster, Ontario, Canada).
- Goat anti-human PS IgG (Affinity Biologicals, Ancaster, Ontario, Canada).

8. Peroxidase substrate is prepared fresh as required by dissolving 10 mg 3,3′-diaminobenzidine tetrahydrochloride hydrate (DAB) in 50 mL TBS-T. 50 μL of 30% $H_2O_2$ are then added immediately before use.
9. Transfer apparatus by the semi-dry method.

### 2.6. Isolation and Culture of Endothelial Cells

1. HUVEC.
2. Dulbecco's modified Eagle's medium high glucose (DMEM/hG), supplemented with 15% of fetal bovine serum (FBS), 2 mM l-glutamine, 100 U/mL penicillin, and 100 μg/mL streptomycin (complete medium).
3. Endothelial cell growth medium (ECGM) plus endothelial cell growth supplement (International PBI, Milan, Italy) supplemented with 10% of FBS, 2 mM l-glutamine, 100 U/mL penicillin, and 100 μg/mL streptomycin.
4. Collagenase I dissolved at 1 mg/mL in DMEM/hG without serum and stored in single aliquots at –80°C. Working solutions at 0.01%.
5. DMEM high glucose containing 300 U/mL penicillin, 300 μg/mL streptomycin, and 150 μg/mL gentamicin supplemented with 0.01% collagenase I solution.
6. PBS: 137 mM NaCl, 2.6 mM KCl, 8 mM $Na_2HPO_4$, 1.4 mM $KH_2PO$. Adjust to pH 7.4 with HCl if necessary and autoclave before storage at 4°C.
7. 0.25% (w/v) trypsin solution and 1 mM EDTA.
8. 1% pig gelatin dissolved in PBS. Autoclave before storage at 4°C. For pig endothelial cells culture coat flask or culture dishes overnight with pig gelatin at room temperature.
9. Trypan blue: dissolve 0.4 g trypan blue in 100 mL of PBS. Pass through a 0.22 μm filter to remove any debris. For stain use 100 μL of cells suspension and 100 μL of 0.4% trypan blue.

10. Hank's balanced salt solution (HBSS): 50 mM Hepes pH 7.4, 150 mM NaCl, 3 mM $CaCl_2$. Pass through a 0.22 μm filter and store at 4°C.
11. HBSS plus 2% bovine serum albumin (BSA).
12. 150 μg/mL gentamicin.
13. Sterile scalpel.
14. Tissue culture incubator at 37°C under 5% CO2 humid atmosphere.

### 2.7. Immunofluorescence Microscopy and Flow Cytometry

1. PBS at room temperature.
2. PBS containing 0.5% BSA.
3. 4% (w/v) paraformaldehyde (PF) (fixing solution): dissolve 4 g of PF powder in 100 mL of PBS and then add 5 μL of 1 M NaOH. Heat to approximately 56°C for 2–4 min. The solution should become clear after the addition of the base. Divide into aliquots and store at −20°C (see Note 3).
4. Quench solution: 50 mM $NH_4Cl$ dissolved in PBS and storage at 4°C.
5. Permeabilization solution: 0.5% (v/v) Triton X-100 in PBS. Store at 4°C.
6. Antibody dilution buffer: 1% BSA in PBS (PBS/BSA). Store at −20°C. Always use fresh solution for antibodies dilution.
7. Nuclear label solution: 1.5 μg/mL Höechst 33258 or 1 μg/mL of 4,6′-diamidino-2-phenylindole (DAPI) in water (see Note 4). Store at −20°C.
8. Mounting medium antifade: Vectashield mounting media.
9. Microscope slides (26 × 76 mm) and microscope sterile circles coverslips.
10. Fluorescence microscope.
11. Flow cytometer.

#### 2.7.1. Primary Antibodies

1. Monoclonal rat anti-human EPCR. Working concentration 5–20 μg/mL. Specific only for human endothelial cells (see Note 5).
2. Sheep anti-human TM. Working concentration 0.1–0.2 μg/mL. Specific for human endothelial cells (see Note 5).
3. Monoclonal mouse anti-human EPCR diluted 1:100 in antibody dilution buffer (see Note 6).
4. Mouse anti-human TM diluted 1:50 in antibody dilution buffer (see Note 6).
5. Mouse anti-pig CD31 fluorescein isothiocyanate (FITC)-conjugated diluted in PBS. Working concentration 5 μg/mL.

6. Mouse anti-human CD62E phycoerythrin (PE)-conjugated. Working concentration 0.1 μg/mL (see Note 6).
7. Monoclonal mouse antibody against human TFPI. Working concentration 5 μg/mL.
8. Mouse IgG1 k PE Isotype control. For cells the suggested use of this reagent is 20 μL per $10^6$ cells in 100 μL PBS.
9. Mouse IgG1 k FITC isotype control. For cells, the suggested use of this reagent is 20 μL per $10^6$ cells in 100 μL PBS.

*2.7.2. Secondary Antibodies*

1. Goat anti-rat IgG FITC conjugated diluted 1:100 in antibody dilution buffer.
2. Rabbit anti-sheep IgG FITC conjugated diluted 1:100 in antibody dilution buffer.
3. Goat anti-mouse IgG FITC conjugated diluted 1:150 in antibody dilution buffer.

### 2.8. Protein C Activation by Endothelial Cells

1. 1 μM human thrombin in water. Working concentration 0.1 U/mL.
2. HBSS containing 2% BSA and 3 mM $CaCl_2$, see above.
3. Purified PC. Working concentration range (0.2–1 μM).
4. Purified APC.
5. 4 U/mL hirudin. Working final concentration 0.1 U/mL.
6. Chromogenic substrate spectrozyme PCa S2366 for determination of activated PC.
7. Multilabel plate reader.

### 2.9. TFPI Coagulation Assay

1. Xa-assay wash buffer: 0.15 mM $NaCl_2$, 4 mM KCl, 11 mM glucose, 10 mM HEPES, 0.5% fatty-acid free BSA, pH 7.5.
2. Xa-assay reaction buffer: Xa-assay wash buffer containing 5 mM $CaCl_2$.
3. Xa-assay stop buffer: Xa-assay wash buffer containing 100 mM EDTA.
4. Human FXa.
5. Human factor X (FX) dissolved in Xa-assay wash buffer.
6. Human TFPI positive control.
7. Recombinant human tissue factor: Innovin dissolved in Xa-assay wash buffer.
8. Factor VIIa dissolved in Xa-assay wash buffer.
9. Spectrozyme FXa chromogenic substrate for the amidolytic assay of FXa.
10. Multilabel plate reader.

## 3. Methods

Platelet and plasma coagulation proteins are isolated from blood of healthy human donors, cynomolgus monkeys (*Macaca fascicularis*) and pigs. Draw venous blood in 3.8% (v/v) sodium citrate (nine parts of blood to one part of sodium citrate) to prevent coagulation and process it within 1–3 h from collection.

### 3.1. Preparation of Platelet Lysate for Detection of Platelet Coagulation FV and PS by Western Blotting

Perform all the following steps at room temperature.

1. Centrifuge blood samples at 200 × *g* for 10 min without brake.
2. Transfer 1 mL of the supernatant platelet-rich plasma (PRP) to a fresh tube, add 0.2 mL of PBS/2% EDTA to prevent platelet activation and centrifuge at 1,000 × *g* for 10–15 s in a microcentrifuge to remove residual red blood cells.
3. Transfer the supernatant to a fresh tube and centrifuge at 14,000 × *g* for 45 s in a microcentrifuge to precipitate platelets.
4. Decant the supernatant, gently resuspend the platelet pellet in 1 mL of PBS/2% EDTA and centrifuge at 14,000 × *g* for 45 s in a microcentrifuge (first washing step).
5. Decant the supernatant and repeat step 4 (second washing step).
6. Decant the supernatant and gently resuspend the pellet in Tyrod's buffer at a final concentration of $1 \times 10^9$/mL. Add to the platelet suspension 1% Triton X-100 (to lyse the platelets) and 5% glycerol, aliquot and store at −80°C until use.
7. Mix the sample with the 4× loading buffer (the ratio of sample/4× loading buffer should be 1:3) and boil at 95–100°C for 5 min (see Note 7).
8. Cool at room temperature for 5 min. Briefly spin to bring down the sample and loading buffer mixture prior to loading gel.
9. Load 20–30 μL of each sample into the wells of a 5–15% gradient SDS-polyacrylamide gel. To reduce disulphide bridges in proteins, prepare a loading buffer containing 100 mM DTT.

### 3.2. Purification of Vitamin K-Dependent Proteins from Plasma by Precipitation with Barium Chloride

All centrifugations are carried out at room temperature.

1. Mix 100 μL of plasma sample with 20 μL of 0.25 M EDTA, 1 μL of 1 M benzamidine hydrochloride hydrate and 1 μL of 10 mM pefabloc (see Note 8).
2. Incubate the mixture with 20 μL of anti-human FV monoclonal antibody coupled to sepharose (3 mg IgG/mL resin), and for 30 min at 4°C with mixing.

3. Centrifuge the mixture at 14,000 × *g* in a microcentrifuge and remove the supernatant.
4. Wash the resin with 200 μL of 20 mM Tris–HCl, pH 7.4, 150 mM NaCl and 5 mM $CaCl_2$.
5. To elute FV bound to the resin, add 10 μL of 0.1 M glycine, pH 2.45 to the washed resin and centrifuge at 14,000 × *g* in a microcentrifuge.
6. Keep the supernatant and add 10 μL of 1 M Tris–HCl, pH 7.4 (to inhibit the effect of acid glycine) and 20 μL of 50% (v/v) glycerol.
7. Mix the samples with the 4× loading buffer (the ratio of sample/4× loading buffer should be 1:3) and boil at 95–100°C for 5 min. After cooling at room temperature, the samples are ready for separation by SDS-PAGE (see Note 7).

### 3.3. Purification of Plasma FV by Immunoadsorption

All centrifugation steps are carried out at room temperature.

1. Mix 100 μL of plasma sample with 20 μL of 0.25 M EDTA, 1 μL of 1 M benzamidine hydrochloride hydrate and 1 μL of 10 mM pefabloc (see Note 8).
2. Incubate the mixture with 20 μL of anti-human FV monoclonal antibody coupled to sepharose (3 mg IgG/mL resin), and for 30 min at 4°C with mixing.
3. Centrifuge the mixture at 14,000 × *g* in a microcentrifuge and remove the supernatant.
4. Wash the resin with 200 μL of 20 mM Tris–HCl, pH 7.4, 150 mM NaCl and 5 mM $CaCl_2$.
5. To elute FV bound to the resin, add 10 μL of 0.1 M glycine, pH 2.45 to the washed resin and centrifuge at 14,000 × *g* in a microcentrifuge.
6. Keep the supernatant and add 10 μL of 1 M Tris–HCl, pH 7.4 (to inhibit the effect of acid glycine) and 20 μL of 50% (v/v) glycerol.
7. Mix the samples with the 4× loading buffer (the ratio of sample/4× loading buffer should be 1:3) and boil at 95–100°C for 5 min. After cooling at room temperature, the samples are ready for separation by SDS-PAGE (see Note 7).

### 3.4. SDS-PAGE for Plasma and Platelet Coagulation Proteins

1. Before preparing the gels, clean the glass plates with a detergent, rinse well with deionized water and then with ethanol.
2. The gradient former consists of a mixing chamber and a reservoir chamber. Add a stir bar in the mixing chamber and place the gradient former on a magnetic stir plate. Connect the gradient former to a peristaltic pump.

3. Prepare the following solutions without catalysts and keep them at room temperature: *Light solution* (5% acrylamide/bis): 2.9 mL of 30% acrylamide/bis solution, 4.4 mL of separating buffer, 175 μL of 10% SDS, 10 mL of distilled water, 87.5 μL of 10% APS and 8.7 μL of TEMED. *Heavy Solution* (15% acrylamide/bis): 8.8 mL of 30% acrylamide/bis solution, 4.4 mL of separating buffer, 175 μL of 10% SDS, 4.1 mL of distilled water, 87.5 μL of 10% APS and 8.7 μL of TEMED.
4. When you are ready to form the gradient, add 87.5 μL of 10% APS and 8.7 μL of TEMED to each solution and mix. Then, pour the light solution in the mixing chamber and the heavy solution in the reservoir chamber.
5. Overlay each gel with a thin layer of water to remove the bubbles and to prevent the gel from drying out. Wait about 30 min for the gel polymerize completely.
6. Pour off the water and prepare the stacking gel by mixing 5 mL of stacking buffer with 1.8 mL of 30% acrylamide/bis solution, 13 mL of distilled water, 200 μL of 10% SDS, 200 μL of APS and 10 μL of TEMED. Pour the stacking gel on top of the separating gel and insert the combs. The stacking gel should be polymerized in about 30 min.
7. Once the stacking gel is set, carefully remove the combs, install the gels into the electrophoresis unit and fill the upper and lower chambers with the running buffer. Load the samples into the wells. Include one well for prestained molecular weight markers.
8. Connect the electrophoresis unit to a power supply and run the gel at 15 mA until the samples reach the resolving gel border, then increase the current at 25 mA. The run is complete when the tracking dye reaches the bottom of the gel (usual running time is 2 h).

#### 3.5. Western Blotting for Plasma and Platelet Coagulation Proteins

1. After the separation of proteins by SDS-PAGE, the gel unit is disconnected from the power supply and disassembled. The stacking gel is removed and discarded except for the coagulation Factor V. The gel is equilibrated in transfer buffer for 5 min (see Note 9).
2. Meanwhile, cut a piece of PVDF membrane and 18 pieces of filter paper to the same size as the gel to be transferred (see Note 10). To avoid membrane contamination, always use tweezers or wear gloves when handling the membrane.
3. Activate PVDF membrane in a small volume of 100% methanol for 30 s, rinse it in deionised water and then equilibrate it in transfer buffer for approximately 5 min.

4. Soak the first layer of nine filter papers in transfer buffer for a few seconds, place them onto the anode electrode plate and roll a pipette over the surface of the papers to eliminate all air bubbles.
5. Place the preequilibrated PVDF membrane on top of the filter papers and remove air bubbles once more.
6. Place the preequilibrated gel on top and on the centre of the membrane, again taking care to avoid trapping air bubbles.
7. Immerse nine additional pieces of filter papers in transfer buffer for a few seconds and place them on top of the gel to complete the transfer sandwich. Roll out air bubbles.
8. Carefully place the cathode electrode plate on top of the transfer sandwich.
9. The blotting unit is then connected to a power supply and proteins are transferred for 1 h at a constant current of 0.8–1.0 mA/cm$^2$ and at room temperature.
10. When the transfer is complete, disconnect the blotting unit from the power supply. Remove the safety lid and the upper cathode electrode. Carefully disassemble the transfer sandwich and remove the membrane for analysis. It is prudent to mark the membrane by notching or clipping a corner to allow its orientation.
11. Incubate the membrane in 50 mL of blocking buffer overnight at 4°C on a rocking platform (see Note 11).
12. Remove the blocking buffer and wash the membrane three times for 10 min each with TBS-T under gentle agitation.
13. Incubate the membrane with HRP-conjugate primary antibody diluted in TBS-T for 2 h at room temperature with shaking. The optimal dilution should be determined for each antibody. Dilution can range from 1:100 to 1:1,000.
14. Wash the membrane three times for 10 min each with TBS-T under gentle agitation to remove residual primary antibody.
15. Immerse the membrane in freshly prepared peroxidase substrate and shake at room temperature until desired band intensity is achieved. Stop the reaction by rinsing the membrane with deionised water. Then, to dry the membrane, press it between two sheets of filter paper. An example of the results produced is shown in Fig. 1.

### 3.6. Isolation and Culture of Endothelial Cells

#### 3.6.1. Isolation of PAEC

1. For PAEC isolation, harvest fresh aortas from large white swine and transport to the laboratory in clean containers.
2. Under a biological laminar-flow cabin, wash the aorta extensively with cold DMEM/hG containing 300 U/mL penicillin, 300 μg/mL streptomycin and gentamycin 150 μg/mL.

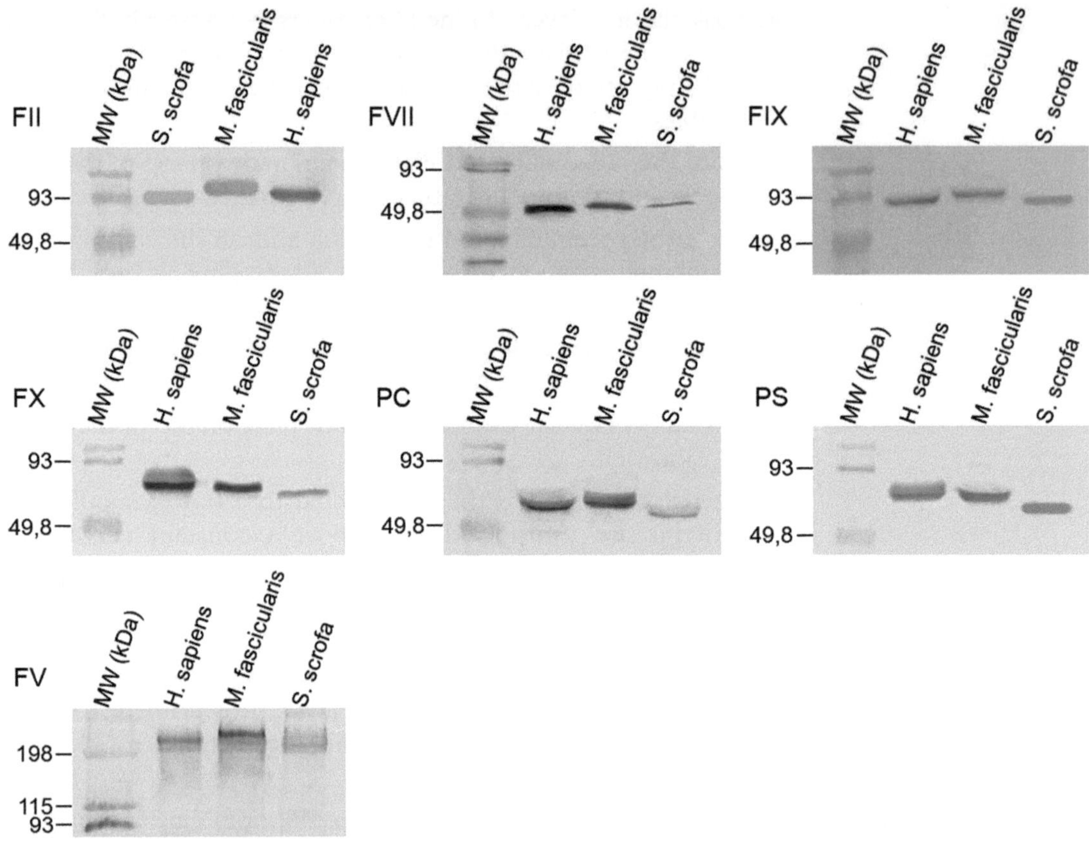


Fig. 1. Vitamin K-dependent proteins (FII, FVII, FIX, FX, PC, and PS) and FV, isolated from citrated blood of human donors (*Homo sapiens, H. sapiens*), cynomolgus monkeys (*Macaca fascicularis, M. fascicularis*), and pigs (*Sus scrofa, S. scrofa*), were analyzed on a 5–15% SDS-PAGE under nonreducing conditions followed by transfer to PVDF membrane. Coagulation factors were then detected with monoclonal or polyclonal antibodies according to the materials and methods described in the Chapter. *MW* molecular weight standards.

3. Dissect the aorta with a sterile scalpel.
4. Open the aorta and wash two times with the same medium above supplemented with 0.01% collagenase I solution.
5. Scrape the aorta with single scalpel strokes.
6. Transfer the cells to a 15-mL centrifuge tube.
7. Incubate the cells for 2 min at 37°C to break up the sheets of tissue in fragments.
8. Add complete medium containing 15% of FBS to block the action of collagenase I.
9. Centrifuge at 400 × *g* for 15 min at 22°C.
10. Discard the supernatant and suspend the cells with 5 mL of complete medium.
11. Seed the cells into a T-25 flask precoated with pig gelatin.

12. Change the culture medium every 2–3 days.
13. To propagate the cells, split the confluent cells 1:4 with trypsin–EDTA into gelatin precoated flasks (see Note 12).

#### *3.6.2. Trypsinization of Pig and Human Endothelial Cells*

1. Examine the flasks of cultured cells under the inverted microscope to check the density and the morphology of the cells.
2. Prewarm all reagents to 37°C in a water bath.
3. Remove the old medium.
4. Wash the monolayer cells with PBS without calcium and magnesium twice.
5. Remove the PBS and add 2–4 mL (for a T-75 flask) of trypsin–EDTA solution which has been prewarmed to 37°C.
6. Incubate the cells in the incubator for about 3–5 min, and check the flask under inverted microscope to make sure the monolayer of cells is lifting off of the flask.
7. Afterward, when a single cell suspension has been obtained, add 5 mL of complete medium to block the action of trypsin.
8. Transfer the cells suspension to a sterile 15 mL conical centrifuge tube.
9. Centrifuge the cells suspension at $400 \times g$ for 10 min.
10. Discard the supernatant carefully.
11. Suspend the cell pellet in about 5 mL of complete medium.
12. Afterward, collect 100 μL to count cell number.
13. Count the total number of cells using a Bürker hemocytometer and evaluate cell viability by trypan blue staining exclusion.
14. Once cells/mL has been determined, seed the appropriate number of cells into gelatin precoated flasks (T25–T75) or 24-well plate (see Note 12).

### 3.7. Immunofluorescence Microscopy and Flow Cytometry

#### *3.7.1. Flow Cytometry*

1. Prepare $0.5\text{–}1 \times 10^6$ cells/mL of PAEC to characterize by flow cytometry for the expression of pig endothelial markers such as CD31, E-selectin, TM, and EPCR.
2. To this end, harvest PAEC cells by trypsin–EDTA as described.
3. Centrifuge at $400 \times g$ for 10 min and resuspend the pellet in 1 mL PBS containing 0.5% BSA.
4. Incubate cells for 30 min in the dark at room temperature with the different antibodies: anti-CD31-FITC, anti-CD62E-PE, anti-TM, and anti-EPCR.
5. When using nonconjugated primary antibodies, incubate the cells with the appropriate secondary antibodies. Incubate secondary antibodies for 30 min in the dark at room temperature.

6. Wash twice with PBS.
7. After centrifugation resuspend the cells in 500 μL of PBS.
8. Analyze by flow cytometry.
9. Use as a negative control the following antibodies: IgG1-PE, clone MOPC-21(BD Biosciences), mouse IgG1-FITC, clone MOPC-21 (BioLegend, Europe).

*3.7.2. Immunofluorescence Analysis*

1. Grow HUVEC and PAEC into gelatin precoated 24 wells plate containing sterile circle coverslips (see Note 13). When the cells are 70% confluent, remove the medium and wash the cells twice in PBS.
2. Add 4% PF solution and incubate for 20 min at room temperature to fix the cells.
3. Discard the PF (into a hazardous waste container) and wash the sample twice for 5 min each with PBS.
4. Afterward, treat the sample with 50 mM $NH_4Cl$ for 10 min at room temperature for quenching residual PF, followed by two more washes with PBS.
5. Permeabilize cells with 0.5% Triton X-100 in PBS for 15 min and then rinse the cells in PBS two times.
6. Incubate HUVEC and PAEC for 1 h at 37°C in a humidified incubator with specific antibodies (anti-TM, anti-EPCR, and anti-TFPI).
7. At the end of the incubation, remove the primary antibody by washing the cells twice with PBS.
8. Incubate the cells for 1 h at 37°C in a humidified incubator with the appropriate secondary antibodies.
9. After the secondary antibody incubation, wash the cells twice for 5 min each at room temperature. To prevent the decay of fluorescence in the samples during the following procedure, cover them with aluminum foil.
10. To identify the nuclei, incubate the cells for 10 min at room temperature in the dark with 1.5 μg/mL Höechst or 1 μg/mL DAPI.
11. Rinse with PBS and mount the coverslips with 5 μL antifading solution on a microscope slide. Be careful to put the coverslip inverted on the drop of mounting medium on the slide. Seal the coverslip with nail polish to prevent drying and movement under the microscope. Store the slides at 4°C in dark for up to a month.
12. In order to assess nonspecific binding of the secondary antibodies, perform the same procedure described above without using primary antibody.

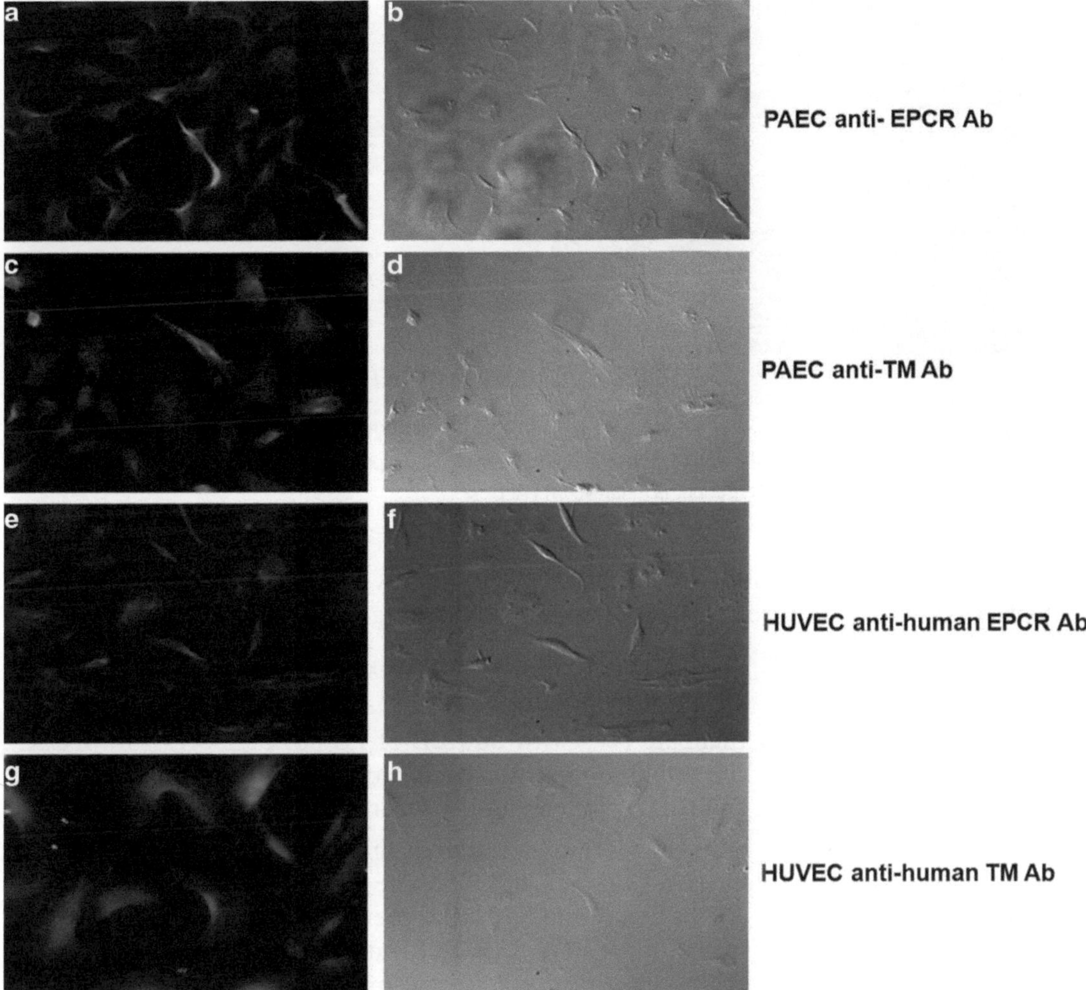


Fig. 2. Immunofluorescence staining of PAEC (**a–d**) and HUVEC (**e–h**) cells. Immunofluorescent labeling of endothelial protein C receptor (EPCR) (**a**, **e**) and thrombomodulin (TM) (**c**, **g**) with antibodies specific for human and pig cells. (**b**, **d**, **f**, **h**) Differential Interference Contrast (DIC). The nucleus was counterstained with Höechst (data not shown). Microscope objective 40×. Images were obtained on a Leica DM 5000 microscope and processing with Leica Application Suite (LAS) Microscope Software. Immunofluorescent staining reveals the presence of TM and EPCR on the membrane surface of cells and in the cytoplasm.

13. Analyze the slides with a fluorescent microscope. Höechst or DAPI fluorescence of nuclei is visualized by excitation at 330–385 nm with a 450 nm barrier filter. FITC fluorescence is visualized by excitation at 475–490 and emission at 530 nm. All samples are analyzed by differential interference contrast (DIC) objective. For this analysis, all images are viewed and captured at 40×, 60×, and 100× (with oil) magnification. The expression of TM and EPCR is clearly visible on the surface of the cells and in the intracellular space in both cell species (Fig. 2). As for TFPI, the expression is localized on the cellular surface in HUVEC and is negative in PAEC (Fig. 3).

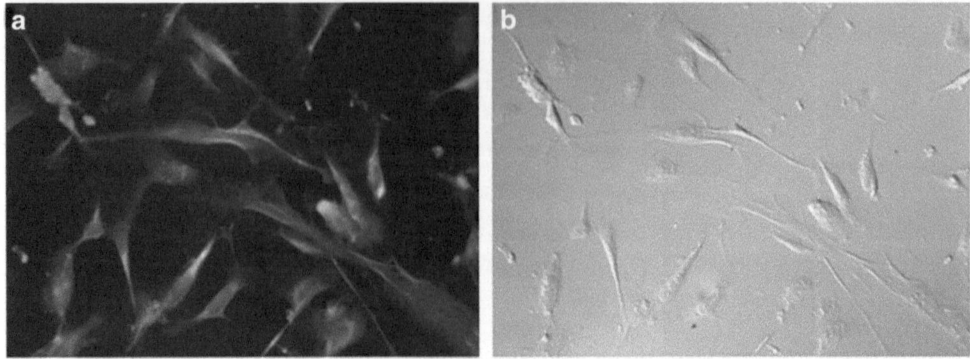


Fig. 3 (**a**) Immunofluorescence staining of HUVEC with anti-human TFPI. (**b**) Differential Interference Contrast (DIC). DNA was labeled with Höechst (data not shown). Microscope objective 40×. The expression of human TFPI is mainly localized on the membrane cells surface on HUVEC, but is negative on PAEC.

### 3.8. Protein C Activation Assay

Monolayers of HUVEC and PAEC are formed by overnight culture in 24-well plates coated with gelatin.

1. Wash monolayer cells once with PBS and then with HBSS containing 2% BSA and 3 mM of $CaCl_2$.
2. Incubate cells with 200 μL of HBSS/2% BSA/3 mM $CaCl_2$ containing different concentrations of human PC (0.2–1 μM) and 0.1 U/mL of human thrombin.
3. Stop the reaction at different time points by adding 0.1 U/mL hirudin for 5 min at 37°C. The amount of activated PC generated is determined by using the specific chromogenic substrate spectrozyme PCa (1 mM). Monitor the absorbances at 405 nm against time using a kinetic microplate reader (Victor). The concentration of APC is calculated by using a standard curve of purified activated PC. As a control, the same experiment is carried out without cells.

### 3.9. TFPI Coagulation Assay

#### 3.9.1. TFPI Activity and Human Factor Xa Binding Assay

1. Seed $1.5 \times 10^5$ HUVEC and PAEC into gelatin precoated 24-well plate and incubate for 24 h.
2. Wash twice with Xa-assay wash buffer.
3. Afterwards, incubate cells for 30 min at 37°C with Xa-assay reaction buffer and different concentrations of FXa (0.3, 0.5, 1 nM/well) in a total final volume of 200 μL.
4. Terminate the reaction by the adding 5 μL of Xa-assay stop buffer.
5. Determine residual human FXa amidolytic activity by using spectrozyme Xa. Spectrozyme FXa chromogenic substrate is added to a final concentration of 1 mM and the absorbance is measured at 405 nM in a multilabel plate reader (Victor). The concentration of unbound FXa is determined by a standard

curve of purified human FXa. The slope of the absorbance curve at each concentration of FXa is compared to that in the absence of cells with and without 12.5 nM recombinant human TFPI.

*3.9.2. Inhibition of Human Factor VIIa/TF Activity*

1. Wash confluent HUVEC and PAEC twice with Xa-assay wash buffer.
2. On ice, mix excess of human FVIIa (200 pM) in FXa-assay reaction buffer with various concentrations of TF (0–32 pM) in a final volume of 200 μL.
3. Incubate the mixture for 15 min at 37°C to form the TF/FVIIa complex.
4. Afterwards, add 200 μL/well of TF/FVIIa complexes to the cells and incubate at 37°C for 15 min.
5. Add 100 μL of 120 nM human FX in Xa-assay reaction buffer and incubate for a further 15 min at 37°C. Terminate the reaction by the addition of 5 μL Xa-assay stop buffer.
6. Measure the decrease in FXa generated by the inhibition of FVIIa/TF activity and compare it to that observed in the presence of recombinant TFPI. To determine the concentration of residual FXa, use spectrozyme FXa chromogenic substrate to a final concentration of 1 mM and measure the absorbance at 405 nm in a Victor multilabel plate reader. As control, determine the inhibition of FVIIa/TF activity or FXa activity in the presence of anti-human TFPI antibody in a saturating concentration.

## 4. Notes

1. Acrylamide is a potent neurotoxin; always wear gloves when handling it.
2. The gel electrophoresis tank must be kept cool in order to prevent a smiling effect of the bands. This is done by running the gel at constant 16°C.
3. Paraformaldehyde is flammable, is a carcinogen, and produces irritant vapor. PF is toxic and the preparation process involving heating results in considerable vaporization, which increases the hazard. It is essential to use appropriate safety procedures, such as working in a fume hood.
4. DAPI and Höechst are known mutagens and should be handled with care. The dyes must be disposed of safely and in accordance with applicable local regulations.
5. The appropriate dilution should be determined for each antibody.

6. The antibody recognizes receptors in both human and pig endothelial cells.
7. During protein sample treatment, the sample should be mixed by vortexing before and after the heating step for best resolution.
8. Benzamidine hydrochloride hydrate and pefabloc are used as inhibitors of serine proteases.
9. Equilibration helps to remove the salts and detergents contained in the electrophoresis buffer. If the salts are not removed, they will increase the conductivity of the transfer buffer and the amount of heat generated during the transfer.
10. To ensure that the current passes through the gel, all components of the transfer sandwich are cut to the same size as the gels to be transferred.
11. This step serves to block unoccupied binding sites on the membrane to prevent nonspecific binding of the primary antibody.
12. Certain cell types are only viable up to a certain passage number. Make sure to check the passage number of cells before splitting beyond the appropriate passage number.
13. If coverslips are not sterilized they must be sterilized before put on the plate. Dip them in 95% ethanol and passing through the flame of a Bunsen burner while holding loosely between forceps and then put on the well (pay attention to ethanol; it is highly inflammable; wait a second so that ethanol drips before flaming the coverslips).

## References


1. Cozzi E, Simioni P, Boldrin M et al (2004) Alterations in the coagulation profile in renal pig-to-monkey xenotransplantation. Am J Transplant 4:335–345
2. Cozzi E, Simioni P, Boldrin M et al (2005) Effects of long-term administration of high-dose recombinant human antithrombin in immunosuppressed primate recipients of porcine xenografts. Transplantation 80:1501–1510
3. Simioni P, Boldrin M, Gavasso S et al (2011) Effects of long-term administration of recombinant human protein C in xenografted primates. Transplantation 91:161–168
4. Robson SC, Cooper DK, d'Apice AJ (2000) Disordered regulation of coagulation and platelet activation in xenotransplantation. Xenotransplantation 7:166–176
5. Chen Y, Qiao J, Tan W et al (2009) Characterization of porcine factor VII, X and comparison with human factor VII, X. Blood Cells Mol Dis 43:111–118
6. Chen Y, Qin S, Tan W et al (2009) Cloning and comparison of factor X from rhesus monkey (Macaca mulatta). Comp Med 59: 476–481
7. Qin SF, Tan WD, Chen YN et al (2008) cDNA cloning, protein structure modeling of rhesus monkey (Macaca mulatta) prothrombin. Transplant Proc 40:603–606
8. Crikis S, Cowan PJ, d d'Apice AJ (2006) Intravascular thrombosis in discordant xenotransplantation. Transplantation 82:1119–1123
9. Bach FH, Winkler H, Ferran C (1996) Delayed xenograft rejection. Immunol Today 17:379
10. Cowan PJ (2007) Coagulation and the xenograft endothelium. Xenotransplantation 14:7–12

11. Robson SC, Schulte am Esch J 2nd, Bach FH (1999) Factors in xenograft rejection. Ann N Y Acad Sci 875:261–276
12. Pierson RN 3rd, Dorling A, Ayares D et al (2009) Current status of xenotransplantation and prospects for clinical application. Xenotransplantation 16:263–280
13. Roussel JC, Morana CJ, Salvaris EJ et al (2008) Pig thrombomodulin binds human thrombin but is a poor cofactor for activation of human protein C and TAFI. Am J Transplant 8: 1101–1112
14. Kopp CW, Grey ST, Siegel JB et al (1998) Expression of human thrombomodulin cofactor activity in porcine endothelial cells. Transplantation 66:244–251
15. Kopp CW, Siegel JB, Hancock WW et al (1997) Effect of porcine endothelial tissue factor pathway inhibitor on human coagulation factors. Transplantation 63:749–758
16. Lee KF, Salvaris EJ, Roussel JC et al (2008) Recombinant pig TFPI efficiently regulates human tissue factor pathways. Xenotransplantation 15:191–197

Chapter 7

# Cellular Studies for In Vitro Modeling of Xenogeneic Immune Responses

**Roberta Sommaggio, Magdiel Pérez-Cruz, and Cristina Costa**

## Abstract

Cellular studies are essential in the xenotransplantation field in order to investigate the cellular immune responses triggered by xenogeneic cells and identify the key molecules involved. A series of functional studies can be conducted with this purpose that include treatment with proinflammatory cytokines and xenogeneic cell-based assays that put together pig cells and human leukocytes such as monocytes, NK cells, and T cells. The choice of the pig cell type is critical to appropriately model the transplant setting of interest. Thus, pig endothelial cells are commonly used for studying the rejection process of vascularized organs. Treatment with cytokines allows studying the regulation of adhesion, costimulatory molecules, and receptors involved in triggering the immune response in an attempt to reproduce the more complex in vivo situation. The adhesion assays are used to determine the capacity of human leukocytes to adhere to porcine cells under various conditions. Furthermore, we describe coculture, costimulatory, and cytotoxicity assays for investigating the cellular and molecular mechanisms that take place during the xenogeneic immune response.

**Key words:** Pig, Human, Monocytes, T cells, NK cells, Cytokines

## 1. Introduction

The cellular assays are used to model the rejection process in a simplified setting. In this manner, they help elucidate the cellular and molecular mechanisms of xenograft rejection and set the bases for the development of strategies that counteract them. Cell-, tissue-, and organ-based xenografts are subjected to distinct immune rejection processes that share humoral and cellular mechanisms. After the introduction of genetic modifications in the donor pig to prevent hyperacute rejection, solid organs obtained from such pigs are mainly rejected by a process called acute xenograft rejection (AXR).

Cristina Costa and Rafael Máñez (eds.), *Xenotransplantation: Methods and Protocols,* Methods in Molecular Biology, vol. 885, DOI 10.1007/978-1-61779-845-0_7, © Springer Science+Business Media, LLC 2012

Acute cellular xenograft rejection (ACXR) corresponds to the cellular component of AXR and is mainly mediated by a T cell response to donor antigens that takes place within days after transplantation. Different studies have demonstrated the presence of a mononuclear cell infiltrate containing primarily CD4+ T cells, CD8+ T cells, macrophages, and some NK cells in solid organs transplanted into nonhuman primates without immunosuppression (1, 2). Nevertheless, current immunosuppression protocols markedly decrease the cellular immune infiltrate and prolong survival of solid organ xenografts, but they do not prevent the strong humoral response that leads to AXR (1, 3). The rejection process of cell-based xenografts also comprises a humoral and a cellular component. However, cell-mediated rejection seems more relevant in this setting as the use of immunotherapies that suppress T cell activation results in long-term survival of porcine islets in diabetic monkeys (<140 days) (4, 5). Thus, understanding the underlying mechanisms of how xenografts trigger cellular rejection is of high interest to develop pharmacological and genetic engineering approaches that promote engraftment.

Most of what we know about the T cell response to xenografts results from cellular studies that coculture porcine aortic endothelial cells (PAEC) with human peripheral blood mononuclear cells (PBMC) or purified lymphocyte populations. Porcine endothelial cells such as PAEC can function as authentic antigen presenting cells promoting direct activation of $CD8^+$ and $CD4^+$ T cells through respective binding of SLA I and II together with costimulatory molecules (6, 7). Thus, engagement of human CD28 by porcine CD86 is a major costimulatory mechanism used by PAEC to activate human T cells (6), and it can be targeted through genetic modification (8).

AXR also comprises an innate immune cellular response that has received much attention because it could be responsible of relevant rejection mechanisms resistant to immunosuppression. In particular, NK cells are known to contribute to solid organ rejection, being their role more prominent in rejection of xenografts. Evidence is in part derived from observations that human NK cells lyse PAEC in vitro (9, 10). The incapacity of SLA class I molecules to signal through human NK cell inhibitory receptors is one of the reasons why pig cells are susceptible to human NK-cell-mediated lysis (9). Furthermore, human NK cells can adhere and lyse porcine endothelial cells both directly and by antibody-dependent cell-mediated cytotoxicity (ADCC) in the presence of human serum (10). The cellular studies have also allowed to identify some key NK cell activating receptors and pig ligands implicated in triggering cytotoxicity (11). Interestingly, selected genetic modifications of the pig cell can dramatically reduce cytolysis (8). Monocytes/macrophages are also important cells involved in AXR. They are present from an early stage in the xenograft of nonhuman primates (1, 2) and contribute to xenograft rejection of solid organs in small animal models (12). Different studies demonstrated that these cells exhibit an activated cytokine profile and mediate their effects through NO and/or TNFα (12). Human monocytes express galectin-3 on their cell surface, a lectin that interacts with the major carbohydrate antigen on pig cells (13). Furthermore, the pig adhesion molecules VCAM-1 and ICAM-1 participate in the binding of PAEC to human monocytes (14). A pathway to take into account as a good target for intervention is that of the signal-regulatory protein-α (SIRPα), a receptor expressed on macrophages which interacts with ubiquitously expressed CD47 and prevents the phagocytosis of cells (15).

In the last decades, major progress has been made to understand the immunobiology of pig-to-primate xenotransplantation. However, the cellular assays remain a key tool to elucidate the molecular mechanisms of rejection and test the strategies to counteract them. In this chapter we describe the methodology of some basic cellular assays that can be easily adapted depending on the purpose of the study. We have used various of these techniques for a recent study characterizing the cellular immune response against porcine chondrocytes (16) and it could be applied to other cells of interest for xenotransplantation. The static adhesion assays are used to assess the capacity of various human immune cells to adhere to porcine cells and determine the adhesion molecules involved. The coculture assays allow to study the response of both human and porcine cells under conditions that favor cell contact. In particular, the costimulation assays are performed to identify the cell surface pig molecules that provide functional costimulatory signals to human T cells. Finally, the cytotoxicity assay is a fundamental tool for the study of NK cell function that can be applied to identify the triggering ligands and receptors.

## 2. Materials

### 2.1. Cell Culture and Cytokine Treatments

1. Porcine cells of interest for xenotransplantation such as PAEC or PK-15 (see Note 1).
2. Cytokines: multiple pig and human cytokines, as well as combinations of them, can be used (see Note 2).
3. Rubbing alcohol.
4. Sterile tissue culture supplies: 15-mL conical tubes, culture flasks, pipettes, Pasteur pipettes and tips.
5. Culture medium appropriate for the chosen cell: Most established cell lines do well in Dulbecco's modified Eagle's medium (DMEM) containing high glucose, 10% fetal bovine serum (FBS) (see Note 3), 200 U/mL penicillin and 200 μg/mL

streptomycin. Primary cultured cells usually have special requirements such as media supplements (see Note 4).

6. Solution of trypsin (0.05%) and ethylenediamine tetraacetic acid (EDTA) (1 mM) for detachment of adherent cells. The reagent TrypLE™ Express (Invitrogen Carlsbad, CA, USA) works well.
7. Tissue culture equipment: Pipettors, laminar-flow cabin, aspiration pump or source of vacuum, 5% $CO_2$ incubator, centrifuge inverted microscope and Neubauer chamber.

### *2.2. Static Adhesion Assays*

1. Porcine cells of interest for xenotransplantation such as PAEC (see Note 1). These can be tested in resting conditions and after cytokine stimulation.
2. Human leukocytes. It is preferable to assess adhesion of a purified population such as NK cells or CD4+ T cells. The activation state of the cells should be taken into account as many can be assayed either freshly isolated or after culture under controlled conditions (e.g., IL-2 activated NK cells, monocyte-derived macrophages). We recommend to assess both settings when possible. A simpler approach is to assess adhesion of an established cell line such as the NK92 cells or the U937 monoblastic cells. The assay we describe here should work for all of them.
3. Sterile tissue culture supplies: 15-mL conical tubes, culture flasks, pipettes, Pasteur pipettes, tips and 96-flat-bottom-well plates.
4. Cell culture medium appropriate for the chosen pig cells: see Subheading 2.1 for description of culture media of pig cells.
5. Phytohemagglutinin (PHA-P). A solution of 100 μg/mL can be prepared and stored at –20°C in aliquots. Do not filter to prevent the reagent loss.
6. Cell culture medium appropriate for the chosen human leukocytes: It has to be adjusted to each cell type. U937 do well in DMEM/10% FBS, but cultured NK cells will need Iscove's modified Dulbecco's medium (IMDM) supplemented with 10% (v/v) human serum, 10%(v/v) nonessential amino acids, 10% (v/v) sodium pyruvate, 200 U/mL of penicillin/streptomycin, and 200 U/mL recombinant human IL-2 (hIL-2). For 3-week culturing, the NK cells are seeded in the same medium in 96-round-bottom-well plates with irradiated JY feeder cells, 0.5 μg/mL PHA-P and 5 ng/mL recombinant human IL-15. Fresh feeder cells are added also on day 5 of culture to restimulate NK cells.
7. TrypLE™ Express or equivalent reagent.
8. Hanks' balanced salt solution (HBSS) 1% BSA for all incubations and washes (see Note 5).

9. Labeling solution: Calcein AM (10 μg/mL, Invitrogen). It has to be prepared fresh for each assay. Prepare a tube with 5 mL of HBSS/1% BSA wrapped in aluminum foil. Dissolve the content of a calcein AM 50 μg vial in 10 μL of dimethylsulfoxide (DMSO), mix well with about 0.5 mL of the HBSS/1% BSA and transfer back to the wrapped tube. This amount usually suffices for an assay, but it can be scaled up if needed. Store at room temperature until use.
10. Equipment: tissue culture equipment as described in Subheading 2.1 and cytofluorometer. The cytofluorometer is to be used at 485 nm excitation filter and 520 emission filter.
11. Optional: multichannel pipettor and basins. For large assays, it can facilitate the task, especially washes.

### 2.3. Coculture Assay

1. Porcine cells of interest for xenotransplantation such as PAEC (see Note 1). These can be tested in resting conditions and after cytokine stimulation.
2. Human leukocytes: preferably a purified population such as monocytes to determine their response in front of pig cells. Nevertheless, more complex settings can be assayed as well such as PBMC or combinations of T cells and monocytes/macrophages. A simpler approach is to use an established cell line such as the U937 monoblastic cells. We describe a basic assay for a single population that can be modified for more complex settings.
3. Sterile tissue culture supplies: 15-mL conical tubes, culture flasks, pipettes, Pasteur pipettes, 1.5-mL tubes, tips, and 96-flat-bottom-well plates. The assay can also be conducted in other multiwell plates such as 24 or 48-well plates depending on the amount of supernatant desired.
4. Cell culture medium: DMEM/10% FBS with antibiotic will work for PAEC and monocyte cocultures.
5. DMEM alone for washes.
6. ELISA kits for detection of pig and human IL-8 and/or IL-6 (see Note 6).
7. Equipment: tissue culture equipment as described in Subheading 2.1, multichannel pipettor, basins, and microplate reader.

### 2.4. Costimulation Assay

1. Porcine cells of interest for xenotransplantation such as PAEC.
2. Human T cells: preferably a purified population such as $CD4^+$ T cells or $CD8^+$ T cells. Nevertheless, a simpler approach is to use an established cell line such as Jurkat T cells that can

provide relevant information about the presence on pig cells of functional costimulatory molecules for human receptors.

3. Sterile tissue culture supplies: 15-mL conical tubes, culture flasks, pipettes, Pasteur pipettes, 1.5-mL tubes, tips, and 96-flat-bottom-well plates.
4. Culture medium for PAEC: DMEM/10% FBS with antibiotic.
5. Culture medium for Jurkat/T cells: RPMI/10% FBS.
6. PHA-P. Remember that a solution of 100 μg/mL can be prepared and stored at −20°C in aliquots. Do not filter to prevent the reagent loss.
7. Purified anti-human CD28 (hCD28) mAb CD28.2 (BD Biosciences, San José, CA, USA). This is an agonist antibody to be used as positive control.
8. ELISA kit for detection of hIL-2.
9. Equipment: tissue culture equipment as described in Subheading 2.1, multichannel pipettor, basins and microplate reader.

### 2.5. NK Cytotoxicity Assay

1. Porcine cells of interest for xenotransplantation such as PAEC.
2. NK cells. These can be tested either freshly isolated or after culture with hIL-2. The NK92 cell line can also be used as parallels well the activity of hIL-2-activated NK cells on pig cells, but it does not reflect the donor variability.
3. Sterile tissue culture supplies: 15-mL conical tubes, culture flasks, pipettes, Pasteur pipettes, tips and round-bottom 96-well plates.
4. Culture medium for PAEC: DMEM/10% FBS with antibiotic. Use the appropriate medium for culturing other pig cells.
5. Culture medium for the cytotoxicity assay: IMDM/10% FBS with 200 U/mL penicillin/streptomycin. The IMDM/10% human serum can only be used when the effect of xenogeneic antibodies is assessed.
6. TripLE Express or cell dissociation buffer.
7. Optional: Blocking antibodies for NK receptors such as NKG2D (clone# 149810 R&D systems, Minneapolis, MN, USA), NKp44 (p44/8, Beckman Coulter GmbH, Krefeld, Germany) and NKp46 (Clone 9E2, Beckman Coulter GmbH).
8. Equipment: tissue culture equipment as described in Subheading 2.1 and a room equipped for radioactive tissue culture with gamma counter.

## 3. Methods

### *3.1. Cell Culture and Cytokine Treatments*

1. Use primary cultured pig cells at low passages with the recommended media and monitor for morphologic and phenotypic changes when necessary. This is not necessary for well established cell lines.
2. Split primary cultured cells such as PAEC when they reach confluence 1–2 or 1–3. Do not further dilute the PAEC for subculture as low density culture causes fast senescence. However, other primary cultured cells may have other particular requirements that need to be taken into account for each specific cell type.
3. Treat PAEC, pig chondrocytes or established cell lines with cytokines after they reach confluence. We routinely pre-incubate with either 10 ng/mL hTNFα for 24 h, or 5–10 ng/mL hIL-1α for 8 h before other assays to activate the cells as a model of a pro-inflammatory setting.

### *3.2. Static Adhesion Assay*

1. Seed the pig cells to reach confluence in 96-flat-bottom-well plates overnight.
2. Leave these cells untreated or pre-incubate with cytokines as described in preparation for the adhesion assay. Prepare triplicates for each assay condition.
3. First, wash the chosen effector cells by centrifugation (see Note 7) and resuspend in 10 μg/mL calcein AM. Incubate for 20 min at 37°C for loading. The amount of cells to be used can be estimated in advance following the calculations described below. Prepare always in excess. Use HBSS/1% BSA for all the assay incubations and washes.
4. In the case of assessing the contribution of specific molecules to adhesion, one good option is the utilization of blocking antibodies just prior to the addition of the human leukocytes. For blocking pig molecules, wash the plate and keep at room temperature for 20 min with the blocking antibody and the isotype control in designated wells at 2× concentration (10 μg/mL is usually sufficient). This procedure can be conducted simultaneously to the labeling of the effector cells.
5. Wash the calcein loaded cells by centrifugation, resuspend in HBSS/1% BSA and count (see Note 8). For blocking human molecules, pre-incubate a sufficient amount of labeled cells (such as NK92) (see Note 7) with the blocking or the isotype control antibody for 10 min at room temperature.

6. Add the effector cells in equal volume (e.g., 100 μL for a 200 μL final volume) to the confluent pig cells at a specific effector/target (E/T) ratio in triplicates (see Note 9). We recommend to work in a range of 1:1 to 5:1 E/T ratio (see Note 10), but it may be of interest to assess higher ratios for establishing the amount needed to saturate the system.
7. Optional: Centrifuge the plate at 16 × *g* for 2 s without brake.
8. Optional: Determine the start fluorescence with cytofluorometer at 485/520 nm.
9. After an incubation of 30–45 min at 37°C, remove nonadherent cells with thorough washes. When testing the effect of antibodies, these can be kept during the coculture time (a final concentration of 5 μg/mL is usually sufficient).
10. Lyse the remaining cells with 1% SDS.
11. Measure fluorescence intensity with a cytofluorometer at 485/520 nm. The percentage of adhesion can be calculated by dividing the end value with its corresponding start values (if taken) and multiplying by 100.

### 3.3. Coculture Assay

1. Seed the pig cells in multiwell plates to reach confluence overnight.
2. Leave them untreated or preincubate with cytokines as described.
3. Wash the cells with DMEM alone just prior to the assay.
4. Harvest and count the effector cells (see Note 8).
5. Initiate the coculture by adding the effector cells to the pig cells at the desired E/T ratio. We recommend to work in a range of 1:1 to 5:1 E/T ratio (see Note 10). Assay each condition at least in duplicate.
6. Maintain the coculture for the desired length of time. The coculture time should be established for each study, but we find that cocultures of 18–24 h usually lead to a clear response. Shorter times such as 8 h can be used for kinetics, but may result in a partial response for changes in protein expression or secretion.
7. Collect and centrifuge the supernatants at 260 × *g* for 5 min.
8. Harvest the cell-free supernatants. These can be stored in 1.5-mL tubes or 96-well plates (depending on the volume) at −20°C until use. Regarding the cellular pellet, it can be discarded or processed for further analyses (see Note 11).
9. Determine the levels of human and porcine IL-8 in supernatants by ELISA using a commercialized kit following the instructions of the manufacturer. Use a microplate reader at 450 nm.

### 3.4. Costimulation Assay

1. Seed the pig cells in 96-flat-bottom-well plates to reach confluence overnight in preparation for the coculture.
2. On the day of the assay, harvest and count the Jurkat or human T cells after some pipetting up and down to disperse the cells (which usually grow in clusters) (see Note 8).
3. At this time point, the media of the pig cells may be removed and changed to RPMI/10% FBS (e.g., 100 μL). In experiments determining the contribution of specific molecules such as pCD86, preincubate (prior to the coculture) selected wells in triplicate with the specific blocking antibody or the isotype-matched control (10 μg/mL in culture media) for 20 min at room temperature. The antibodies can be conveniently maintained during the coculture (5 μg/mL).
4. Add the effector cells in RPMI/10% FBS in equal volume (e.g., 100 μL for a 200 μL final volume) to the confluent pig cells. Prepare the first cocultures with at least two different E/T ratios such as 2:1 and 5:1 in duplicates or triplicates. A multichannel pipettor and sterile basins can be used for this purpose.
5. Add the PHA to selected wells to a final concentration of 10 μg/mL at the time of initiating the cocultures and incubate at 37°C for the desired times (see Note 12) and conditions. All experiments should include a positive control such as T cells incubated with PHA and the anti-hCD28 agonist antibody (5 μg/mL).
6. Collect cell-free supernatants at the end of the assay by transferring a fixed amount of media to 96-V-bottom-well plates and centrifuging at 260 ×*g*. The harvested supernatants can be assayed instantly and/or frozen for storage at −20°C.
7. Assay for hIL-2 by ELISA following the instructions of the manufacturer. Final determinations are done in a microplate reader.

### 3.5. NK Cytotoxicity Assay

1. Detach the adherent pig cells from the culture flask using TripLE Express or cell dissociation buffer (see Note 13) and wash by centrifugation. Resuspend the cells in IMDM/10% FBS with antibiotic and count (see Note 8).
2. Incubate $5 \times 10^5$ target (pig) cells with 100 μCi (3.7 MBq) $^{51}Cr$ as sodium chromate in a 15-mL tube for 1 h at 37°C. The tube is placed on a rotating system (in a way that the sample cannot be spilled) inside the 5% $CO_2$ incubator with the tap loose.
3. Harvest and wash the NK cells to avoid any traces of human serum and resuspend in IMDM/10% FBS with antibiotic. When multiple E/T ratios are used, the effector cells are plated in serial dilution before the labeled target cells are added.

The E/T ratios 5:1, 10:1, 20:1 and 40:1 are usually tested for this purpose. Have ready the appropriate amount of cells in 100 μL/well in a round-bottom 96-well plate.

4. Wash the labeled pig cells with IMDM/10% FBS with antibiotic twice and resuspend in the appropriate amount of media to have $5 \times 10^4$ cells/mL. Load 100 μL per well for the assay.
5. For experiments assaying the contribution of specific molecules with blocking antibodies, pre-incubate the NK cells for 30 min at room temperature with 10 μg/mL blocking antibodies specific for NK cell receptors such as NKG2D, NKp30, NKp44 and NKp46. The antibodies are later left for the duration of the coculture at a concentration of 5 μg/mL.
6. Add the labeled target cells to the NK cells into the round-bottom 96-well plate to have a final volume of 200 μL/well and coculture at 37°C and 5% $CO_2$. Triplicate each assay condition.
7. Stop the assay after 4 h of coculture and collect the supernatants.
8. Determine the maximum release by incubating the labeled target cells with 1% Triton X-100 and the spontaneous release by incubating the labeled target with medium alone.
9. Analyze the $^{51}$Cr release on a gamma counter. To this end, thick filters are placed on each well to absorb all the supernatant and transferred to the appropriate tubes for measures in the gamma counter.
10. Calculate the percentage of specific lysis as [(experimental release – spontaneous release)/(maximum release – spontaneous release)] × 100.

## 4. Notes

1. There are many pig cells that are of interest for xenotransplantation and can be studied in cellular assays such as chondrocytes, hepatocytes, and fetal mesencephalon cells between others. In general, primary cultured cells such as PAEC are better models for cellular assays. Nevertheless, it is also common to use immortalized and established cell lines because they are easier to grow in culture and genetically engineer.
2. In xenogeneic cellular models, it is especially interesting to assess the effect of human cytokines. However, at least one of them has been reported not to be functional on pig cells, that is human IFNγ (17). On the contrary, human TNFα displays a very strong effect on pig cells (14, 16).

3. Remember to heat-inactivate the FBS before use to prevent undesired complement activation. We usually place a bottle with 1 L FBS recently thawed in a 56°C-water bath for 1 h for heat inactivation. Then, let it cool down and prepare 50-mL aliquots for storage at −20°C. For growing primary cultures, it is advisable to test different batches and providers of FBS and reserve various bottles of a good batch.
4. To enhance the growth and survival of primary cultured pig cells, we supplement the basal media DMEM/10% FBS with a bovine hypothalamus extract such as Endothelial Mitogen (Biomedical Technologies, Stoughton, MA, USA). We use 50 mg/L for PAEC and 25 mg/L for chondrocytes with good results. We remove this supplement at least 24 h before initiating any assay.
5. Note that HBSS can loose its properties when stored for a long time. Renew your stock often as calcium/magnesium is needed for good results.
6. Note that some ELISA antibodies cross-react with homologues of other species. Thus, it is advisable to include always controls of cells without coculture in the same quantities assessed in the coculture. Nevertheless, we have good experience with kits from R&D Systems to detect the pig cytokines and from Raybiotech for the human counterparts. Other cytokines could be tested as well taking into account this limitation and the capabilities of the effector cell chosen.
7. Human NK cells, monocytes and macrophages have Fc receptors. In the case of assaying NK cells cultured with human serum, these should be washed at least twice before the assays to avoid any traces of the human serum that could affect adhesion. In the case of using blocking antibodies in the assay, optimize the conditions to minimize the effect of Fc receptors unless these are the subject of study. For instance, assess the effect of molecules with NK92 cells that lack CD16 or use mouse IgG1. In any case, we advise to include an isotype-matched control antibody when including antibodies in the assay.
8. The cells can be simply counted with the microscope using a Neubauer chamber and an appropriate dilution. It is especially advisable to also assess viability (with trypan blue for instance) to make sure that the cells are in good condition for the assay.
9. Be especially careful during washes and the removal of the media. We recommend inverting the plate over a container for waste collection to avoid touching the cells or leaving media. The remaining drops are removed by patting the plate on a clean paper towel. Be also very thorough with the pipetting when adding the effector cells. To determine the number of target (pig) cells, it can be appropriate to count cells after

harvest from a larger culture surface such as a T25 flask in the previous culture step. The number of confluent cells per well in a 96-well plate can then be estimated by correcting for culture surface (e.g., $y = 0.32 \times (n/25)$ in which $n$ is the number of cells isolated from a confluent T25).

10. We recommend to work at low E/T ratios such as 2:1 for assessing the effect of specific molecules.
11. It may be possible to determine changes in the cells after coculture at the level of mRNA or protein that could be assessed respectively by RT-PCR and western blotting.
12. Cocultures of 18–24 h should suffice to detect the T cell response. Evaporation of the media may be a problem for longer cocultures such as 3–5 days.
13. The choice of buffer for harvesting the adherent pig cells can be important for determining their susceptibility to NK cell-mediated lysis, as some ligands can be destroyed by enzymatic treatments. Thus, this should be evaluated in the first experiments if it has not been previously reported in the literature.

## Acknowledgments

This work was supported by Ministerio de Educación y Ciencia (SAF2008-00499) and a grant from Fundación de Investigación Médica Mútua Madrileña (338/05) to C.C. R.S. was supported by a 1-year fellowship from Fundació catalana de trasplantament and M.P.-C. by an IDIBELL fellowship.

### References

1. Ashton-Chess J, Roussel JC, Manez R et al (2003) Cellular participation in delayed xenograft rejection of hCD55 transgenic pig hearts by baboons. Xenotransplantation 10:446–453
2. Davila E, Byrne GW, LaBreche PT et al (2006) T-cell responses during pig-to-primate xenotransplantation. Xenotransplantation 13:31–40
3. Tseng YL, Kuwaki K, Dor FJ et al (2005) alpha1,3-Galactosyltransferase gene-knockout pig heart transplantation in baboons with survival approaching 6 months. Transplantation 80:1493–1500
4. Cardona K, Korbutt GS, Milas Z et al (2006) Long-term survival of neonatal porcine islets in nonhuman primates by targeting costimulation pathways. Nat Med 12:304–306
5. Hering BJ, Wijkstrom M, Graham ML et al (2006) Prolonged diabetes reversal after intraportal xenotransplantation of wild-type porcine islets in immunosuppressed nonhuman primates. Nat Med 12:301–303
6. Murray AG, Khodadoust MM, Pober JS, Bothwell AL (1994) Porcine aortic endothelial cells activate human T cells: direct presentation of MHC antigens and costimulation by ligands for human CD2 and CD28. Immunity 1:57–63
7. Rollins SA, Kennedy SP, Chodera AJ et al (1994) Evidence that activation of human T cells by porcine endothelium involves direct recognition of porcine SLA and costimulation by porcine ligands for LFA-1 and CD2. Transplantation 57:1709–1716
8. Costa C, Barber DF, Fodor WL (2002) Human NK cell-mediated cytotoxicity triggered by CD86 and Galα1,3-Gal is inhibited in genetically modified porcine cells. J Immunol 168: 3808–3816

9. Sullivan JA, Oettinger HF, Sachs DH, Edge AS (1997) Analysis of polymorphism in porcine MHC class I genes: alterations in signals recognized by human cytotoxic lymphocytes. J Immunol 159:2318–2326
10. Seebach JD, Comrack C, Germana S, LeGuern C, Sachs DH, DerSimonian H (1997) HLA-Cw3 expression on porcine endothelial cells protects against xenogeneic cytotoxicity mediated by a subset of human NK cells. J Immunol 159:3655–3661
11. Forte P, Lilienfeld BG, Baumann BC, Seebach JD (2005) Human NK cytotoxicity against porcine cells is triggered by NKp44 and NKG2D. J Immunol 175:5463–5470
12. Li S, Waer M, Billiau AD (2009) Xenotransplantation: role of natural immunity. Transpl Immunol 21:70–74
13. Jin R, Greenwald A, Peterson MD, Waddell TK (2006) Human monocytes recognize porcine endothelium via the interaction of galectin 3 and α-GAL. J Immunol 177:1289–1295
14. Hauzenberger E, Hauzenberger D, Hultenby K, Holgersson J (2000) Porcine endothelium supports transendothelial migration of human leukocyte subpopulations: anti-porcine vascular cell adhesion molecule antibodies as species-specific blockers of transendothelial monocyte and natural killer cell migration. Transplantation 69:1837–1849
15. Ide K, Wang H, Tahara H et al (2007) Role for CD47-SIRPα signaling in xenograft rejection by macrophages. Proc Natl Acad Sci USA 104:5062–5066
16. Sommaggio R, Máñez R, Costa C (2009) TNF, pig CD86 and VCAM-1 identified as potential targets for intervention in xenotransplantation of pig chondrocytes. Cell Transplant 18:1381–1393
17. Sultan P, Murray AG, McNiff JM et al (1997) Pig but not human interferon-gamma initiates human cell-mediated rejection of pig tissue in vivo. Proc Natl Acad Sci USA 94: 8767–8772

Chapter 8

# Production of Transgenic and Knockout Pigs by Somatic Cell Nuclear Transfer

Angelica M. Giraldo, Suyapa Ball, and Kenneth R. Bondioli


## Abstract

Xenotransplantation is one alternative to transplantation of human organs which has been investigated. It is generally accepted that the pig represents the most logical choice of animals to serve as organ donors for xenotransplantation. Moreover, the implementation of cloning by somatic cell nuclear transfer (SCNT) and transgenic techniques have resulted in the production of numerous transgenic pigs than can be used for xenotransplantation purposes as well as models for human diseases.


**Key words:** Cloning, Somatic cell nuclear transfer, Transgenesis, Knockout, Pig engineering

## 1. Introduction

The shortage of human organs for transplantation is well documented, and few practical solutions have been offered. Xenotransplantation is one alternative to transplantation of human organs which has been investigated. It is generally accepted that the pig represents the most logical choice of animals to serve as organ donors for xenotransplantation (1). This consensus has been arrived at because of physiological and genetic similarities between humans and pigs and the long history of safe utilization of porcine tissues and proteins in human medicine. There are many biological barriers to successfully transplanting animal organs into humans, particularly those involving the inflammatory and immune reactions induced by the xenograft. A number of transgenic strategies have been investigated and proposed to overcome these reactions. These include, but are not limited to, elimination of complement-mediated hyperacute rejection and stimulation of cell-mediated immune rejection and are reviewed elsewhere in this volume. Many of these strategies are best accomplished through a gene targeting

Cristina Costa and Rafael Máñez (eds.), *Xenotransplantation: Methods and Protocols,* Methods in Molecular Biology, vol. 885,
DOI 10.1007/978-1-61779-845-0_8, © Springer Science+Business Media, LLC 2012

process resulting in a functional removal of a gene function, "gene knockout," or a replacement of part of a gene sequence, "gene knock in." Such gene targeting reactions can be accomplished in cultured cells by homologous recombination and subsequent selection of the targeted cells by a variety of strategies. In mice, these procedures are performed in embryonic stem (ES) cells that readily support homologous recombination and are readily propagated in culture (2). Genetically modified and selected murine ES cells can be combined with normal embryos and have the capability of transferring the modified ES cell genome through the germ line of resulting animals. Unfortunately, numerous attempts to isolate ES cells from pigs have failed to produce cells with this germ line transfer capability. Somatic cell nuclear transfer (SCNT) has provided a means to create live animals capable of germ line transfer from cells genetically modified in culture from non-rodent species such as the pig. Several recent reports have demonstrated the feasibility of these types of genetic manipulations utilizing homologous recombination in somatic cells and nuclear transfer to produce live pigs (3).

Strategies for the design of targeting constructs have been described in the literature for a number of years. There are no indications that the design principles for these constructs are any different for targeting in porcine fetal fibroblasts. The acquisition of isogenic DNA to include as the homologous arms in a targeting construct is not a practical reality in this application and successful targeting with non-isogenic DNA has shown this to be a moot point. A strategy that has been successful in porcine fetal fibroblasts when the gene to be targeted is expressed consists of a single positive selection strategy followed by screening for targeted insertions by PCR. The positive selection strategy consists of a promoter trap design with a promoterless Neomycin resistance gene targeted to be inserted in frame behind or to replace the ATG of the gene to be disrupted. Cloned construct DNA is prepared for transfection in the same manner as DNA prepared for nuclear injection. In this chapter, we describe the procedures for transfection and selection of porcine somatic cells and the production of live animals from these cells by SCNT.

## 2. Materials

### 2.1. Collection of Porcine Oocytes

#### 2.1.1. Collection and Transport of Ovaries

1. Disinfectants: 70% alcohol and Nolvasan.
2. Clean plastic container with lid and colander.
3. Washing solution: 0.9% saline solution (autoclaved) supplemented with 10 μg/mL of gentamicin.
4. Optional: Betadine (see Note 1).

*2.1.2. Aspiration of Follicles*

1. 5-mL air syringes, 50-mL conical tubes, and 20–22 G needles.
2. Optional: Regulated vacuum pump (Cook, Indiana, USA).

*2.1.3. Collection and In Vitro Maturation of Oocytes*

1. Oocyte manipulation supplies: Pasteur pipettes, pulled capillaries of about 400–500 μm in diameter, mouth pipette, suction bulb, or Drummond pipette.
2. Culture supplies: Square grid dishes, Petri dishes, four-well dishes.
3. Holding medium: M199 with Hank's salts, L-glutamine and 25 mM HEPES supplemented with 0.1% bovine serum albumin (BSA), 100 U/mL of penicillin, and 100 μg/mL of streptomycin. Filter and store for up 1 month at 4°C. Bring the medium at 38.5°C before use.
4. Maturation medium: The medium of choice should be placed in the incubator at least 1 h before use.
5. Equipment: Dissection microscope with under-stage illumination, dry bath and $CO_2$ incubator at 38.5°C.

**2.2. Establishment of Cell Lines and Preparation of Donor Cells**

*2.2.1. Establishment and Maintenance of a Cell Line*

1. Rubbing alcohol.
2. Collection and dissection tools: Razors, biopsy punches, scissors, scalpels, blades.
3. Culture supplies: 15- or 50-mL conical tubes, culture dishes, Pasteur pipettes, pipettors, tips.
4. Washing medium: Dulbecco's phosphate-buffered saline (PBS, calcium and magnesium free) containing 100 U/mL of penicillin and 100 μg/mL of streptomycin.
5. Cell culture medium: Dulbecco's modified Eagle's medium (DMEM) high glucose containing 10% of fetal bovine serum, 100 U/mL of penicillin, and 100 μg/mL of streptomycin.
6. Trypsin, 0.05% with ethylenediamine tetraacetic acid (EDTA).
7. Equipment: Inverted microscope, aspiration pump, $CO_2$ incubator, centrifuge, and hemocytometer.

*2.2.2. Electroporation and Selection of Donor Cells*

1. Cytosalts: 120 mM KCl, 0.15 mM $CaCl_2 \cdot 2H_2O$, 10 mM $K_2HPO_4$, 5 mM $MgCl_2 \cdot 5H_2O$. Adjust pH to 7.6. Sterilize by filtration and store at room temperature (see Note 2).
2. Opti-MEM (Invitrogen, Carlsbad, CA, USA).
3. Electroporation medium: 75% cytosalts, 25% Opti-MEM mixed fresh daily.
4. Electroporation cuvette with 4-mm gap.
5. Cell culture supplies: 15-mL conical tubes, cryovials, fine tip transfer pipettes, Pasteur pipettes, pipettors, tips, 25-cm$^2$ tissue culture flasks, 100-mm tissue culture plates, 24- and 48-well tissue culture plates, sterile cloning rings.

6. Washing medium: see Subheading 2.2.1.
7. Cell culture medium: see Subheading 2.2.1.
8. Trypsin, 0.05% with EDTA.
9. Geneticin (G418).
10. Freezing medium: 10% DMSO in fetal bovine serum.
11. Equipment: Cell electroporator.

#### *2.2.3. Preparation of Donor Cells for Cloning*

1. Culture supplies: 15-mL conical tubes, 1.5 mL vials, cryovials, culture dishes, Pasteur pipettes, pipettors, tips.
2. Washing medium: see Subheading 2.2.1.
3. Cell culture medium: see Subheading 2.2.1.
4. Trypsin, 0.05% with EDTA.
5. Equipment: Liquid nitrogen tank, inverted microscope, aspiration pump, centrifuge, hemocytometer, humidified $CO_2$ incubator, and water bath at 38°C.

### 2.3. Production of Cloned Embryos

#### *2.3.1. Removal of Cumulus Cells and Selection of Oocytes*

1. Glass pipettes: Pasteur pipettes with polished end, glass capillaries with internal diameter of 180–200 and 250–300 μm.
2. Culture supplies: four-well dishes, Petri dishes, and 15-mL conical tubes.
3. Holding medium: M199 with Hank's salts, L-glutamine, and 25 mM HEPES supplemented with 0.4% BSA and 50 μg/mL of gentamicin. Filter and store for up to 1 month at 4°C. Bring the medium at 38.5°C before use.
4. Hyaluronidase solution: 0.3 mg/mL of hyaluronidase in holding medium (see Subheading 2.3.1). Bring the solution at 38.5°C before use.

#### *2.3.2. Enucleation of Mature Oocytes*

1. Mineral oil.
2. Depression slide.
3. Micromanipulation tools: Holding pipette with an internal diameter of 80–100 μm, and enucleation pipette of 18–20 μm in diameter.
4. Holding medium: see Subheading 2.3.1.
5. Enucleation medium: 7.5 μg/mL of Hoechst and 7.5 μg/mL of cytochalasin B in holding medium (see Subheading 2.3.1). Bring the solution at 38.5°C before use. It is recommended to prepare 100× stock solutions of Hoechst and cytochalasin B. Cytochalasin B is soluble in DMSO. The stocks can be aliquoted and store at −20°C.
6. Equipment: Inverted microscope equipped with UV light, Hoffman or phase-contrast system, stage warmer; and a set microinjectors and micromanipulators.

#### 2.3.3. Reconstruction of Embryos

1. See Subheading 2.3.2.
2. Four-well dishes.
3. Injection pipette: Injection needle of 20–25 μm in diameter.
4. Fusion medium: 260 mM mannitol, 0.1 mM $MgSO_4$, 0.001 mM $CaCl_2$, and 0.1% polyvinyl alcohol (PVA) in water. Add the PVA to prewarmed water and stir. Once the PVA dissolves, add the mannitol. Dissolve the $MgSO_4$ and $CaCl_2$ in separate tubes and then add them to the solution. Filter and store at 4°C for up to 6 months.
5. Equipment: Electro cell manipulator (BTX Inc., San Diego, CA), fusion/activation chamber with a 3.2-mm gap (BTX Inc.) and dissection microscope with under-stage illumination.

#### 2.3.4. Activation of Fused Oocytes

1. See Subheading 2.3.3.
2. Activation medium: 260 mM mannitol, 0.1 mM $MgSO_4$, 0.1 mM $CaCl_2$, and 0.1% PVA in water. Add the PVA to prewarmed water and stir. Once the PVA dissolves, add the mannitol. Dissolve the $MgSO_4$ and $CaCl_2$ in separate tubes and then add to the solution. Filter and store at 4°C for up to 6 months.

### 2.4. Embryo Culture

1. Petri dishes and mineral oil.
2. Embryo culture medium.
3. Equipment: Humidifier incubator.

### 2.5. Synchronization of Recipients

1. Gilts between 280 and 400 lb in good body condition and health.
2. Altrenogest Solution 0.22%, 2.2 mg/mL (Matrix®, Millsboro, DE, USA).
3. Human chorionic gonadotropin (hCG), 1,000 IU/mL.
4. Syringes and needles.

### 2.6. Loading of Embryos in the Transfer Tubing

1. Petri dish and 1-mL air syringe.
2. Sterile saline solution.
3. Holding medium: see Subheading 2.3.1.
4. Transfer tubing: 20-cm long polyethylene tube (PE90 tube, Becton Dickinson, Franklin Lakes, NJ, USA). Immerse the tubing on 70% alcohol and rinse it with holding medium before use.

### 2.7. Surgical Embryo Transfer

1. Recipient animals.
2. Flunixin meglumine injection (50 mg/mL), antibiotic and sterile saline solution (0.9%).
3. Ethanol and Nolvasan.

4. Drape, gauze and catgut chromic suture.
5. 1-mL syringe.
6. Pack of surgical instruments.

### 2.8. Post-Surgical Care

1. Pregnant mare serum gonadotropin (PMSG; 1,000 IU/mL) and hCG (1,000 IU/mL).
2. Syringes and needles.

### 2.9. Vaccination and Deworming Protocols Prior to Farrowing

1. Dewormer of choice.
2. Vaccine 1—Bordetella bronchiseptica-Erysipelothrix rhusiopathiae-Pasteurella multocida bacterin-toxoid-rhinitis bacterin-toxoid.
3. Vaccine 2—Mycoplasma hyopneumoniae bacterin.
4. Vaccine 3—Clostridium perfringens type C and enterotoxigenic strains of Escherichia coli.
5. Syringes and needles.

### 2.10. Induction of Labor

1. Dexamethasone (2 mg/mL).
2. Cloprostenol sodium (263 μg/mL).
3. Syringes and needles.

### 2.11. Post-Natal Care of Piglets

1. Vaccines: Parvovirus vaccine-killed virus-Erysipelothrix rhusiopathiae-Leptospira canicola-grippotyphosa-hardjo-icterohaemorrhagiae-pomona bacterin, Bordetella bronchiseptica-Erysipelothrix rhusiopathiae-Pasteurella multocida bacterin-toxoid-rhinitis bacterin-toxoid and Mycoplasma hyopneumoniae bacterin.
2. Dewormer of choice.
3. Iron dextran complex injection (200 mg/mL), antibiotic and vitamin A and D.
4. Iodine solution and Nolvasan solution.
5. Scale, Pig tooth nipper, small V ear notcher, tail docker.
6. Syringes and needles.

## 3. Methods

### 3.1. Collection of Porcine Oocytes

#### 3.1.1. Collection and Transport of Ovaries

1. All reusable materials used to collect ovaries and oocytes should be disinfected with 70% alcohol and/or Nolvasan solution.
2. Collect the ovaries in a clean container containing warmed washing solution as soon as the tissue is harvested from the carcasses.

3. Once enough number of ovaries is collected, discard the liquid from the container and add enough washing solution to cover the ovaries. Stir the ovaries, discard the liquid and wash ovaries with saline solution again. Repeat this process until the washing solution has a clear appearance. Alternatively, place the ovaries in a clean colander and rinse them thoroughly with washing solution until blood and small debris are no longer observed. This method is especially useful when a large amount of ovaries is processed (see Note 1).
4. Transport ovaries to the laboratory. Transport temperature and time of ovary collection to oocyte aspiration may influence the quality of oocytes. Thus, ovaries can be held in washing solution between 25 and 39°C, while time between collection and aspiration of ovaries should not exceed 6 h.

*3.1.2. Aspiration of Follicles*

1. Use a 20–22 G needle attached to a 5-mL air syringe to aspirate every follicle in the size range of 3–8 mm (see Note 3). Alternatively, a vacuum pump set at 40–50 mmHg can be used to aspirate the follicles. Direct the bevel of the needle towards the ovary and try to aspirate as many follicles as possible using a single point of entry. Apply only enough suction to empty the follicles and avoid spillage of follicular fluid.
2. Deposit the follicular fluid in a 50-mL conical tube. After aspirating all the ovaries, allow the oocytes to settle. Keep the follicular fluid at 38°C.

*3.1.3. Collection and In Vitro Maturation of Oocytes*

1. Using a Pasteur pipette attached to a suction bulb, slowly aspirate a portion of the pellet from the bottom of the 50-mL tube. Be careful not to disturb the pellet.
2. Place the pellet in a grid dish containing holding medium. Do not overload the search dish with a large volume of pellet. Otherwise, the medium will turn cloudy and it will impair the identification of good quality oocytes. Instead, use as many grid dishes as needed to search throughout the pellet (see Note 4).
3. Collect the oocytes as quick as possible to avoid a drastic drop in temperature. Porcine oocytes are very sensitive to cold or drastic changes in temperature.
4. Using a pulled glass capillary mouth pipette attached to suction bulb or a Drummond pipette, select oocytes with at least three layers of cumulus cells, and dark and homogeneous ooplasm.
5. Transfer the selected oocytes to a Petri dish containing clean holding medium. Place the dish on a dry bath to keep the oocytes warm.
6. Wash the oocytes three additional times with holding medium. At this point the oocytes should be free of debris and possible contaminants.

7. Rinse the oocytes in a dish containing equilibrated maturation medium (see Note 5).
8. Transfer 50 oocytes to each of the wells of a four-well dish containing 500 μL of maturation medium.
9. Incubate the oocytes at 38°C and 5% $CO_2$ in 90% humidity for 38–42 h (see Note 6).

### 3.2. Establishment of Cell Lines and Preparation of Donor Cells

#### 3.2.1. Establishment and Maintenance of Cell Lines

1. Fetuses should be decapitated and eviscerated before processing. If adult animals are used to establish the cell culture, the skin should be shaved and disinfected prior to the collection of the sample. The skin sample, of approximately 10-mm in diameter, can be collected using a biopsy punch or a scalpel. All the instruments used during the procedure should be sterile.
2. Place the skin sample or fetus in a sterile tube containing holding medium. The sample can be processed immediately or refrigerated. If kept at 4°C, biopsy samples will survive for at least 24 h and even up to 3–4 days, although the longer the time from collection to culture, the lower the likelihood of establishing a healthy cell line.
3. Transfer the sample to a culture dish. Finely chop the tissue using scissors or scalpels to about 1-mm cubes. Rinse the pieces with holding medium.
4. Place the pieces onto a tissue culture dish and add just enough culture medium to cover the culture surface of the dish and hold the tissue pieces in place.
5. Culture the explants under 5% $CO_2$ in air and 90% humidity at 38°C.
6. Check the culture dish 3–5 days later. If the pieces have adhered, add enough culture medium to cover the explants. Cells will start to migrate from the explants 5–10 days after seeding.
7. After 5–7 days in culture, the medium should be changed every 3 days. With a sterile Pasteur pipette attached to a vacuum pump, remove the medium from the dish and add fresh medium.
8. After the primary culture is established and the cells reach 80–100% of confluence, the cell line can be sub-cultured. Discard the culture medium of the culture dish.
9. Add enough PBS to cover the surface of the dish and dilute the remaining culture medium. Gently, agitate the dish in an effort to rinse the cell culture.
10. Remove the PBS and the explants that come lose during the process. Repeat this process two more times.

11. Add enough prewarmed trypsin to cover the cell culture. Incubate the dish at 38°C for 5 min.
12. Check the dish under an inverted microscope to verify that the cells have detached from the bottom of the dish. Cells in primary cultures sometimes require longer incubation time in trypsin. If cells are not completely detached after 10 min, tap the sides of the dish vigorously.
13. Transfer the cell suspension into a 15-mL tube containing prewarmed culture medium.
14. Centrifuge the cells at 350×*g* for 5 min and discard the supernatant.
15. Add enough culture medium to resuspend the cell pellet. Pipette up and down several times to obtain a cell suspension free of cell aggregates.
16. Count cells using a hemocytometer. Seed the desired number of cells into a new culture dish containing fresh medium. Generally, skin fibroblasts can be sub-cultured with a split ratio of 1:5 or 1:10. However, some cell lines may require higher or lower initial seeding densities to obtain a satisfactory growth curve.
17. Steps 8–16 can be repeated every time the cells reach 80–100% of confluence.

#### 3.2.2. Electroporation and Selection of Donor Cells

1. Aseptic cell culture techniques should be utilized throughout the electroporation and selection procedure.
2. Cells from Subheading 3.2.1 can be used directly following removal from culture dish with trypsin. Alternatively, cells from Subheading 3.2.1 which have been cryopreserved at early passage can be used for electroporation immediately after thawing (see Note 7).
3. Prepare electroporation medium by mixing 75% cytosalts and 25% Opti-MEM medium.
4. Pellet 1–2×$10^6$ cultured cells released by trypsin or cryopreserved cells after thawing in a 38°C water bath by centrifugation at 350×*g* for 5 min.
5. Resuspend cells in 5 mL of electroporation medium and centrifuge at 350×*g* for 5 min.
6. Resuspend cells in 800 mL of electroporation medium containing 20 μg of linearized DNA. Let cells incubate at room temperature in electroporation medium containing DNA for approximately 10 min, then load cells into electroporation cuvette.
7. Apply an electroporation pulse of 450 V and 350 μF capacitance.

8. Remove cells from the electroporation cuvette with a fine-tipped transfer pipette and place in a 25-cm² tissue culture flask with culture medium. Culture for 48 h without selection to allow expression of the antibiotic resistance gene.
9. Release cells with trypsin and centrifuge at 350×*g* for 5 min. Resuspend in selection medium consisting of culture medium plus 600 μg/mL of G418. Plate cells in 100-mm tissue culture plates at $2\times10^5$ cells per plate in selection medium (see Note 8).
10. Culture under selection with medium changes every 3–4 days for approximately 10 days until colonies of 100–200 cells appear.
11. Isolate individual colonies by trypsinization using an appropriate diameter cloning ring for the size of the colony. Passage isolated colonies in wells of a 48-well plate with selection medium.
12. When surviving colonies reach 80–100% confluency, trypsinize and place the cells in a 24-well plate while utilizing approximately 10% of the cells for initial screening by PCR (see Note 9).
13. When colonies selected from the initial PCR screen reach 80–100% confluency, passage by trypsin release to a new well of a 24 well plate while cryopreserving approximately half of the cells in small aliquots of $1–2\times10^5$ cells in 100 μL of freezing medium. Repeat this step until sufficient cell aliquots for nuclear transfer have been cryopreserved (see Note 10).
14. Expand colonies to larger tissue culture plates to obtain enough DNA for final screening by PCR and/or Southern blot if desired.

*3.2.3. Preparation of Donor Cells for Cloning*

1. Cells successfully transfected with the gene(s) of interest can be used fresh or stored in liquid nitrogen. It is recommended to freeze the cells to avoid prolonged time in culture and sequential passages.
2. Remove the vial containing the transgenic cells from the liquid nitrogen tank and hold it at room temperature for a couple of seconds.
3. Place the vial in a water bath until the medium thaws out.
4. Transfer the thawed cell suspension to a 15-mL tube containing cell culture medium.
5. Centrifuge the cell suspension at 350×*g* for 5 min and discard the supernatant.
6. Add enough culture medium to resuspend the cell pellet. Pipette up and down several times to obtain a cell suspension free of cell aggregates (see Note 11).
7. Seed cells at an appropriate dilution such that the cells reach confluency 24–48 h prior to nuclear transfer.

8. Culture the cells until in a humidifier incubator at 5% $O_2$, 5% $CO_2$ balanced with $N_2$.
9. Approximately 1–2 days after cells reach confluency, discard the culture medium from the dish.
10. Add enough PBS to cover the surface of the dish. Gently, agitate the dish in an effort to rinse the cell culture.
11. Remove the PBS from the dish.
12. Add enough prewarmed trypsin to cover the cell culture. Incubate the dish at 38°C for 5 min.
13. Check the dish under an inverted microscope to verify that the cells have detached from the bottom of the dish and transfer the cell suspension into a 15-mL tube containing prewarmed culture medium.
14. Centrifuge the cells at 350×*g* for 5 min and discard the supernatant.
15. Resuspend the cells in 1 mL of holding medium and transfer the suspension to a 1.5-mL vial. The cells can be held at 38°C or room temperature for a couple of hours.

### 3.3. Production of Cloned Embryos

#### 3.3.1. Removal of Cumulus Cells and Selection of Oocytes

1. Using a Pasteur pipette (with polished opening of ~400 μm), transfer ≤300 expanded cumulus oocytes complexes (COC's) in each well of a four-well dish containing 500 μL of hyaluronidase solution.
2. Pipette the COC's up and down to remove some of the cumulus cells surrounding the oocytes.
3. Transfer the oocytes of each well into a 15-mL tube containing 2–3 mL of hyaluronidase solution.
4. Vortex the tube vigorously for 4–5 min. Use 5 mL of holding medium to rinse the walls of the tube forcing the oocytes adhered to the wall to settle in the medium.
5. Centrifuge the oocytes at 250×*g* for 1 min and transfer the pellet to a dish containing holding medium.
6. Use a dissection microscope to select oocytes with intact membrane and transfer them to a new dish containing holding medium (see Note 12). Use a glass capillary of 250–300 μm of intern diameter to move the oocytes.
7. Wash the oocytes two more times in holding medium.
8. Place pools of 50–100 denuded oocytes in 50 μL drops of holding medium covered with mineral oil. Keep the oocytes at 38°C.

#### 3.3.2. Enucleation of Mature Oocytes

1. Prepare the micromanipulation chamber by placing ~100 μL of enucleation medium in the center of a depression slide (see Note 13) and cover the medium with mineral oil.

2. Under the microscope, place a pool of oocytes (see Note 14) on the depression slide containing enucleation medium. Incubate the oocytes in this medium for 10 min before exposure to UV light.
3. Use the holding pipette to hold an oocyte by applying pressure on the holding system and the aspiration needle to rotate the oocyte and search for the first polar body (PB). Discard all immature oocytes.
4. Place the PB at around 4–5 o'clock. Usually, the metaphase II (MII) is in the vicinity of the PB; however, it is recommended to verify the localization of the metaphase plate using UV light.
5. With the bevel facing down, insert the enucleation needle throughout the zona pellucida and remove the PB and metaphase plate.
6. Verify the removal of the MII by exposing the aspirated karyoplast to UV light. If the metaphase is still present, rotate the oocyte and relocate the MII around 4–5 o'clock. Remove the metaphase.
7. Discard the PB and cytoplasm from the pipette.
8. Place pools of 50–100 enucleated oocytes in 50 μL drops of holding medium covered with mineral oil.

#### 3.3.3. Reconstruction of Embryos

1. Prepare the micromanipulation chamber by placing ~100 μL of holding medium in the center of a depression slide and cover the medium with mineral oil.
2. Place the enucleated oocytes on the left side of the micromanipulation chamber and a pool of donor cells on the right hand corner of the depression slide (see Subheading 3.2). Allow the cells to settle.
3. Focus the microscope on the bottom of the chamber to visualize the cells. Aspirate a cell into the injection pipette (see Note 15).
4. Select an oocyte and keep it in place using the holding pipette. Insert the reconstruction pipette through the zona pellucida and deposit the donor cell into the perivitelline space. Remove the pipette out of the oocyte.
5. Place pools of 50–100 reconstructed oocytes in 50 μL drops of holding medium covered with mineral oil.
6. In a four-well dish, place 500 μL of holding medium in well one; 250 μL of holding and 250 μL of fusion medium in well two, 500 μL of fusion medium in well three and 500 μL of holding medium in well four. Place pools of 50–100 oocytes in well one of the fusion dish.

7. Transfer 10–15 oocytes to the well two and allow the oocytes to settle. Then, transfer the oocytes to the well three and allow them to settle.
8. Place the fusion chamber onto the stage of a dissection microscope and connect the wires to a BTX fusion machine.
9. Place the oocytes into the fusion chamber containing fusion medium. Move each oocyte until the plane of contact between the cell and the ooplasm is parallel to the electric bars of the fusion chamber (see Note 16).
10. Fusion is induced with an AC pulse of 5 V for 5 s followed by two DC pulses of 1.5 kV/cm for 60 μs (see Note 17).
11. Wash the oocytes in the well four of the fusion dish and then transfer the couplets to 50 μL drops of holding medium covered with mineral oil.
12. Repeat steps 7–11 for all the couplets.
13. Check fusion visually approximately 30 min after the fusion procedure. Discard the unfused oocytes.

#### *3.3.4. Activation of Fused Oocytes*

1. Fused oocytes can be activated 1–4 h post-fusion.
2. In a four-well dish, place 500 μL of holding medium in well one; 250 μL of holding and 250 μL of activation medium in well two, 500 μL of activation medium in well three and 500 μL of holding medium in well four. Place pools of 50–100 oocytes in well one of the activation dish.
3. Transfer 20–30 oocytes to the well two and allow the oocytes to settle. Then, transfer the oocytes to the well three and allow them to settle.
4. Place the activation chamber onto the stage of a dissection microscope and connect the wires to a BTX fusion machine.
5. Place the oocytes into the activation chamber containing 700 μL of activation medium.
6. Activate the oocytes with an AC pulse of 5 V for 5 s followed by two DC pulses of 1.25 kV/cm for 60 μs.
7. Wash the oocytes in the well four of the activation dish and then transfer the couplets to 50 μL drops of holding medium covered with mineral oil.

### *3.4. Embryo Culture*

Activated oocytes should be transferred into a recipient animal as soon as possible. However, the presumptive zygotes can be held overnight or for a couple of days in embryo culture medium.

1. Place pools of 10–15 activated couplets in 25 μL drops of embryo culture medium covered with mineral oil (see Note 18).
2. Incubate the embryos at 38°C in a humidifier incubator at 5% $O_2$, 5% $CO_2$, and 90% $N_2$.

### 3.5. Synchronization of Recipients

The recipient animals should be synchronized and show signs of heat the day that cloning is performed, regardless of the day of embryo transfer (ET). If the embryos are to be transferred after some days in culture, the recipient gilts should have shown heat the day that cloning was performed.

1. House the gilts individually in synchronization crates or pens.
2. Deworm gilt at the same time as synchronization begins.
3. Days 1 through 14, feed recipients with 8 mL of altrenogest solution (18 mg/day) with ration once a day. Draw out with a 12-mL syringe, 8-mL of altrenogest solution and administer directly into the feed once per day. Repeat this process daily for 14 days. Wear protective gloves while handling the altrenogest solution and bottle.
4. At day 18, administer 500 IU of hCG intramuscularly in the neck, behind the ear or ham with a 3-mL syringe and 18 G 1 ½-in. needle.
5. At day 19, transfer 1-cell embryos.

### 3.6. Loading of the Embryo in the Transfer Tubing

1. After activation, place the reconstructed embryos in a dish containing holding medium.
2. Group the activated oocytes very tightly in the center of the dish.
3. Attach the transfer tubing to a 1-mL air syringe and aspirate enough holding medium to create a column of liquid of approximately 0.5 cm.
4. Then, create a column of about 0.5 cm of air in the tubing.
5. Place the tip of the transfer tubing directly on top of the cloned embryos and load them with the minimum possible amount of holding medium.
6. After all the embryos have been loaded (see Note 19), create a column of air followed by one of medium.
7. Carefully, detach the syringe from the transfer tubing and check that the embryos are still in the tubing.
8. The embryos should be held and transported at 38°C.

### 3.7. Surgical Embryo Transfer

1. The pigs are denied access to food or water the morning before surgery. One antibiotic treatment should be performed the day prior to surgery.
2. Sedate and anesthetize the recipient gilts. Follow the protocols recommended by the veterinarian in charge.
3. Place the pig on a dorsal position and disinfect the abdomen.
4. Administer 2 mL of flunixin meglumine and give one more antibiotic treatment prior to surgery.

5. Clean the abdominal area to remove any loose dirt and debris. Clip the ventral abdomen and clean and disinfect the skin using gauze soaked in Nolvasan scrub and gauze soaked in ethanol.
6. Cover the abdomen with a sterile drape and secure in place with towel clips. Cut an oval shape hole 7-cm wide by 20-cm long with operating scissors. Make a midline incision about 5–10-cm long through the abdominal wall between the last and second to last teats.
7. Carefully exteriorize one of the uterine horns. Insert the transfer tubing through the ostium of the infundibulum so that the end of the tubing is in the region of the ampullary-isthmic junction. Embryos are expelled from the tubing using the 1-mL syringe.
8. Rinse the uterus with 35°C sterile saline solution before returning it to the abdomen. Gently pour remaining sterile saline solution into the abdominal cavity to prevent possible adhesions.
9. Close the abdomen using an appropriate suture material and technique. The gilt is then taken to a recovery room until consciousness is regained.
10. Check on recipient gilt 2–4 h after surgery. Transport recipient gilt to gestation housing 12 h after surgery.

### 3.8. Post-Surgical Care

Several laboratories use a combination of PMSG and hCG injections after embryo transfer in an attempt to help to maintain pregnancy.

1. Recipient gilt must have a suture check at 24 and 48 h post-ET. Check for excess bleeding and swelling of abdominal area.
2. At day 11 post-ET, administer 1,000 IU of PMSG intramuscularly in the neck, behind the ear or ham.
3. At day 14 post-ET, administer 500 IU of hCG intramuscularly in the neck, behind the ear or ham.
4. An ultrasound should be performed at day 28 post-ET to confirm pregnancy. Weekly ultrasounds can be scheduled to verify the viability of the fetuses.

### 3.9. Vaccination and Deworming Protocols Prior to Farrowing

1. Five weeks prior to farrowing, administer vaccines 1 and 3 intramuscularly in the neck, behind the ear or ham.
2. Two weeks prior to farrowing, administer vaccines 1, 2, and 3 intramuscularly in the neck, behind the ear or ham.
3. Deworm accordingly to the product directions.
4. The vaccination schedule and type of vaccine used depends on the herd and health status.

5. Ten days prior to farrowing, transport recipient gilt to farrowing facility. Recipient gilt must be washed with soap or disinfectant upon arrival.

### *3.10. Induction of Labor*

Labor should be induced when pregnancy has reached 120–121 days and no signs of parturition are apparent.

1. At day 1, administer 12 mL of dexamethasone and 2 mL of cloprostenol (see Note 20) intramuscularly in the neck, behind the ear or ham.
2. A day later, administer 1 mL of cloprostenol intramuscularly.
3. Pig should start showing signs of parturition 24–36 h after the first set on injections. Farrowing should occur no later than 48 h after initial treatment.

### *3.11. Post-Natal Care of Piglets*

1. Dip umbilical cord in iodine solution as soon as possible after birth.
2. Process newborn litter 24–48 h after birth or when suitable for piglets health.
3. Assign litter and pig number at day 1. Notch the right ear with the litter number and the left ear with the individual number. Never repeat a litter/piglet number, it can be very confusing to track specific animals.
4. Place the piglet in lateral recumbency between arm and side. Restrain the head with your hand. Locate the long, sharp teeth on both sides of the upper and lower jaw. Cut teeth off slightly above the gum line.
5. Hold the piglet by one rear leg so that it is suspended in the air upside down, place tail docker around tail so that approximately 3/4 in. will remain, and with the crimped side facing the piglet, dock the tail. Rinse instruments with Nolvasan solution between piglets.
6. With piglet in same position, administer 1 mL of iron dextran complex injection, antibiotic, and vitamin A and D. Deliver these injections intramuscularly in the hams-two injections per side. Withdraw needle slowly and place finger firmly on injection sites for several seconds so as to prevent medications from coming out.
7. Place piglet on scale to determine weight and record on sheet.
8. At day 21, administer the vaccines intramuscularly into the ham and deworm.
9. Wean 1–2 weeks after vaccination to minimize the stress on the piglet.
10. All instruments must be cleaned and autoclave prior to their next use.

## 4. Notes

1. Chance of contamination generally increases in abattoirs with poor sterilization techniques or when a large quantity of ovaries is collected. To prevent contamination carried over from the surface of biological tissues, ovaries can be immersed for 3–5 min in washing solution containing 1–2% of Betadine. Rinse the ovaries thoroughly with washing solution to remove any traces of the antiseptic.
2. The cytosalts formulation (4) is intended to more closely mimic intracellular ionic conditions rather than extracellular conditions and thus limit loss of viability upon electroporation. Alternatively, electroporation can be carried out in Opti-MEM.
3. Substances present in the rubber plunger tips of syringes have been reported to have detrimental effects on oocytes and embryos. Only air syringes should be used during aspiration of oocytes, preparation of culture media, and manipulation of embryos (5, 6).
4. Alternatively, remove the supernatant from the 50-mL tube and add 50 mL of Ringer lactate or PBS supplemented with 0.1% of BSA. Invert the tube carefully to resuspend the pellet. Place the content of the tube in an embryo filter with a 75-μm mesh nylon membrane and filter the cell suspension. Rinse the filter with abundant medium to eliminate blood and other small particles. Do not allow the membrane to dry out at any time. Place the cell suspension remaining on top of the membrane on the search dish.
5. Successful in vitro maturation of porcine oocytes has been achieved using a variety of culture media types (simple or complex) containing fetal calf serum (7) or follicular fluid (8) and other supplements, such as gonadotropins and growth factors (9). Additionally, serum-free media have been successfully used for the in vitro maturation of pig oocytes (10).
6. Some protocols describe a sequential maturation system. In these protocols, the oocytes are cultured in maturation medium containing gonadotropins for the first 20–22 h before being changed to a gonadotropin-free medium for additional 18–20 h.
7. Cells can be expanded post thaw prior to electroporation if needed. This is not recommended unless necessary. The overall goal is to have cells for nuclear transfer with a minimum number of population doublings and expansion will only increase the number of population doublings.
8. The seeding density of $2 \times 10^5$ is a starting point and may need to be adjusted for different DNA constructs or donor cell lines. The goal is to form single cell colonies that are sufficiently separated to allow isolation. Too low of a seeding density may decrease cell viability.

9. One of the direct PCR reagents sold by several molecular biology reagents suppliers that eliminate the need to purify DNA can be beneficial at this step.
10. Three or four cryopreserved cell aliquots should be sufficient for any single colony. If the nuclear transfer procedures described here are followed and a pregnancy does not occur after four embryo transfers, the investigator is advised to try a different cell line.
11. It is recommended to culture the cells for a couple of days to allow them to recover from the freezing–thawing procedure. However, frozen/thawed cells can be used directly as donors without intermediate culture.
12. Remaining cumulus cells after the vortex procedure can be stripped by pipetting the oocytes up and down with a glass capillary of an internal diameter of 180–200 μm.
13. Alternatively, a microscope slide surrounded by a thick wall of silicone or a 100-mm dish can be used as micromanipulation chamber.
14. Place as many oocytes on the micromanipulation medium as you can enucleate in about 30–60 min. Incubation of the oocytes in Hoechst for extended periods of time may have detrimental effects on embryo development.
15. Insert as many cells as possible in the injection pipette without compromising the suction control.
16. If the membranes of the ooplasm and the somatic cell are not in contact, you can reduce the osmolarity of the fusion medium by adding water. An osmolarity below 260 mM will induce a rapid increase in the volume of the cytoplasms, better contact between cells and consequently higher fusion rates. However, dramatic reduction in the osmolarity will cause lysis of the oocyte.
17. BTX offers fusion/activation chambers with different gap sizes. The voltage settings and the volume of medium should be adjusted accordingly to the gap size. Additionally, the number of oocytes that can be efficiently aligned at once depends of the space between gaps.
18. A variety of culture media have been used to successfully culture porcine oocytes. Common culture media include but are not limited to NCSU-23 (11), BECM-3 (12), and PZM-3 (13).
19. The appropriate number of embryos that should be transferred per recipient animal depends of factors such as the cell line used as donor, the transgene and the quality of oocytes. Under optimal conditions, transfer of 100–200 cloned embryos into a single recipient should be sufficient to establishing a pregnancy.
20. Estrumate tends to work better that Lutalyse for induction of parturition.

## References


1. Lunney JK (2007) Advances in swine biomedical model genomics. Int J Biol Sci 3:179–184
2. Te Riele H, Maandag ER, Berns A (1992) Highly efficient gene targeting in embryonic stem cells through homologous recombination with isogenic DNA constructs. Proc Natl Acad Sci USA 89:5128–5132
3. Ramsoondar JJ, Machaty Z, Costa C, Williams BL, Fodor WL, Bondioli KR (2003) Production of α1,3-Galactosyltransferase-Knockout cloned pigs expressing human α1,2-Fucosylosyltransferase. Biol Reprod 69:437–445
4. Van den Hoff MJB, Moorman AFM, Lamers WH (1992) Electroporation in 'intracellular' buffer increases cell survival. Nucleic Acids Res 20:2902–2905
5. Takeda T, Hasler JF (1986) Effect of plastic disposable syringes on development of mouse embryos in culture. Theriogenology 25:205
6. Bondioli KR, Hill KG (1986) The effect of exposing media to syringes on the viability of bovine embryos. Theriogenology 25:142
7. Zheng YS, Sirard MA (1992) The effect of sera, bovine serum albumin and follicular cells on in vitro maturation and fertilization of porcine oocytes. Theriogenology 37:779–790
8. Yoshida M, Ishizaki Y, Kawagishi H, Bamba K, Kojima Y (1992) Effects of pig follicular fluid on maturation of pig oocytes in vitro and on their subsequent fertilizing and developmental capacity in vitro. J Reprod Fertil 95: 481–488
9. Wang W, Niwa K (1995) Synergetic effects of epidermal growth factor and gonadotropins on the cytoplasmic maturation of pig oocytes in a serum-free medium. Zygote 3:345–350
10. Abeydeera LR, Wang WH, Prather RS, Day BN (1998) Maturation in vitro of pig oocytes in protein-free culture media: fertilization and subsequent embryo development in vitro. Biol Reprod 58:1316–1320
11. Petters RM, Wells KD (1993) Culture of pig embryos. J Reprod Fertil 48:61–73
12. Dobrinsky JR, Johnson LA, Rath D (1996) Development of a culture medium (BECM-3) for porcine embryos: effects of bovine serum albumin and fetal bovine serum on embryo development. Biol Reprod 55:1069–1074
13. Yoshioka K, Suzuki C, Tanaka A, Anas IM, Iwamura S (2002) Birth of piglets derived from porcine zygotes cultured in a chemically defined medium. Biol Reprod 66:112–119

# Chapter 9

# Small Animal Models of Xenotransplantation

**Hao Wang**


## Abstract

Organ transplantation has become a successful and acceptable treatment for end-stage organ failure. Such success has allowed transplant patients to resume a normal lifestyle. The demands for transplantation have been steadily increasing, as more patients and new diseases are being deemed eligible for treatment via transplantation. However, it is clear that human organs will never meet the increasing demand of transplantation. Therefore, scientists must continue to pursue alternative therapies and explore new treatments to meet the growing demand for the limited number of organs available. Transplanting organs from animals into humans (xenotransplantation) is one such therapy. The observed enthusiasm for xenotransplantation, irrespective of the severe shortage of human organs and tissues available for transplantation, can be said to stem from at least two factors. First, there is the possibility that animal organs and tissues might be less susceptible than those of humans to the recurrence of disease processes. Second, a xenograft might be used as a vehicle for introducing novel genes or biochemical processes which could be of therapeutic value for the transplant recipient.

To date, millions of lives have been saved by organ transplantation. These remarkable achievements would have been impossible without experimental transplantation research in animal models. Presently, more than 95% of organ transplantation research projects are carried out using rodents, such as rats and mice. The key factor to ensure the success of these experiments lies in state-of-the art experimental surgery. Small animal models offer unique advantages for the mechanistic study of xenotransplantation rejection. Currently, multiple models have been developed for investigating the different stages of immunological barriers in xenotransplantation. In this chapter, we describe six valuable small animal models that have been used in xenotransplantation research. The methodology for the small animal model establishment includes animal selection, preoperative care, anesthesia, postoperative care, and detailed procedures.


**Key words:** Xenotransplantation, Animal models, Microsurgery, Rodents, Heart transplantation, Kidney transplantation, Small bowel transplantation, Vascularized thymus transplantation, Vascularized ear transplantation, Limb transplantation

## 1. Introduction

Xenotransplantation is defined as the transplantation of organs or tissues between members of different species (1). Based on the severity and pattern of the graft rejection, xenotransplantation can

Cristina Costa and Rafael Máñez (eds.), *Xenotransplantation: Methods and Protocols,* Methods in Molecular Biology, vol. 885, DOI 10.1007/978-1-61779-845-0_9, © Springer Science+Business Media, LLC 2012

be divided into two categories: concordant and discordant (2). Organs transplanted between closelyrelated species, such as the combination of pig-to-goat, hamster-to-rat, rat-to-mouse, or chimpanzee-to-human, generally survive for a period of days before succumbing to rejection, and are referred to as concordant grafts. Thus, organ xenografts performed between species that are phylogenetically distant, such as pig-to-human, guinea pig-to-rat, or rabbit-to-pig, are known as discordant grafts. These grafts are rejected within minutes by preformed antibodies and/or complement within the recipient (3, 4). Before xenotransplantation can be offered to patients, a number of hurdles must be first overcome which include immunologic barriers, potential xenozoonosis, physiological incompatibility, and ethical concerns. The first step is to overcome xenogeneic immunologic responses so that the remaining barriers can be systematically characterized and addressed. The immunological barriers of xenotransplantation can be divided into four stages: hyperacute rejection (HAR), acute vascular rejection (AVR), acute cellular rejection (ACR), and chronic rejection (CR) (5–8). Since the long-term survival of solid organ xenografts is hard to achieve due to the limitation of current immunosuppressive agents, the study of the immune mechanisms tends to be focused mainly on HAR, AVR, and ACR (although considerable overlap among these phases in both clinical features and pathogenesis is often observed).

Vigorous rejection remains a major barrier for successful clinical xenotransplantation, owing to the widely disparate genetic background between donor and recipient. Like many other medical breakthroughs, animal research continues to make enormous contributions to the eventual success of transplantation. Suitable animal models that can mimic clinical xenotransplantation are needed to study immunological mechanisms. Small animal xenotransplantation models tend to offer certain advantages. First, the availability of inbred and transgenic strains allows the mechanisms of rejection to be studied at the molecular level. Second, a greater genetic difference tends to exist between smaller animals than between other xenotransplantation models. For example, the nucleotide difference is 33.4% between rat and mouse, but only 2.6% between macaque monkey and baboon (9). Thirdly, when using rodent models, experiments can be performed in a much more cost-effective manner than when using larger animals. Although there are limitations in the extrapolation of data generated from rodents to humans, tremendous contributions in elucidating the role of humoral and cellular factors in the pathogenesis of xenograft rejection have already been made. Advances in microsurgery over the last decade have permitted the use of the rodent model in transplantation research, thereby facilitating the careful dissection of rejection mechanisms at the molecular level (10–15).

### 1.1. Small Animal Models for HAR Study

In discordant models of xenotransplantation, such as the transplantation of pig organs to human or nonhuman primates, vascularized grafts are rapidly rejected within minutes to hours after transplantation by a process now recognized as HAR (16). HAR is the first immunological barrier to future clinical xenotransplantation using pig organs. Traditionally, the guinea pig-to-rat heterotopic heart transplant model has been the classic model of study for HAR in rodents (13, 17–19). However, the precise role of xenoantibodies in this model remains controversial. Accumulating evidence indicates that tissue rejection in this model is both T-cell and B-cell independent, and that the alternative complement pathway plays an important role (20, 21). The clinical relevance of this model is questionable owing to the fact that activation of the classical complement pathway will most likely maintain a primarily role in future clinical xenotransplantation. As a result, several other animal models have been developed to study HAR mechanisms. We demonstrated a novel guinea pig-to-mouse heart transplant model in 2000 (22). The heart from a 1-day-old guinea pig was heterotopically transplanted in an adult BALB/c mouse. The grafts were rejected within 24 h with typical pathological features of HAR without treatment. Due to the wide availability of transgenic and knockout mice, this model may provide a powerful tool for further study of HAR mechanisms in a molecular basis. Moreover, a concordant model can artificially be switched to a "discordant" model by presensitizing the recipients with donor antigen prior to transplantation. In this way, HAR can be reliably observed in presensitized recipients. Our laboratory developed a HAR model by presensitizing recipients with donor splenocytes in a rat-to-mouse heart transplant procedure (23–25). Hearts from 2-week-old Lewis rats were transplanted into adult BALB/c recipients that were presensitized with $2 \times 10^7$ donor-type splenocytes through intravenous infusion 14 days before transplantation. The rat heart graft turned black immediately after reperfusion and was rejected in 14 min (Fig. 1a) with typical pathological features of HAR (Fig. 1b). In addition, the rat-to-presensitized mouse small bowel transplant model offers the unique advantage of permitting serial biopsies in rapid succession, thus enabling the analysis of sequential rejection events in a single animal during the first few minutes of reperfusion (26). This rat-to-presensitized mouse transplant model can be utilized in the mechanistic study of HAR using a variety of transgenic or knockout mice, thereby aiding in the search for an effective therapeutic approach to HAR progression.

### 1.2. Small Animal Models for AVR and ACR Study

If HAR is blocked or attenuated, xenogeneic organs will undergo AVR within days after transplantation. Currently, AVR represents the main obstacle in preventing the use of xenogeneic donor organs for routine transplantation, as modern immunosuppressive agents, such as cyclosporine (CsA), fail to prevent AVR (6). The role of

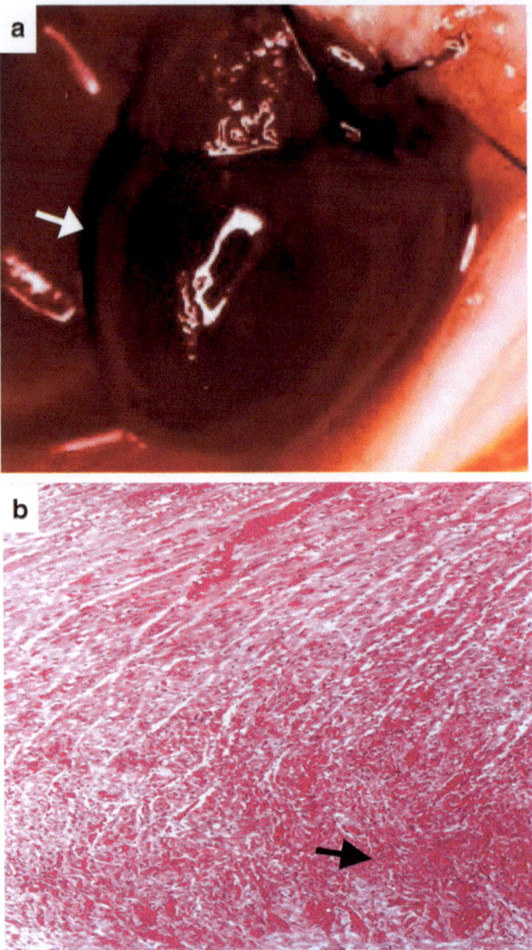


Fig. 1. (**a**) Heart from 2-week-old Lewis rats was transplanted into adult BALB/c recipients that was presensitized with $2 \times 10^7$ donor type splenocytes through intravenous infusion 14 days before transplantation. The rat heart graft turned black immediately after reperfusion, and was rejected in 14 min (*white arrows*: rejected heart xenograft); (**b**) Transplanted heart showed a typical pathological feature of HAR: massive interstitial hemorrhage without cellular infiltration (*black arrows*: massive interstitial hemorrhage).

T cells in xenotransplantation is not well demonstrated, but ACR is predicted to be the third immunological barrier following AVR. Models commonly used in AVR and ACR mechanism research often include hamster-to-rat and rat-to-mouse transplants. The hamster-to-rat transplant model has been widely used to study AVR. Hamster hearts are rejected in 3–4 days with typical pathological features of AVR (27). Our laboratory demonstrated that hamster hearts survived more than 2 months in anti-μ treated nude rats (28). These data indicate that antibodies play a predominant role in the pathogenesis of AVR. The hamster-to-rat kidney transplant model has been attempted by several groups with less favorable results than those established in the heart model (29, 30).

It has been suspected that the small size of the hamster kidney is not able to maintain sufficient renal function in bilaterally nephrectomized rats.

Owing to the emergence of genetically modified mice, as well as advances in microsurgery techniques during the past decade, the rat-to-mouse combination serves as an excellent xenotransplant model when using mice as recipients. Heart, kidney, small bowel, limb, ear, and vascularized thymus models in this combination have been well established in our centre and in others (22–26, 31–36) (Table 1). The surgical techniques for mouse xenografting are similar to those of mouse allografting. The rat-to-mouse small bowel xenotransplantation model is an appropriate experimental model to study humoral rejection (26). In addition, as intestine has abundant vessels of varying size, this unique anatomy makes possible the observation of vascular rejection.

Xenograft rejection patterns vary with different recipient strains (33, 37). BALB/c and C57BL/6 are two immunologically well-characterized mouse strains which develop opposing immunological responses when challenged with pathological organisms. For example, when challenged with *Leishmania major*, C57BL/6 mice respond by developing a vigorous cell-mediated immune response dominated by the production of classic type 1 T-helper-cell cytokines (IL-12, IFN-γ and IL-18), and subsequently are able to fend off infection by this organism (38). In contrast, BALB/c mice develop an immune response dominated by type 2 T-helper-cell cytokines (IL-4 and IL-10), and later succumb to infection. We have further demonstrated that the contrasting patterns of rejection between these two strains are attributable to the different cytokine profiles and antibody levels between BABL/c mice and C57BL/6 mice (33). Reactions in C57BL/6 mice are predominantly Thl cytokine (IFN-γ and IL-12) dependant, which may inhibit AVR. Alternately, Th2 cytokines, such as IL-6, may be

**Table 1**
**Rat-to-mouse xenotransplant models: experience from our centre**

| Transplant model | Median graft survival (days) | Pathology |
|---|---|---|
| Heart transplantation | 13 | AVR+ACR |
| Kidney transplantation | 5.5 | AVR+ACR |
| Small bowel transplantation | 7 | AVR |
| Ear transplantation | 6 | AVR |
| Limb transplantation | 9 | AVR |

*AVR* acute vascular rejection, *ACR* acute cellular rejection

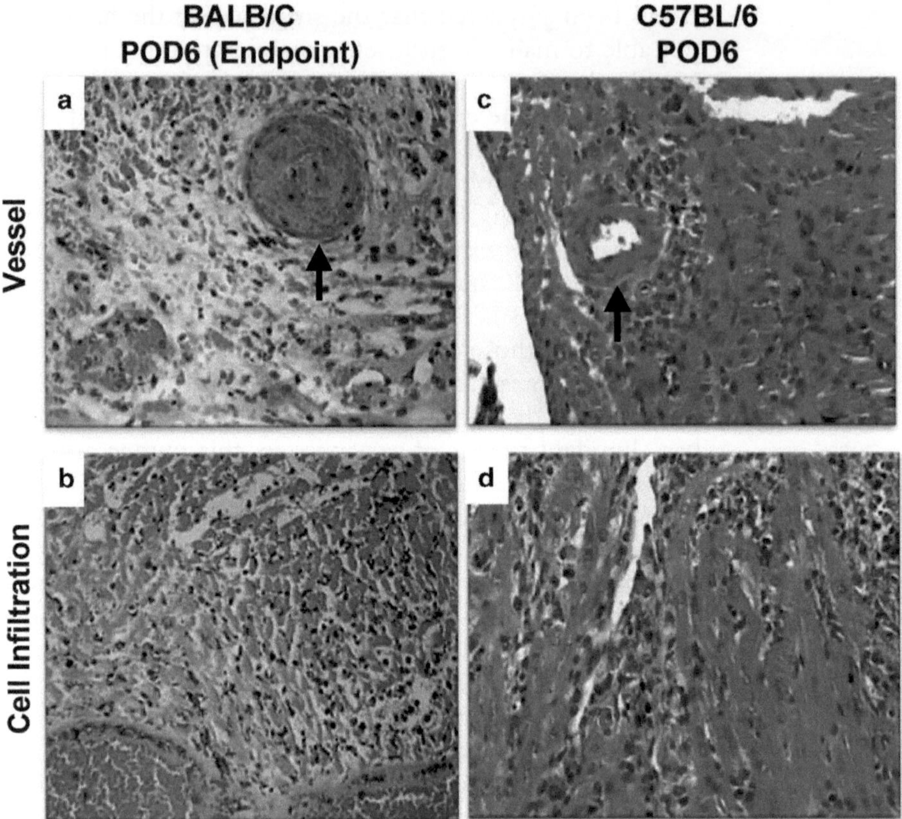


Fig. 2. (**a**, **b**) At the endpoint of rejection (approximately day 6) in BALB/c recipients, transplanted hearts showed a typical AVR profile, with prominent thrombosis (*black arrow*: thrombosis in vessel), interstitial hemorrhage, edema, endothelial damage, fibrin deposition, and few infiltrating cells; (**c**, **d**) on day 6 in C57BL/6 recipients, transplanted rat hearts showed no AVR, had "clear" vessels, but showed heavy cellular infiltrates.

responsible for facilitating AVR in BALB/c mice. Xenoreactive antibody levels are also different between these two strains. BALB/c mice have substantially higher levels of antibodies against rat antigen than do C57BL/6 recipient mice. Xenograft immunostain in BALB/c mice also show positive staining around the vessels, whereas C57BL/6 mice show almost no antibody deposition. We heterotopically transplanted newborn Lewis rat hearts into C57BL/6 or BALB/c recipient mice, which then rejected the cardiac transplants with significantly different kinetics. Most saliently, Lewis rat hearts had a mean survival time of 6.0 ± 0.6 days when heterotopically transplanted into BALB/c mice, but had a significantly increased survival time in C57BL/6 mice (mean survival of 20.6 ± 4.9 days). At the endpoint of rejection (approximately day 6) in BALB/c recipients, transplanted hearts showed a typical AVR profile with prominent thrombosis, interstitial hemorrhage, edema, endothelial damage, fibrin deposition, and cell infiltration (Fig. 2a, b). In contrast, on day 6 in C57BL/6 recipients, transplanted rat hearts showed no AVR, had "clear" vessels (Fig. 2c), and heavy $CD4^+$ and $CD8^+$ infiltrates (Fig. 2d). Kidney xenotransplantation showed similar survival trends when Lewis rat kidneys were transplanted into C57BL/6 and BALB/c mice recipients. Using these two mouse strains and cytokine mutant mice as recipients can help us to investigate how cytokine network responses regulate AVR, as well as the precise role of cytokines in xenotransplantation. It may also be helpful to use immunological prescreening measures to determine whether individuals are high or low producers of these cytokines, as a predictive tool in assessing the outcome of xenotransplants.

In this chapter, we introduce several widely used rodent xenotransplantation models for studying the immune response.

## 2. Materials

### 2.1. Preoperative Care and Anesthesia

1. Atropine.
2. Ketoprofen.
3. Xylazine.
4. Ketamine.
5. Oxygen.
6. Isoflurane.
7. Warming blanket and heating lamp.

### 2.2. Postoperative Care

1. Warm water blanket.
2. Heating lamp.
3. Ketoprofen.
4. Penicillin G.

### 2.3. Surgical Procedures

*Animals* (see Note 1)

1. Male mice weighing between 25 and 30 g. A variety of inbred strains including congenic, transgenic and knockout mice can be selected as either the donor or the recipient depending on the experimental design.
2. Two-week-old male Lewis rats (25–30 g).
3. Three-week-old Golden Syrian hamsters (30–35 g).
4. One-day-old guinea pigs (70–100 g).

*Instruments*

1. Operating Microscope. Small animal xenotransplantations can be performed under 4–25× magnification.
2. Plastic operating board.

3. Surgical Instruments: Microscissors (curved, straight), microneedle holder (curved), microforceps (curved, straight), microvascular clips, halsted-mosquito clamp, ophthalmic iris scissors and needle holder.
4. Sutures: 11-0 and 10-0 nylon sutures; 8-0, 7-0, 6-0, 5-0 and 3-0 silk sutures.

## 3. Methods

### 3.1. Preoperative Care and Anesthesia

1. Do not restrict food and water for donors or recipients.
2. Give atropine (0.04 mg/kg, subcutaneously) and ketoprofen (5 mg/kg, IM) before anesthesia.
3. Anesthetize rodents with an intramuscular injection of mixture of ketamine (100 mg/kg) and xylazine (5 mg/kg).
4. Maintain anesthesia by 2–3 mL/min oxygen with a fraction of 0.5% isoflurane.
5. Keep recipients on a warming blanket and under a heating lamp throughout the procedure.

### 3.2. Postoperative Care (See Note 2)

1. Follow the animal care guideline for postoperative care.
2. Keep the animals on a warm water blanket and under a heat lamp for 24 h after surgery.
3. Make available food and water ad libitum after all the surgeries.
4. Give the first dose of ketoprofen (5 mg/kg, I.M.) immediately preoperatively and a second dose 24 h later.
5. Give penicillin G 500 U/10 g/day I.M. for 7 days to mice after transplantation.

### 3.3. Abdominal Heterotopic Heart Xenotransplantation

Surgical techniques for performing the abdominal heterotopic heart xenotransplantation were based on the previous publications from our research group (28, 33).

#### 3.3.1. Donor Procedure (See Note 3)

1. Separate the anterior chest wall of the donor from the diaphragm, and divide the anterior rib cage, using a pair of scissors, on both sides of the sternum from the lower rib margin to the clavicle.
2. Perfuse the heart graft slowly in situ with 1.0 mL of cold heparinized lactated Ringer's solution through the inferior vena cava (IVC) and aorta.
3. Ligate the superior vena cava and pulmonary veins with 6-0 silk near the atrium and divide distally.

4. Mobilize the ascending aorta and transect close to the innominate artery before the bifurcation.
5. Mobilize the main pulmonary artery and transect at the point of bifurcation.
6. Remove the heart and its vascular supply en bloc and store in lactated Ringer's solution at 4°C.

*3.3.2. Recipient Preparation Procedure*

1. Make a long midline abdominal incision in the recipient.
2. Pull the gut carefully to the left side and cover with moist gauze.
3. Expose the abdominal aorta and IVC, and then mobilize for a short segment from bifurcation of renal vessels to bifurcation of common iliac vessels.
4. Ligate all bifurcations from this segment and place two micro-clips to stop blood flow.

*3.3.3. Venous Anastomosis*

1. Make a longitudinal incision (1.4–1.6 mm) on the IVC.
2. Anastomose the main pulmonary artery of the donor heart end-to-side to the recipient IVC with 11-0 sutures. Continuous suturing using 10–12 stitches is recommended.

*3.3.4. Arterial Anastomosis*

1. Make a longitudinal incision (0.8–1.0 mm) on the recipient aorta.
2. Anastomose the aorta of the donor heart end-to-side to the recipient abdominal aorta with 11-0 sutures. Usually, 8–10 stitches are required to complete the anastomosis.
3. Place a small quantity of microfibrillar collagen (Avitene) around the arterial anastomosis before releasing the clamps.
4. Release the clips first from the distal end.
5. If there is no persistent bleeding, release slowly the proximal clips.
6. Apply gentle pressure to the anastomotic site with a dry cotton swab for 1–2 min after reperfusion. With successful anastomosis, the heart graft is perfused instantly and starts beating (see Note 4).
7. Place the gut back and allow resuming its normal position.
8. Close the abdominal wall using continuous 5-0 sutures.

### 3.4. Orthotopic Kidney Xenotransplantation (See Notes 5 and 6)

Surgical techniques for performing the orthotopic kidney xenotransplantation were based on our previous publications from the research group (28, 39, 40).

*3.4.1. Donor Procedure (See Note 7)*

1. Open the abdomen via a long midline incision.
2. Expose the left kidney by moving the intestine laterally to the right side and using a mosquito clamp to retract the stomach.

3. Isolate the left kidney by ligating and dividing the adrenal and testicular vessels with a 10-0 nylon suture.
4. Mobilize the aorta and IVC at their junction with the left renal artery and vein after ligating and dividing a few lumbar branches.
5. Tie four 8-0 silk sutures around the aorta and IVC above and below the renal artery and vein.
6. Dissect the left ureter free from the renal hilum to the bladder (see Note 8).
7. To minimize bleeding, ligate the aorta first below the renal vessel.
8. Expose the ureterovesical junction using caudad retraction on the bladder dome.
9. Excise a small, elliptical patch of bladder containing the left ureterovesical junction.
10. After ligating the aorta above the renal artery, introduce a 30-gauge needle into the infrarenal aorta, and perfuse the graft slowly in situ with 0.2–0.4 mL of cold, heparinized, lactated Ringer's solution.
11. Transect the renal vein at its junction with the IVC.
12. Divide the aorta obliquely, approximately 2 mm below the renal artery (Fig. 3).
13. Remove en bloc the kidney and its vascular supply, with the ureter attached to the bladder patch, and store in lactated Ringer's solution at 4°C.

*3.4.2. Recipient Preparation Procedure*

1. Make a long midline abdominal incision in the recipient.
2. Pull the gut of the recipient carefully to the left side and cover with moist gauze.
3. Remove the recipient's native right kidney first.
4. After ligating the lumbar branches, isolate infrarenal aorta and IVC carefully and cross clamp with two microvascular clamps (Fig. 4).
5. While retracting with a 11-0 nylon suture placed through a full thickness of aorta, made an elliptical aortotomy (approximately one fifth the diameter of the vessel) with a single cut using iris scissor (Fig. 5). Avoid larger openings as they would invariably cause narrowing at the anastomotic site, resulting in an anastomotic thrombosis.
6. Make a longitudinal venotomy in the IVC by first puncturing it with a 30-gauge needle and then snipping it with iris scissors parallel to the aortotomy.

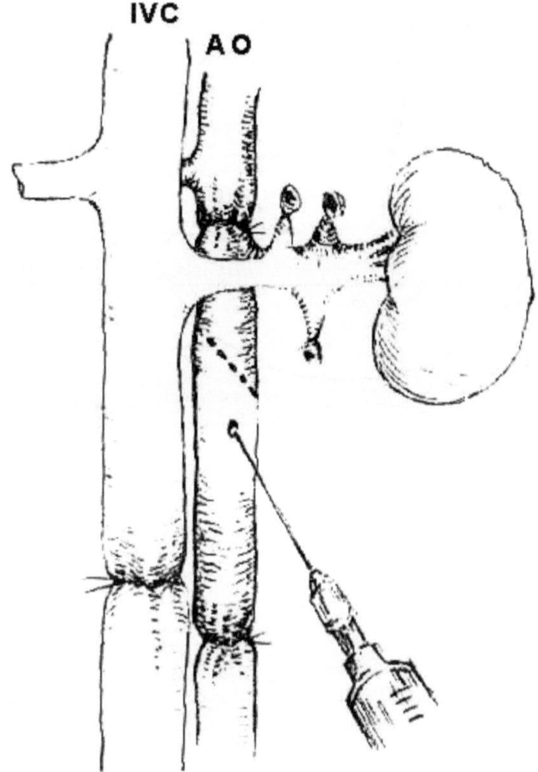


Fig. 3. Preparation and perfusion of the donor kidney with cold Ringer's solution through the aorta prior to graft removal (*AO* aorta, *IVC* inferior vena cava).

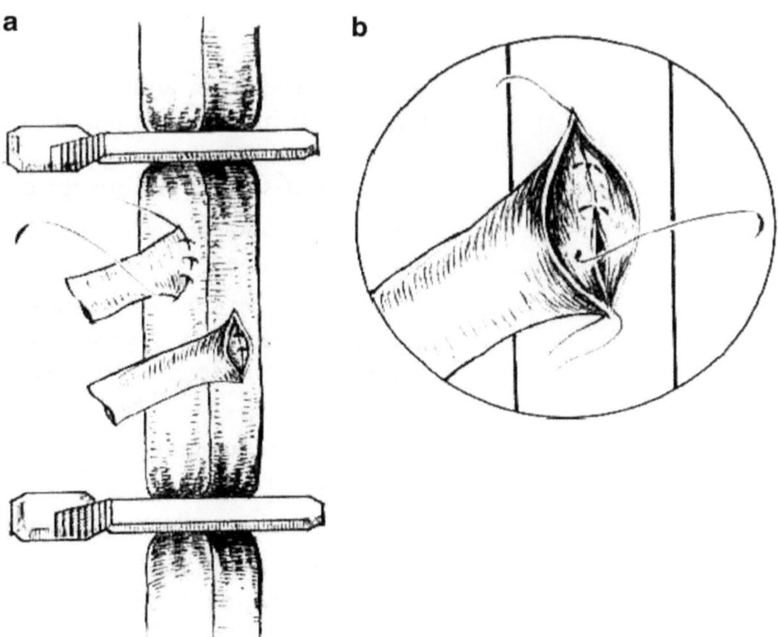


Fig. 4. (**a**) Cross clamping the aorta and inferior vena cava using two microvascular clamps; (**b**) the posterior wall is sutured within the vessel lumen during vascular anastomosis.

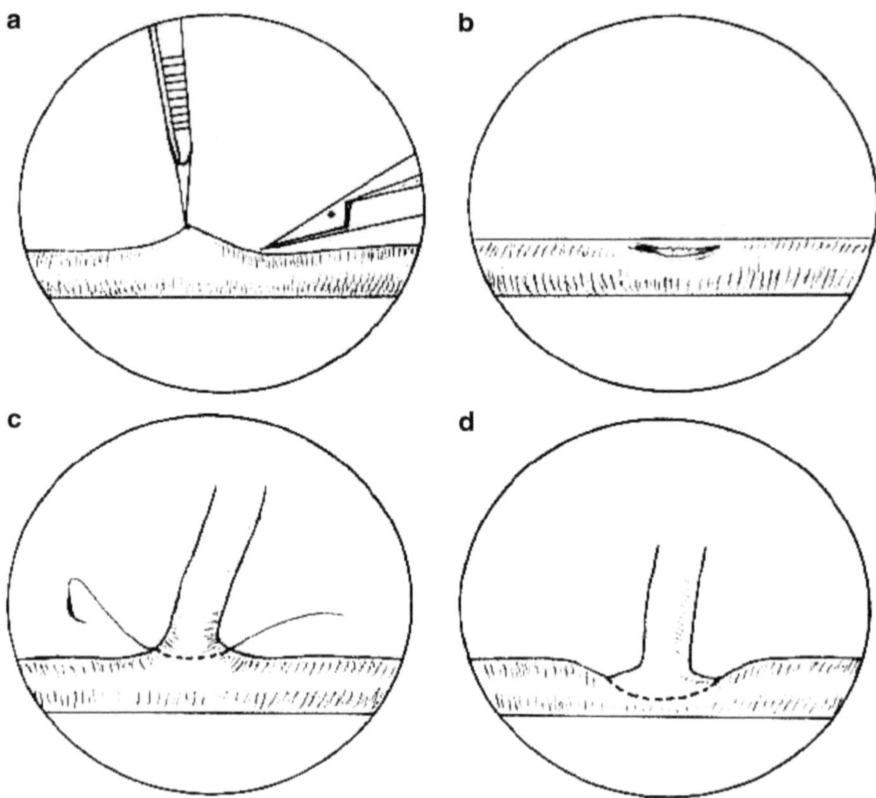


Fig. 5. The steps for creating an aortotomy and performing the arterial anastomosis. (**a**) Elliptical aortotomy made by a single cut with retraction by an 11-0 nylon suture; (**b**) completed aortotomy; (**c**) completed anastomosis with donor aortic cuff; (**d**) incorrect arterial anastomosis (narrowed anastomotic site due to large aortotomy).

7. Flush both the aorta and IVC thoroughly with heparinized saline to clear intraluminal blood or clots.

*3.4.3. Venous Anastomosis*

1. Place two stay sutures at both apices of the venotomy.
2. Remove the donor kidney from the ice and place in the right flank of the recipient.
3. After ensuring that the orientation of the donor renal vein is correct, perform an end-to-side anastomosis between donor renal vein and recipient IVC using continuous 11-0 nylon sutures.
4. Suture the posterior wall within the vessel lumen without repositioning the graft.
5. Close the anterior wall externally using the same suture.
6. Once the venous anastomosis is completed, stretch the vein gently before tying the sutures to avoid narrowing at the anastomotic site. The venous anastomosis is completed in less than 10 min using only four or five sutures for each side of the anastomosis (see Note 9).

*3.4.4. Arterial Anastomosis*

1. Perform the arterial anastomosis between the donor aortic cuff and recipient aorta in the same fashion as the venous anastomosis except that only two or three stitches are required for each side.
2. Place a small quantity of microfibrillar collagen (Avitene) around the arterial anastomosis before releasing the clamps.
3. Apply gentle pressure to the anastomotic site with a dry cotton swab for 1–2 min after revascularization. With successful anastomosis, the kidney graft is perfused instantly (see Note 10).

*3.4.5. Urinary Tract Reconstruction*

1. After ensuring correct orientation of the ureter, divide the dome of the recipient bladder. Bleeding is controlled by cautery.
2. Place two stay sutures of 10-0 nylon 180° apart through the recipient bladder and donor bladder patch, and hold by mosquito clamps.
3. Anastomose each side of the bladder with four or five interrupted 10-0 sutures (Fig. 6).
4. Perform a contralateral nephrectomy following the bladder anastomosis in the one-stage procedure, or in the two-stage procedure on the seventh postoperative day (Fig. 7).
5. Close the abdomen in a single layer with a continuous 5-0 silk suture (see Note 11).

### *3.5. Small Bowel Xenotransplantation (See Notes 12 and 13)*

Surgical techniques for performing the small bowel xenotransplantation were based on the previous publications from our research group (41, 42).

*3.5.1. Donor Procedure*

1. Open the abdomen via a midline incision.
2. Resect first the entire colon and ileum, identify a segment of proximal jejunum (50% of the total intestine), and ligate mesenteric vessels are ligated with 8-0 silk sutures.
3. Separate carefully the portal vein from the pancreas.
4. Following exposure of the aorta and ligation of the celiac trunk, ligate the pyloric and splenic veins separately using 8-0 silk sutures.
5. Introduce a 30-gauge needle into the infrarenal aorta, and perfuse the graft in situ with 0.2–0.4 mL of cold, heparinized Ringer's lactate solution.
6. After dividing the portal vein close to the hilus of the liver, remove the graft with a Carrel patch of aorta, surrounded by a gauze sponge packed with crushed ice, and store in Ringer's lactate solution at 4°C (see Notes 14 and 15).

*3.5.2. Recipient Operation*

1. Open the abdomen through a midline incision.
2. Cover the native intestine with wet gauze and carefully retract to the left side.

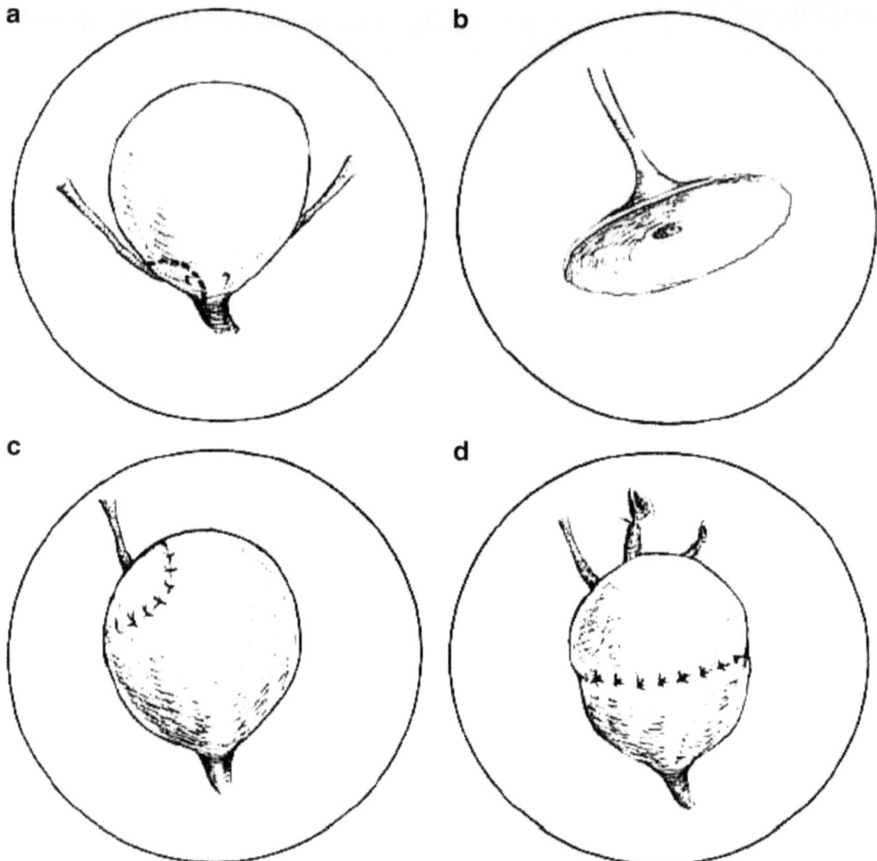


Fig. 6. The steps for urinary tract reconstruction. (**a**) Resection of the bladder patch with the ureter during the donor surgery; (**b**) A small elliptical patch of bladder containing the left ureterovesical junction; (**c**) Completed bladder-to-bladder anastomosis; (**d**) Incorrect urinary tract reconstruction (a eurogenic bladder due to large donor bladder patch).

3. Mobilize and isolate the recipient's infrarenal aorta and IVC.
4. Block blood flow of the infrarenal aorta and IVC by placing two 4-mm microvascular clamps about 1 cm apart.

*3.5.3. Venous Anastomosis*

1. Make a longitudinal venotomy in the IVC using a pair of iris scissors following puncture by a 30-gauge needle.
2. Place two 11-0 nylon stay sutures at both apices of the venotomy.
3. After ensuring that the donor portal vein is not twisted, perform an end-to-side anastomosis using a continuous 11-0 nylon suture.
4. Anastomose the posterior wall from within the vessel lumen without repositioning the graft.

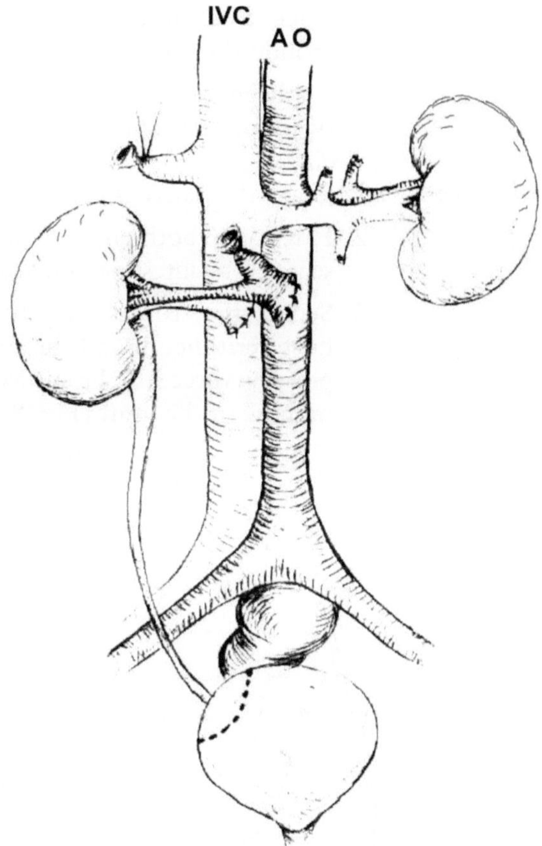


Fig. 7. A completed recipient surgery (*AO* aorta, *IVC* inferior vena cava).

5. Anastomose the anterior wall of the portal vein externally using the same suture. Cold saline irrigation helps keep the vessel walls apart during the anastomosis.
6. Before tying the sutures, pull apart gently the vein to avoid collapse and narrowing of the vessel at the anastomotic site. Only 4-5 sutures are needed for each side of the anastomosis.

#### *3.5.4. Arterial Anastomosis*

1. Make an elliptically shaped aortotomy by either directly grasping the adventitia of the aorta and gently lifting vertically with #5 jewelers forceps, or penetrating the aortic wall vertically with an 11-0 suture, then using a pair of iris scissors to make a single cut at a 30° angle from the most distal part of the aorta. This cut is approximately one fifth the diameter of the vessel (see Note 16).
2. Join the front and back wall of the arterial anastomosis end-to-side with 11-0 nylon suture.
3. After completing the front wall, rotate the intestinal graft to the left to facilitate suturing of the back wall of the artery.
4. Rinse the graft with cold saline several times during the anastomosis.

5. Place a small quantity of microfibrillar collagen (Avitene) around the arterial anastomosis before releasing the clamps.
6. Apply gentle pressure to the anastomotic site with a dry cotton swab for 1–2 min after revascularization (see Note 17).

*3.5.5. Graft Exteriorization*

1. Leave the native intestine intact.
2. Exteriorize both ends of the graft as stomas to allow the mucus secreted by intestinal mucosa to drain outside of the abdomen.
3. Secure the stomas with two 9-0 nylon sutures between the host peritoneum and the sero-muscular layer of the graft, and position three 7-0 silk sutures between the skin and the reverted mucosa of the graft (Fig. 8).
4. Close the abdomen in a single layer with a continuous 5-0 silk suture (see Notes 18 and 19).

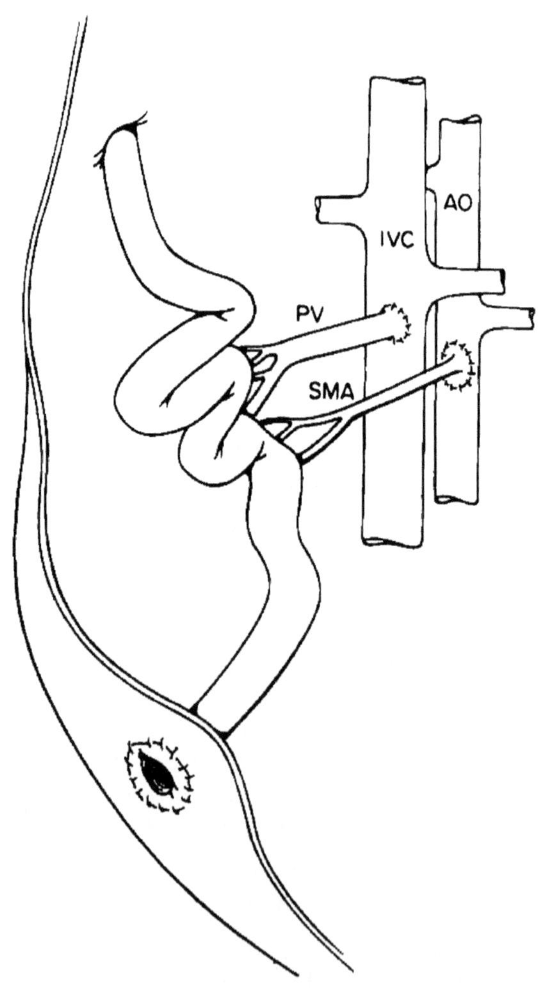


Fig. 8. Completed vascular anastomosis and graft exteriorization (*AO* aorta, *PV* portal vein, *IVC* inferior vena cava, *SMA* superior mesenteric artery).

### 3.6. Vascularized Thymus Xenotransplantation (See Note 20)

Surgical techniques for performing vascularized thymus xenotransplantation were based on the previous publication from our research group (35).

#### 3.6.1. Donor Procedure

1. Make a midline abdominal incision in the donor animal.
2. After ligating the aorta and IVC below the renal artery and vein, introduce a 30-gauge needle into the IVC.
3. Perfuse the graft slowly in situ with 0.4–0.6 mL of cold, heparinized Ringer's lactate solution.
4. Remove the diaphragm and open the thoracic cavity by cutting the ribs on both sides of the spine.
5. Flip backward the anterior chest wall over the animal's head and fix it in that position
6. Rinse continuously the thymus graft with cold Ringer's lactate solution to prevent damage.
7. Remove a piece of right clavicle to expose subclavian vessels.
8. After ligating and cutting the subclavian and internal thoracic arteries and veins (as well as their branches), create the single vascular pathway of the donor thymus. The arterial blood supply of the right thymus lobe comes from the brachycephalic artery and is perfused into the thymic artery. The venous route is from the thymic vein to the superior cava vein (Fig. 9).
9. Separate the right lobe of the thymus from the left lobe, and remove the surrounding tissues by blunt dissection.
10. After cutting the brachycephalic artery and the superior cava vein, remove the thymus graft, along with its vascular supply, and store in Ringer's lactate solution at 4°C.

#### 3.6.2. Recipient Preparation Procedure

Working through an incision at the right mid–clavicular line in the neck of the recipient, expose the right external jugular vein and common carotid artery (CCa) and dissect bluntly.

#### 3.6.3. Arterial Anastomosis

1. After isolating and clamping the right CCa, make a longitudinal incision with a single cut using iris scissors.
2. Flush the right CCa thoroughly with cold, heparinized Ringer's solution to prevent air and blood emboli.
3. Place two stay sutures at both apices of the incision.
4. Remove the donor thymus from the ice, place in the right side of the neck, and cover with wet gauze.
5. Perform end-to-side anastomosis between the donor brachycephalic artery and recipient right CCa using 11-0 interrupted sutures.
6. After suturing the anterior wall, close the posterior wall without repositioning the graft.

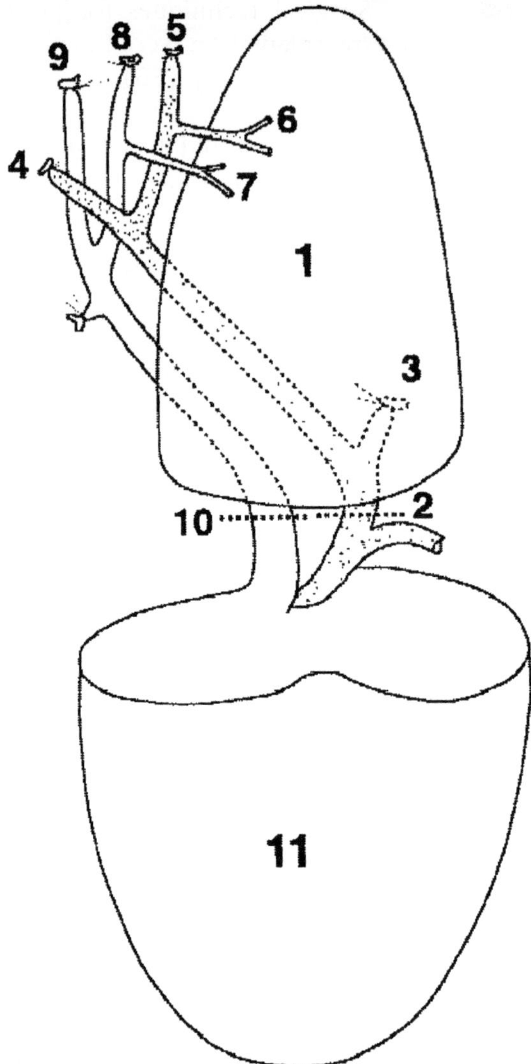


Fig. 9. The donor operation. The *dotted line* indicates the level of resection: (1) right thymus; (2) branchiocephalic artery; (3) common carotid artery; (4) subclavian artery; (5) internal thoracic artery; (6) thymic artery; (7) thymic vein; (8) internal thoracic vein; (9) subclavian vein; (10) superior cava vein; (11) heart.

7. Complete the arterial anastomosis using five or six sutures for each side of the anastomosis (see Note 21).

*3.6.4. Venous Anastomosis*

1. After clamping the right jugular vein, make a longitudinal venotomy.
2. Flush the right jugular vein is then flushed with cold, heparinized Ringer's solution.
3. Perform the venous anastomosis between the donor superior cava vein and recipient right external jugular vein in the same

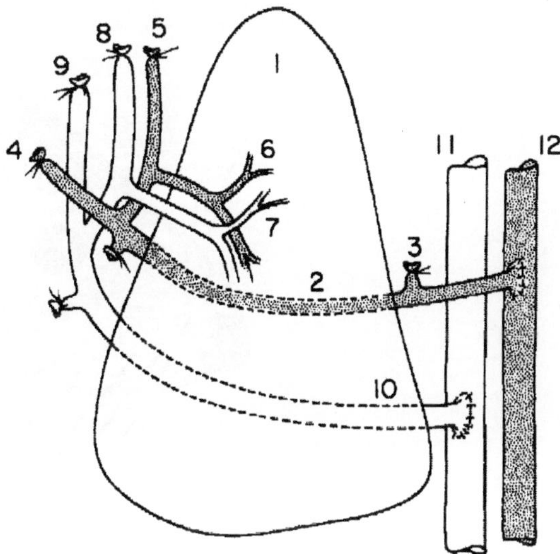


Fig. 10. A completed recipient surgery: (1) right thymus; (2) branchiocephalic artery; (3) common carotid artery; (4) subclavian artery; (5) internal thoracic artery; (6) thymic artery; (7) thymic vein; (8) internal thoracic vein; (9) subclavian vein; (10) superior cava vein; (11) recipient's right external jugular vein; (12) recipient's right common carotid artery.

fashion as the arterial anastomosis, except only four or five stitches are required for each side.

4. Sprinkle the graft with cold Ringer's solution during the transplant procedure.
5. Place a small quantity of microfibrillar collagen (Avitene) around the arterial anastomosis before opening the vessel clamps. With successful anastomosis, the blood supply to the donor thymus results in a pink color of the thymus graft.
6. After covering the graft with the salivary gland, close the skin in a single layer with interrupted 5-0 silk sutures. A completed recipient surgery is shown in Fig. 10.

### *3.7. Vascularized Ear xenotransplantation (See Note 22)*

Surgical techniques for performing vascularized ear xenotransplantation were based on the previous publication from our research group (36).

#### *3.7.1. Donor Procedures*

1. Induce anesthesia.
2. Shave the region of the neck and left ear and prepare in a sterile fashion.
3. Tape the animal in a supine position, but the left forelimb can be moved from left to right, allowing the donor to be brought from a supine to a right decubitus position. The latter is especially useful when dissection is performed around the posterior aspect of the ear.

Fig. 11. Drawing to illustrate the incision in the donor animal. An elliptical incision circumscribing the auricular base is made and extended toward a midline neck incision.

4. Make an elliptical incision circumscribing the left auricular base and extend toward a midline neck incision (Fig. 11).
5. Excise the left lobe of the thymus gland to facilitate the exposure of the neck vessels.
6. Start the dissection by first developing the arterial pedicle based on the CCa. Identify this vessel in the lower neck and carry dissection in a cranial fashion, individually ligating the branches as they are identified.
7. Divide the posterior belly of the digastric muscle to allow for better exposure of the external carotid artery (ECa).
8. Ligate the superior thyroid, as well as the lingual and external maxillary arteries, in turn until the terminal branches of the ECa are reached. Preserve the auricular artery but divide the internal maxillary and temporal.
9. Start the development of the venous pedicle by exposing the posterior facial vein under the superficial muscles. Follow the vein caudally to its junction with the jugular vein and further still to the subclavian. Ligate branches encountered individually.
10. Divide the medially located internal maxillary vein, terminal branches, and lacrimal and palpebral veins, which lie cranially. Leave the venous drainage of the ear intact by carefully preserving the anterior and posterior auricular veins (see Note 23).
11. At this point, turn the animal to the decubitus position and continue dissection from the posterior aspect of the ear.
12. Cut the tubal cartilage at the base of the auricle giving close attention to the developed pedicles, which lie just anteriorly at this point.

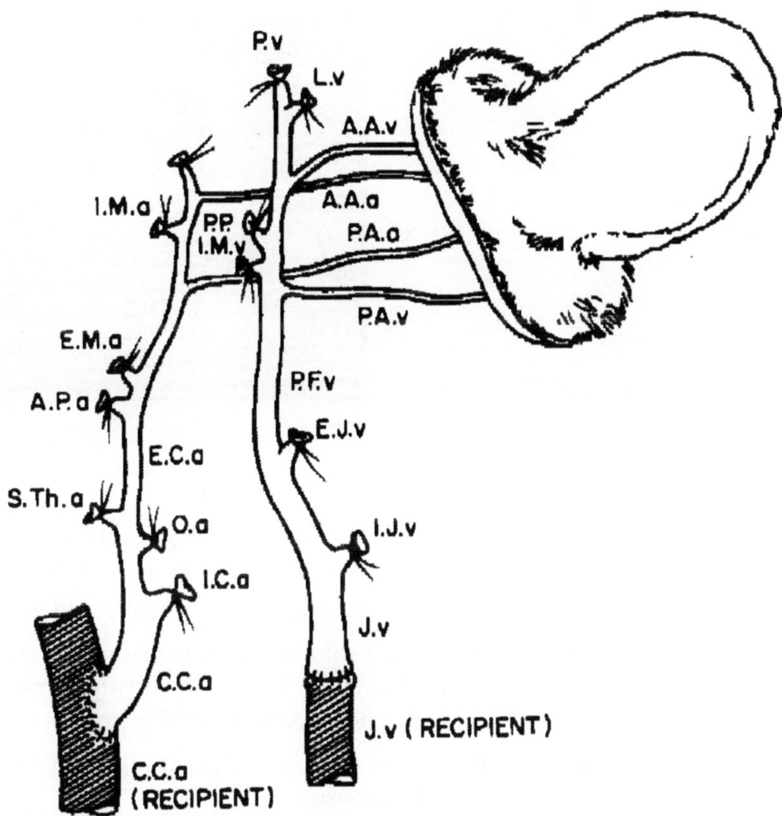


Fig. 12. Donor ear with its vascular pedicle. The vessels shown are arteries, including *CCa* common carotid artery, *ICa* internal carotid artery, *Oa* occipital artery, *STha* superior thyroid artery, *ECa* external carotid artery, *APa* ascending pharyngeal artery, *EMa* external maxillary artery, *IMa* internal maxillary artery, *AAa* anterior auricular artery, and *PAa* posterior auricular artery; and veins, including *Jv* jugular vein, *IJv* internal jugular vein, *EJv* external jugular vein, *PFv* posterior facial vein, *PAv* posterior auricular vein, *IMv* internal maxillary vein, *PPv* pterygoid plexus vein, *AAv* anterior auricular vein, *Pv* palpebral vein, and *Lv* lacrimal vein.

13. Amputate the ear at its base with the exception of the connecting vasculature.
14. Divide the arterial pedicle proximally on the CCa and the venous pedicle close to the jugular–subclavian junction. This gives the necessary length for the two pedicles (Fig. 12).
15. Flush the artery with heparin solution and place the ear in cold saline solution while preparing the recipient.

*3.7.2. Recipient Procedures (See Note 24)*

1. Prepare the recipient initially in the same manner as the donor.
2. Make an elliptical incision around the base of the auricle chosen for the graft site.

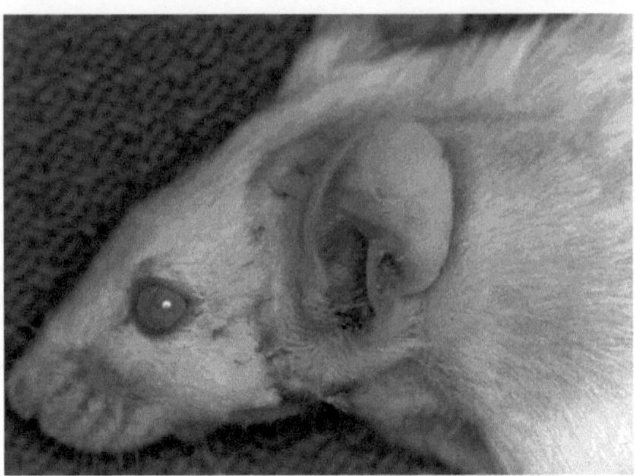

Fig. 13 Normal gross appearance of an ear graft on postoperative day 30.

3. Divide the tubal cartilage, and ligate or burn the auricular vessels with a hand cautery as they are encountered.
4. Amputate the ear completely and achieve hemostasis.
5. Make a midline neck incision and connect to the graft site via a subcutaneous tunnel (through this would traverse the arterial and venous pedicles of the graft).
6. Dissect out the recipient jugular vein and CCa and prepare for anastomosis.
7. Place the graft in position, taking care to ensure that there is no twisting of the vessels.
8. Perform the arterial anastomosis first using 11-0 nylon to create an end to the side anastomosis.
9. Divide the recipient jugular vein with a ligature distally and a microvascular clamp proximally.
10. Perform the venous anastomosis in an end-to-end fashion.
11. Remove all clamps and perfuse the graft.
12. Approximate the tubal cartilage of the graft to that of the recipient using interrupted 7.0 prolene, and finally close the skin incision in a similar manner (Fig. 13).

### 3.8. Limb Xenotransplantation

Surgical techniques for performing the orthotopic or heterotopic limb xenotransplantation were modified from the mouse limb transplantation model described by Zhang et al. (43) and by Tung et al. (44), as well as based on our previous publication (34).

#### 3.8.1. Donor Procedure

1. A skin incision was made parallel to and 1 cm distal to the groin crease.
2. In the groove between medial thigh and lateral thigh musculature, the femoral vessels were dissected from the femoral sheath.

3. Femoral vessels were traced back to the origin of the iliac vessels and small proximal branches were taken down with a hand-held cautery.
4. The femoral vessels below the inguinal ligament were ligated and divided, preserving maximum vessel length.
5. The muscles were divided sharply down to the femur at the level of the mid-femur.
6. Posteriorly, the sciatic nerve was divided and the femur was divided at the level of the mid-thigh.
7. The prepared limb was then wrapped in saline-dampened gauze and stored in cold Ringer's solution. The donor animal was then euthanized with overdose of xylazine and ketamine.

*3.8.2. Recipient Procedure for Orthotopic Limb Transplant Model*

1. An incision was made parallel to and 5 mm distal to the inguinal crease.
2. Femoral vessels were dissected using microsurgical techniques.
3. The femoral vein and artery were divided proximal to the superficial epigastric branches to maximize their stump length.
4. The sciatic nerve was transected, and the femur was divided at the mid-thigh.
5. The graft was brought into the surgical field and positioned so that it took on a normal anatomical configuration.
6. The donor limb was attached to the recipient limb stump by nonrigid bony fixation, using a portion of 25-gauge needle (2.5 mm in length) as an intramedullary pin.
7. Muscles on the medial and lateral thigh were approximated with 6-0 silk sutures.
8. The venous anastomosis was performed in an end-to-side manner with 11-0 nylon sutures;
9. An end-to-end arterial anastomosis was performed using 11-0 nylon sutures. The sciatic nerve was repaired using 11-0 nylon.
10. Lastly, skin was closed using 5-0 silk sutures, which were removed on POD15.

*3.8.3. Recipient Procedure for Heterotopic Limb Transplant Model*

1. The femoral vessels were dissected using microscissors and then clamped.
2. Care was taken to preserve the branches of the femoral artery in the proximal thigh in order to maintain perfusion to the native limb after division of the femoral artery.
3. The graft was brought into the surgical field.

4. The donor limb was not attached to the recipient's femur, but was secured through muscle and skin closure.
5. The donor limb graft was inset heterotopically at a position either medial or lateral to the recipient's proximal thigh, depending on the configuration and reach of the donor vessels.
6. The venous and arterial anastomoses were then performed in the same manner as in the orthotopic model.
7. A layer of muscle plication sutures using 5-0 silk was used to secure the donor limb musculature to the medial thigh muscles of the recipient animal.
8. Another single skin layer suture using 5-0 silk was used to close the donor proximal thigh skin to the medial thigh skin on the recipient.

## 4. Notes

1. Experiments involving animals must be carried out according to institutional and national animal care guidelines.
2. The animals will be sacrificed immediately under the following conditions: (1) any evidence of surgical complications (e.g., lethargy, 15% of weight loss, dehydration, etc.), (2) any evidence of organ failure (e.g., nonbeating heterotopic heart grafts; renal or liver graft failure, development of necrosis in transplanted limb or ear, etc.). Animals free of complications will be sacrificed after surgery based on experimental designs. At sacrifice, a study of graft function and histology will be performed.
3. Young rats or hamsters (2–3 weeks old) can be used as almost size-matched donors.
4. The beating of the graft is monitored by daily direct abdominal palpation. The degree of pulsation is scored as (a) beating strongly, (b) noticeable decline in the intensity of pulsation, or (c) complete cessation of cardiac impulses.
5. Intraoperative fluid replacement to recipient includes 0.4 mL warm saline during surgery by intermittent injections via the dorsal penile vein. Saline (0.2 mL) is given before and after vascular clamping. After closing the abdomen 1.5 mL of saline is given subcutaneously.
6. Renal graft function is evaluated by: (a) serum blood creatinine levels or urea nitrogen levels at a designated time after transplant and; (b) insulin and para-aminohippuric acid clearances.

It is suggested that changes in class II antigen expression be detected by radioimmune scintigraphy. This form of scintigraphy targets MHC class II antigens on donor organ cells and acts a sensitive and noninvasive method for detecting kidney xenograft rejection.

7. Avoid manipulation of the donor kidney and apply 0.5% xylocaine around the renal pedicle to prevent renal vasospasm.
8. A generous amount of connective tissue is left around the donor ureter and bladder patch to minimize risk of compromising the blood supply which nourishes these organs.
9. The venous wall is extremely thin and fragile when using mice as recipients; it usually collapses after cutting. Thus, saline irrigation is useful during anastomosis to keep the vessel walls apart and to provide better visualization for suturing.
10. Since the aortic diameter is smaller than the vein, there is a high risk of thrombosis. Techniques used to avoid this complication included making a small elliptical aortotomy, penetrating the full thickness of the aortic wall, using as few stitches as possible and atraumatically handling the arterial wall. Adventitial stripping is avoided since full thickness sutures of the aorta are required to ensure a leak-proof anastomosis.
11. The ureter implantation (Fig. 14), introduced in 1999 by Han et al., serves as an alternative simple technique for urinary tract reconstruction, and can save on average, 25 min of operation time compared to classic bladder-to-bladder anastomosis (45).
12. Intraoperatively recipients receive a total of 0.8 mL of fluid replacement normally by intermittent injection in the dorsal penile vein during the procedure, including 0.3 mL of saline given before and after vascular clamping. The rate of infusion is very slow to avoid congestive heart failure due to sudden fluid overload. A dose of 1.5 mL of saline is given subcutaneously after closing the abdomen.
13. Heterotopic intestinal transplant graft survival is defined as the end-point, but rejection does not necessarily affect the recipient's survival. Hence, observing clinical signs from the intestinal graft is very important when assessing rejection. The commonly used criteria include: (a) Increasing output mucus from stomas; (b) necrosis or closure of stomas and; (c) development of a palpable mass. Clinical signs of intestinal rejection should be monitored daily by two independent observers who are blinded to the experimental design. Sign of rejection can be scored as (a) no change, (b) mild change, and (c) marked change. Two or more signs showing marked changes are considered end-stage rejection.

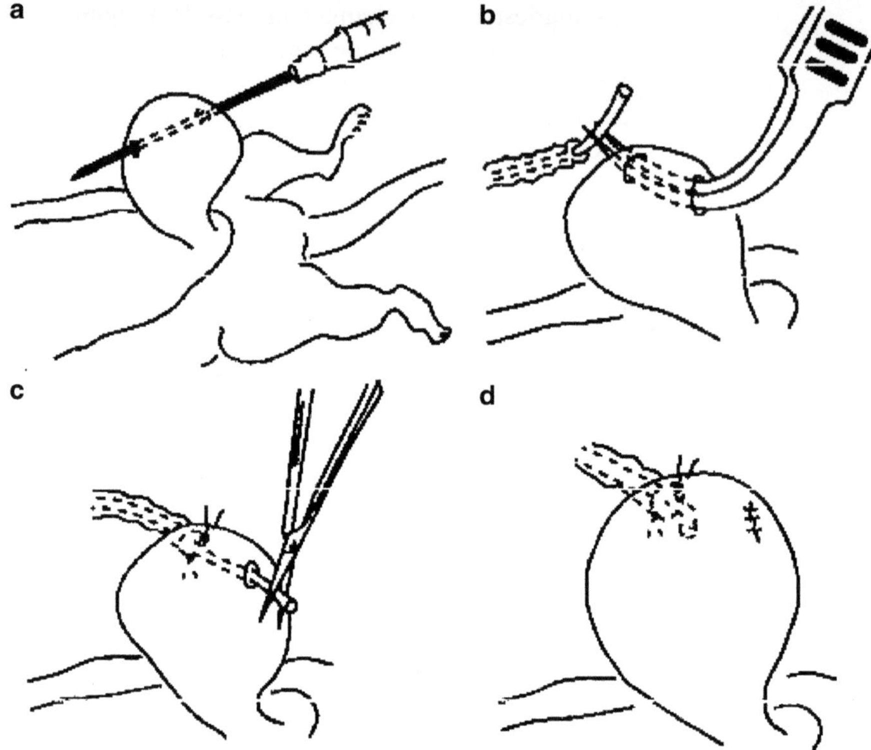


Fig. 14. Ureter implantation technique described by Han et al. (45). (**a**) A 25-gauge needle is used to make diagonally opposing holes in the recipient bladder; (**b**) A pair of curved forceps is passed through the holes gripping the distal end of the ureter and pulling it through both holes; (**c**) The ureter is fixed to the bladder by stitching the periureteral tissue to the outside bladder wall at the edges of the hole using 10-0 nylon (two stitches). Once this is complete, the ureter is trimmed such that 2–3 mm is left inside the bladder; (**d**) The second bladder hole is closed (10-0 nylon suture, 1–2 stitches).

14. The donor surgery takes approximately 45 min. If surgery lasts longer, or there is blood loss during the procedure, normal saline (0.4–0.6 mL) is given intravenously.
15. A wet swab is used when necessary to handle the graft. Manipulation of the intestine is minimized.
16. A longitudinal venotomy in the recipient IVC should be made just above the aortotomy in aorta.
17. Since the aorta's diameter (approximately 0.4 mm) is much smaller than the vein, the following precautions should be taken to minimize the risk of an arterial anastomotic complication: (a) High magnification (×25) and good lighting are essential; (b) each stitch is passed through all layers of the arterial wall,

with bites that are equidistant from the edge of the vessel and are never larger than the thickness of the vessel wall; (c) only the adventitia is grasped to perform the anastomosis, the intima is never manipulated or touched; (d) the mural surfaces of the anastomosis should be in gentle contact with each other after tying to avoid kicking and strangulation; (e) the anastomostic site is repeatedly flushed with heparinized saline during anastomosis.

18. An alternative approach is to create one stoma for the proximal end of the graft and connect the distal end of the graft with the recipient's jejunum by end-to-side anastomosis using 9-0 interrupted sutures. This method may minimize the incidence of graft obstruction by providing better drainage of mucus from the graft. However, this procedure is time consuming with a potential risk of intestinal anastomotic leakage.
19. Stomas are checked daily and kept open in order to facilitate mucus draining.
20. This vascular thymus transplant model can be used to study T-cell development and activation in the thymus, as well as the tolerogenic potential of vascularized thymus xenografts.
21. Vascular thrombosis is minimized by avoiding a larger aortotomy. The caliber of the right carotid artery is extremely small when using mice as recipients as well as when using interrupted sutures to avoid purse string, all of which can cause anastomotic thrombosis.
22. Clinical features of rejection include redness, edema, and development of ulceration and necrosis in the transplanted ear.
23. Crucial to success is the careful procurement of the ear graft with its vascular pedicles intact.
24. The procedure is technically demanding when using mice as recipients.

## Acknowledgments

I wish to acknowledge my past mentor, Dr. Robert Zhong who passed away in 2006. He was a pioneer in experimental microsurgery and transplantation and will be remembered for his gentle spirit as well as his unwavering focus on translational transplant research. Since his death, we strive to carry on not only his work but also his dreams and his spirit. I also acknowledge Dr. Shuhua Luo for his excellent assistance in preparation of this chapter.

## References


1. Auchincloss H Jr (1988) Xenogeneic transplantation. A review. Transplantation 46:1–20
2. Calne RY (1970) Organ transplantation between widely disparate species. Transplant Proc 2: 550–556
3. Auchincloss H Jr, Sachs DH (1998) Xenogeneic transplantation. Annu Rev Immunol 16:433–470
4. van den Bogaerde J, White DJ (1997) Xenogeneic transplantation. Br Med Bull 53: 904–920
5. Leventhal JR, Matas AJ, Sun LH, Reif S, Bolman RM 3rd, Dalmasso AP, Platt JL (1993) The immunopathology of cardiac xenograft rejection in the guinea pig-to-rat model. Transplantation 56:1–8
6. Zaidi A, Schmoeckel M, Bhatti F, Waterworth P, Tolan M, Cozzi E, Chavez G, Langford G, Thiru S, Wallwork J et al (1998) Life-supporting pig-to-primate renal xenotransplantation using genetically modified donors. Transplantation 65:1584–1590
7. Lin SS, Weidner BC, Byrne GW, Diamond LE, Lawson JH, Hoopes CW, Daniels LJ, Daggett CW, Parker W, Harland RC et al (1998) The role of antibodies in acute vascular rejection of pig-to-baboon cardiac transplants. J Clin Invest 101:1745–1756
8. Platt JL, Bach FH (1991) The barrier to xenotransplantation. Transplantation 52:937–947
9. Goodman M (1970) Molecular-genetic approaches to the classification of animals. Transplant Proc 2:432–437
10. Chong AS, Shen J, Xiao F, Blinder L, Wei L, Sankary H, Foster P, Williams J (1996) Delayed xenograft rejection in the concordant hamster heart into Lewis rat model. Transplantation 62:90–96
11. Gorczynski RM, Fu XM, Chung S, Sullivan B, Chen Z (1995) Manipulation of xenogeneic skin and/or renal graft survival in the rat-mouse concordant combination by portal vein pretransplant transfusion. Transpl Immunol 3:321–329
12. Miyagawa S, Hirose H, Shirakura R, Naka Y, Nakata S, Kawashima Y, Seya T, Matsumoto M, Uenaka A, Kitamura H (1988) The mechanism of discordant xenograft rejection. Transplantation 46:825–830
13. Pruitt SK, Baldwin WM 3rd, Barth RN, Sanfilippo F (1993) The effect of xenoreactive antibody and B cell depletion on hyperacute rejection of guinea pig-to-rat cardiac xenografts. Transplantation 56:1318–1324
14. Matsumiya G, Shirakura R, Miyagawa S, Izutani H, Sawa Y, Nakata S, Matsuda H (1994) Analysis of rejection mechanism in the rat to mouse cardiac xenotransplantation. Role and characteristics of anti-endothelial cell antibodies. Transplantation 57:1653–1660
15. Zhang Z, Zhong R, Jiang J, Wang J, Garcia B, Le Feuvre C, White M, Stiller C, Lazarovits A (1997) Prevention of heart allograft and kidney xenograft rejection by monoclonal antibody to CD45RB. Transplant Proc 29:1253
16. Platt JL, Parker W (1995) Another step towards xenotransplantation. Nat Med 1:1248–1250
17. Salame E, Chereau C, Calmus Y, Ayani E, Houssin D, Weill B (1992) Superacute xenogenic rejection: attempted treatment with antiidiotypic antibodies. Presse Med 21:1945–1946
18. Grimm H, Mages P, Lindemann G, Potthoff M, Bohnet U, Korom S, Ermert L (2000) Evidence against a pivotal role of preformed antibodies in delayed rejection of a guinea pig-to-rat heart xenograft. J Thorac Cardiovasc Surg 119:477–487
19. Tanaka M, Murase N, Ye Q, Miyazaki W, Nomoto M, Miyazawa H, Manez R, Toyama Y, Demetris AJ, Todo S et al (1996) Effect of anticomplement agent K76 COOH on hamster-to-rat and guinea pig-to-rat heart xenotransplantation. Transplantation 62:681–688
20. Urbani L, Fabre M, Cardoso J, Lambin P, Devillier P, Soubrane O, Houssin D, Gautreau C (1998) Predominant role of the Fab fragment in delaying hyperacute rejection in guinea pig-to-rat xenotransplantation. Transplantation 66:395–397
21. Alwayn IP, van Bockel HJ, Daha MR, Scheringa M (1999) Hyperacute rejection in the guinea pig-to-rat model without formation of the membrane attack complex. Transpl Immunol 7:177–182
22. Kiyochi H, Zhang Z, Jiang J, Wang H, Garcia B, Kellersmann R, Blomer A, Zhong R, Grant D (2000) Histologic comparison of small bowel, heart, and kidney xenografts in a rat to mouse model. Transplant Proc 32:964
23. Wang H, Rollins SA, Gao Z, Garcia B, Zhang Z, Xing J, Li L, Kellersmann R, Matis LA, Zhong R (1999) Complement inhibition with an anti-C5 monoclonal antibody prevents hyperacute rejection in a xenograft heart transplantation model. Transplantation 68:1643–1651
24. Wang H, Hosiawa KA, Garcia B, Shum JB, Dutartre P, Kelvin DJ, Zhong R (2003) Treatment with a short course of LF 15-0195 and continuous cyclosporin A attenuates acute

xenograft rejection in a rat-to-mouse cardiac transplantation model. Xenotransplantation 10: 325–336

25. Wang H, Hosiawa KA, Garcia B, Shum JB, Dutartre P, Kelvin DJ, Zhong R (2003) Attenuation of acute xenograft rejection by short-term treatment with LF15-0195 and monoclonal antibody against CD45RB in a rat-to-mouse cardiac transplantation model. Transplantation 75:1475–1481
26. Kiyochi H, Grant D, Zhang Z, Garcia B, Kellersmann R, Wang H, Zhong R (1998) Rat-to-mouse small bowel xenotransplantation: novel models to study hyperacute and acute humoral rejection. Transplant Proc 30:2589
27. Yin D, Ma LL, Blinder L, Shen J, Sankary H, Williams JW, Chong AS (1998) Induction of species-specific host accommodation in the hamster-to-rat xenotransplantation model. J Immunol 161:2044–2051
28. Zhang Z, Bedard E, Luo Y, Wang H, Deng S, Kelvin D, Zhong R (2000) Animal models in xenotransplantation. Expert Opin Investig Drugs 9:2051–2068
29. Salomon S, Steinbruchel D, Nielsen B, Kemp E (1996) Hamster to rat kidney transplantation: technique, functional outcome and complications. Urol Res 24:211–216
30. Miyazawa H, Murase N, Demetris AJ, Matsumoto K, Nakamura K, Ye Q, Manez R, Todo S, Starzl TE (1995) Hamster to rat kidney xenotransplantation. Effects of FK 506, cyclophosphamide, organ perfusion, and complement inhibition. Transplantation 59: 1183–1188
31. Wang H, Arp J, Huang X, Liu W, Ramcharran S, Jiang J, Garcia B, Kanai N, Min W, O'Connell PJ et al (2006) Distinct subsets of dendritic cells regulate the pattern of acute xenograft rejection and susceptibility to cyclosporine therapy. J Immunol 176:3525–3535
32. Hosiawa KA, Wang H, DeVries ME, Garcia B, Jiang J, Zhou D, Cameron MJ, Zhong R, Kelvin DJ (2006) Regulation of B- and T-cell mediated xenogeneic transplant rejection by interleukin 12. Transplantation 81:265–272
33. Wang H, DeVries ME, Deng S, Khandaker MH, Pickering JG, Chow LH, Garcia B, Kelvin DJ, Zhong R (2000) The axis of interleukin 12 and gamma interferon regulates acute vascular xenogeneic rejection. Nat Med 6:549–555
34. Zhong T, Liu Y, Jiang J, Wang H, Temple CL, Sun H, Garcia B, Zhong R, Ross DC (2007) Long-term limb allograft survival using a short course of anti-CD45RB monoclonal antibody, LF 15-0195, and rapamycin in a mouse model. Transplantation 84:1636–1643
35. Jiang J, Wang H, Madrenas J, Zhong R (1999) Surgical technique for vascularized thymus transplantation in mice. Microsurgery 19:56–60
36. Jiang J, Humar A, Gracia B, Zhong R (1998) Surgical technique for vascularized ear transplantation in mice. Microsurgery 18:42–46
37. Wang H, Hosiawa KA, Min W, Yang J, Zhang X, Garcia B, Ichim TE, Zhou D, Lian D, Kelvin DJ et al (2003) Cytokines regulate the pattern of rejection and susceptibility to cyclosporine therapy in different mouse recipient strains after cardiac allografting. J Immunol 171:3823–3836
38. Heinzel FP, Sadick MD, Holaday BJ, Coffman RL, Locksley RM (1989) Reciprocal expression of interferon gamma or interleukin 4 during the resolution or progression of murine leishmaniasis. Evidence for expansion of distinct helper T cell subsets. J Exp Med 169:59–72
39. Zhang Z, Lazarovits A, Gao Z, Garcia B, Jiang J, Wang J, Xing JJ, White M, Zhong R (2000) Prolongation of xenograft survival using monoclonal antibody CD45RB and cyclophosphamide in rat-to-mouse kidney and heart transplant models. Transplantation 69:1137–1146
40. Zhang Z, Schlachta C, Duff J, Stiller C, Grant D, Zhong R (1995) Improved techniques for kidney transplantation in mice. Microsurgery 16:103–109
41. Zhong R, Zhang Z, Quan D, Duff J, Stiller C, Grant D (1993) Development of a mouse intestinal transplantation model. Microsurgery 14:141–145
42. Zhong R, Zhang Z, Quan D, Garcia B, Duff J, Stiller C, Grant D (1993) Intestinal transplantation in the mouse. Transplantation 56: 1034–1037
43. Zhang F, Shi DY, Kryger Z, Moon W, Lineaweaver WC, Buncke HJ (1999) Development of a mouse limb transplantation model. Microsurgery 19:209–213
44. Tung TH, Mohanakumar T, Mackinnon SE (2001) Development of a mouse model for heterotopic limb and composite-tissue transplantation. J Reconstr Microsurg 17: 267–273
45. Han WR, Murray-Segal LJ, Mottram PL (1999) Modified technique for kidney transplantation in mice. Microsurgery 19:272–274

# Chapter 10

## Heart Xenotransplantation in Primate Models

Johannes Postrach, Andreas Bauer, Michael Schmoeckel, Bruno Reichart, and Paolo Brenner

### Abstract

Xenotransplantation is a potential solution for the worldwide persisting donor organ shortage. However, immunological and physiological barriers need to be overcome before the first clinical trials can be started. Nonhuman primates are considered the most suitable recipients in preclinical xenotransplantation models. Heterotopic abdominal cardiac xenotransplantation is a well-established nonworking heart model for immunological and biological studies on acute and delayed xenograft rejection and xenograft survival. Nevertheless, orthotopic life-supporting pig-to-baboon heart transplantation is the only accepted model for future cardiac xenotransplantation in humans so far. Survival times of 3 months in at least 60% of consecutive experiments have to be achieved and a minimum number of ten nonhuman primates have to survive for this period of time before clinical transplantation may be started. We recently introduced the heterotopic thoracic technique of pig-to-baboon heart transplantation. We believe that this technique combines the advantages of a working heart model with the safety of heterotopic transplantation. We describe the technical procedure of the three different pig-to-baboon models and give detailed information on perioperative care of the recipients.

**Key words:** Xenotransplantation, Pig, Baboon, Heterotopic abdominal, Heterotopic thoracic, Orthotopic, Heart transplantation

## 1. Introduction

Transplantation of solid organs is the gold standard therapy for end-stage heart failure so far. This procedure is limited by a persisting disparity between the number of heart donors and potential recipients, therefore the transplantation of genetically modified pig hearts into humans could be the ideal solution (1). Since the mid 90s the pig-to-baboon model for cardiac xenotransplantation was established in Cambridge/UK and Munich (2, 3). Since then,

Cristina Costa and Rafael Máñez (eds.), *Xenotransplantation: Methods and Protocols,* Methods in Molecular Biology, vol. 885, DOI 10.1007/978-1-61779-845-0_10, 

wild-type, single- and multitransgenic pig hearts were transplanted in the heterotopic abdominal (max. xenograft survival of 99 days (4)) and orthotopic position (max. survival of 25 days (5)) into baboons by our group.

Recent progress in the production of transgenic pigs and the improvement of immunosuppressive strategies make the first clinical applications of cardiac xenotransplantation likely in the near future (6). Genetically engineered knockout pigs lacking α(1,3)-galactosyl-transferase (GalT-KO) (7, 8) prolonged survival of hearts (max. xenograft survival of 179 days) and kidneys (max. xenograft survival of 82 days) transplanted into primates (9, 10). However, antibody responses against non-Gal epitopes were observed (11), often associated with consumptive coagulopathy; underscoring the need to control non-Gal antibody production (12). Heterotopic (i.e., non-life-supporting) transplantation of pig hearts transgenic for hCD46 into immunosuppressed baboons achieved a maximum xenograft survival of 137 days and a median xenograft survival of 96 days (13). If the same immunosuppressive protocol was applied in the orthotopic position, survival still reached a maximum period of 57 days (14). The worldwide first experiments of xenogeneic heterotopic thoracic heart transplantation (max. survival of 50 days (15)) in combination with more efficient immunosuppressive strategies and new multitransgenic donor pigs may offer the opportunity for a further step towards clinically applicable xenotransplantation. However, GalT-KO pigs which are also transgenic for a human complement regulator are needed for a successful outcome, and further transgenes, e.g., to overcome coagulation disorders (such as human thrombomodulin), have to be considered.

## 2. Materials

### *2.1. Donor Animals*

Genetically modified donor piglets are needed to overcome hyperacute rejection enabling a xenograft survival of more than a few hours. Different single or multitransgenic donor pigs can be used:

1. Piglets overexpressing human decay accelerating factor (hDAF, Imutran Novartis Pharma, Harlan, Correzzana, Italy).
2. Donor pigs transgenic for the human complement regulator protein CD46 (William J. von Liebig Transplant Center, Mayo Clinic, Rochester, MN, USA).
3. GalT-KO pigs (Laboratory of Developmental Engineering, Maiji University, Japan and Institute of Farm Animal Genetics, Department of Biotechnology, Mariensee, Germany).

4. GalT-KO pigs expressing the human heme oxygenase-1 (hHO-1) transgene (Institute of Farm Animal Genetics, Department of Biotechnology, Mariensee, Germany).
5. GalT-KO pigs transgenic for hCD46 (Revivicor, Blacksburg, VA, USA and Institute for Molecular Animal Breeding and Biotechnology, Gene Centre, Munich, Germany).
6. GaT-KO piglets transgenic for both hCD46 and human thrombomodulin (hTM) (Revivicor, Blacksburg, VA, USA and Institute for Molecular Animal Breeding and Biotechnology, Gene Centre, Munich, Germany).

### 2.2. Donor Anesthesia and Heart Explantation

1. Reagents for sedation: Midazolam and ketamine for intramuscular administration.
2. Reagents for anesthesia: Propofol, fentanyl, and isoflurane for induction and maintenance of anesthesia.
3. Standard monitorization devices: Sirecust 960 monitoring system (Siemens, Erlangen, Germany) and pulse oxymeter N-20 PA (Covidien-Nellcor, Boulder, CO, USA) for continuous monitorization of heart rate, electrocardiogram (ECG), invasive blood pressure, and oxygen saturation.
4. Servo ventilator 900 (Siemens, Erlangen, Germany) for pigs to be mechanically ventilated after endotracheal intubation.
5. Standard instruments for heart surgery.
6. Sutures: Ethibond Excel 2 (Ethicon, Norderstedt, Germany) for ligations and double-armed polypropylene sutures (Prolene 5-0 RB-2, Ethicon).
7. Histidine–tryptophan–ketoglutarate cardioplegic solution (HTK Custodiol, Dr. Franz Köhler Chemie GmbH, Bensheim, Germany) cooled to 4°C.
8. Aortic root cannula (DLP 10016, 5°F, Medtronic, Düsseldorf, Germany) for administration of HTK Custodiol.
9. Ringer's solution for topical cooling.

### 2.3. Recipient Animals and Perioperative Care

1. Captive-bred baboons (papio anubis and papio hamadryas) of both genders (10–30 kg body weight) (German Primate Center, Leibniz Institute for Primate Research, Göttingen, Germany) to be used as heart recipients (see Note 1).
2. Same reagents and equipment described for sedation, anesthesia, monitorization, and ventilation of the donor animal.
3. Tunneled double-lumen intravenous catheter with silver coating (Logicath AgTive, 8.5°F, length 20 cm, Smiths Medicals, Grasbrunn, Germany) to be placed in the internal or external jugular vein for continuous venous access.

4. Continuous infusion system (Large Animal Infusion Set, Lomir Biomedical Inc., Quebec, Canada). It consists of a special baboon jacket and a metal tube (length 3 m) connected to the cage by a 3-lumen swivel and to the jacket by an end plate (see Note 2).
5. Immunosuppressants: Methylprednisolone, tacrolimus, sirolimus, and mycophenolate mofetile. To be used in the basic immunosuppressive protocol: methylprednisolone (IV 10–0.3 mg/kg tapering); tacrolimus (0.01–0.05 mg/kg IV; blood concentration. 20–30 ng/mL), and sirolimus (1.0 mg/kg IM; blood level: 10–20 ng/mL) or mycophenolate mofetile (50–100 mg/kg IV; blood level: 2–3 μg/mL).
6. Rejection therapy: Methylprednisolone and antithymocyte globulin (ATG). At rejection, a high dose methylprednisolone (10 mg/kg) is given for 3 days and ATG (1.5 mg/kg IV) for 5 days (see Note 3).
7. Analgesics: Ketamine, fentanyl, lidocaine (optional), metamizol, buprenorphine, or piritramid. Postoperative analgesia is maintained by an infusion therapy (1 mL/kg/h) of a double (ketamine: 60 mg/500 mL and fentanyl: 0.6 mg/500 mL) or triple (additionally 2% lidocaine: 500 mg/500 mL) combination with Ringer's solution and further continued with metamizol and buprenorphine or piritramid for at least 7 days.

### 2.4. Immunoadsorption

1. Life 18 apheresis unit (Miltenyi Biotec Company, Teterow, Germany). Consists of a pair of reusable Ig-TheraSorbTM columns, a special tubing set (Life 18—Theraline, Miltenyi) and a plasma disk separator (Life 18—Disk Separator, Milthenyi).
2. Solutions: TheraSorb NaCl (pH 4.5–7.0; Miltenyi), TheraSorb Gly-HCl (0.2 M, pH 2.8; Miltenyi), and TheraSorb PBS (30 mM $PO_4$, Miltenyi). Before immunoadsorption is started and while the treatment is performed, the tubing set is alternatively filled with TheraSorb NaCl, TheraSorb Gly-HCl, and TheraSorb PBS.
3. Anticoagulants: IV heparin and citrate dextrose solution A (ACD-A) (500 mL Baxter, Maurepas, France). Anticoagulation is initiated with IV heparin (60 IU/kg) and maintained with ACD-A, which is applied into the apheresis system.
4. Therasorb-PBS acid solution (1,000 mL, Miltenyi) to wash and preserve the Ig-TheraSorbTM columns after immunoadsorption.

### 2.5. Xenogeneic, Heterotopic Abdominal Heart Transplantation (According to Ono and Lindsey)

1. Standard surgical instruments (e.g., De Bakey atraumatic dissection forceps 200 mm, Metzenbaum dissection scissor 180 mm, both Aesculap (Tuttlingen, Germany) to be used for the implantation procedure.

2. Microsurgical scissors and needle holder (e.g., microspring scissors with flat handle 45° and 125°, micro-needle holder 200 mm, both Aesculap).
3. Abdominal retractor (Balfour 200 × 240 mm, Aesculap), to be used for optimal exposure of the recipient's abdominal vessels.
4. Cardiovascular forceps (Cooley 165 mm, Aesculap), to be used for the clamping of abdominal vessels (abdominal aorta and IVC).
5. Sutures: Double-armed polypropylene (Prolene 6-0 ACC-1, Ethicon) for vascular anastomoses, poligalactin suture (Vicryl 1 CTX plus, Ethicon) for peritoneum closure, Vicryl 3-0 SH plus suture for the subcutaneous layer, and the absorbable poliglecaprone suture (Monocryl 3-0, Ethicon) to perform an intracutaneous skin suture.

***2.6. Xenogeneic, Orthotopic Heart Transplantation (According to Lower and Shumway)***

1. Conventional and microsurgical instruments (see Subheading 2.5).
2. Oscillating saw (System 6, Stryker, Duisburg, Germany) for sternum opening.
3. Thoracic retractor (thoracic retractor 245 × 180 mm, Asculap) for good exposure of the heart.
4. Continuous autotransfusion system (CATS) to minimize blood loss.
5. Closed cardiopulmonary bypass (CPB) system (animal experimental set, model Großhadern, Medos, Stolberg, Germany) (see Note 4).
6. Two straight venous polyurethane-cannulas (Paediatric Cannula 16–20°F, Stöckert/Sorin Group, Mirandola, MO, Italy) and one aortic cannula (Fem Flex II Femoral Aterial Cannula 12–16°F, Edwards Lifescience, Irvine, CA, USA), which are secured by polypropylene suture (Prolene 5-0 RB-2, Ethicon) and polyester tapes (Mersilene 4 mm, Ethicon), to be used for extracorporeal circulation. The cannulas are connected to the closed CPB system with small priming volume (see Note 4).
7. HTK solution to be used for cardiac arrest of the recipient's heart.
8. Aortic root cannula (DLP10016 5°F, Medtronic, Düsseldorf, Germany) to administer the HTK solution.
9. Double-armed polypropylene (Prolene 5-0 RB-2 or 6-0 ACC-1, Ethicon) for all anastomoses (left and right atrium, pulmonary trunk and aorta).
10. Sutures: Steel wires 5 CPX in combination with Vicryl 1 CTX plus (both Ethicon) to close the sternum, Vicryl 3-0 SH plus (Ethicon) for the subcutaneous layer and Monocryl 3-0, (Ethicon) for the intracutaneous suture.
11. Plaster spray to cover the wound.

#### 2.7. Xenogeneic, Heterotopic Thoracic Heart Transplantation (According to Barnard and Losman)

1. Materials and instruments as in the orthotopic operation.
2. Polyester vascular prosthesis (Gelweave, 10 mm, Vascutek, Hamburg, Germany) for the connection of the grafts pulmonary trunk to the pulmonary trunk of the recipient.
3. Implantable telemetry system [PhysioTel D70-PCTP, Data sciences international (DSI), St. Paul, MN, USA] for postoperative monitorization of the xenograft function and hemodynamic conditions of the recipient.

## 3. Methods

#### 3.1. Donor Animals

An important factor for successful transplantation is the size matching of the donor and recipient with a view to the operation technique (see Note 5). Therefore, the body weight of the donor pigs is determined weekly before the day of surgery. To reduce preoperative stress of the donor animals, an adequate equilibration period is needed and the donor pigs are transferred to the animal facility at least 1 week before transplantation. Preoperatively the animals are also examined for potential infections and heart defects. Blood is taken by a puncture of the internal jugular vein. Afterwards, the lungs and the heart are auscultated and echocardiography can be performed.

#### 3.2. Heart Explantation and Donor Anesthesia

1. Sedate donor pigs with IM 7–10 mg/kg ketamine and 0.5–1.0 mg/kg midazolam via an injection line. Induce anesthesia with IV administration of propofol 1.5–2.0 mg/kg and fentanyl 2.5–5.0 μg/kg.
2. Endotracheal intubation and maintenance of anesthesia by a continuous administration of 0.1–0.2 mg/kg/min propofol and repeated injections of 2.5–5 μg/kg fentanyl (every 30–60 min).
3. Keep a continuous mechanical ventilation and monitoring (ECG, oxygen saturation and invasive arterial pressure) of the animal.
4. After median sternotomy and opening of the pericardium, separate the ascending aorta from the pulmonary trunk. Expose the superior and inferior vena cava (SVC and IVC) and the course of the pulmonary veins.
5. Administer heparin IV (400 IU/kg); afterwards, insert the needle-vent cannula into the aortic root. Divide the SVC between two ligations as close as possible to the junction with the azygos vein. Clamp the IVC at the diaphragmatic level and incise proximally to allow unloading of the right heart.

6. Cross-clamp the ascending aorta just before the junction of the brachiocephalic trunk and apply 30 mL/kg body weight cardioplegic solution (e.g., HTK solution) through the aortic root cannula (see Note 6). An incision into the pulmonary veins unloads the left heart. Cool the heart topically with ice-cold Ringer's solution.
7. After cardioplegic arrest, perform the excision of the donor heart from caudal to cranial. First divide the IVC, followed by a high transection of the ascending aorta and the pulmonary trunk near the bifurcation.
8. Submerge the heart in a bowl filled with ice-cold Ringer's solution and prepare for implantation (see Subheadings 3.4–3.6).

### *3.3. Recipient Anesthesia, Peri- and Postoperative Care*

1. Sedate baboons with IM 0.5–1.0 mg/kg midazolam and IM 7–10 mg/kg ketamine. Use a blow pipe or a squeeze back system in the cages for intramuscular injection.
2. Insert a peripheral intravenous catheter (vena saphena lateralis).
3. Induce anesthesia in animals with an intravenous bolus of 1.5–2.0 mg/kg propofol and fentanyl 6–8 μg/kg. Maintain anesthesia with 0.1–0.2 mg/kg/min of propofol and repetitive intravenous bolus injections of 5 μg/kg of fentanyl every 30–60 min.
4. Place a venous catheter (double lumen) into the left internal or external jugular vein. Dissect the jugular veins and tunnel the catheter subcutaneously from the anterior chest wall. Insert the IV line into the vein and fixe with a ligation. Postoperatively, secure the central venous catheter with a primate jacket and a flexible metal tube (Large Animal Infusion System, Lomir Biomedical Inc.).
5. After the transplantation surgery, bring back the baboons to the primate facility, and give oxygen and water. Assess daily the general condition of the recipient. Monitor xenograft function with continuous ECG and blood pressure monitoring (PhysioTel D70-PCTP, DSI). Perform echocardiography of the xenograft one to two times per week.
6. Apply heparin continuously via the tethering system (target value: PTT 60–80 s).
7. Maintain postoperative analgesia by an infusion therapy (1 mL/kg/h) of a double (ketamine: 60 mg/500 mL and fentanyl: 0.6 mg/500 mL) or triple (additionally lidocaine 2%: 500 mg/500 mL) combination with Ringer's solution and continue further with metamizol and buprenorphine or piritramid for at least 7 days.

### 3.4. Immunoadsorption

1. Perform immunoadsorption (IA) preoperatively and when xenoreactive antibody levels increase. Anesthetize animals before treatment as previously described. Use two IV lines for an efficient treatment, therefore employ two permeable catheters with a minimum lumen of 6°F. Mainly, use the permanent central venous double lumen catheter; otherwise insert a peripheral catheter.
2. Before starting immunoadsorption and during treatment, fill alternatively the tubing set with TheraSorb NaCl, TheraSorb Gly-HCl, and TheraSorb PBS.
3. Start the treatment and assess hemodynamic and blood parameters (e.g., blood count, albumin). Assess immunoglobulin (IgG and IgM) before, during (at the half of cycles), 1 h and 1 day after immunoadsorption procedure. The reduction of the immunoglobulins and immune complexes (about 80%) need between 21 and 38 cycles. Blood flow ranges between 5 and 120 mL/min with an extracorporeal blood volume about 80 mL.
4. Initiate anticoagulation with IV heparin (60 IU/kg) and maintain with anticoagulant ACD-A, which is applied into the apheresis system.
5. At the end, reinject the all blood and preserve the Ig-TheraSorbTM columns with the Therasorb-PBS acid solution.

### 3.5. Xenogeneic, Heterotopic Abdominal Heart Transplantation (According to Ono and Lindsey)

1. Before implanting the donor heart heterotopically into the abdomen of the recipient, prepare the xenograft: perform a ligation of the SVC and IVC and the six pulmonary veins of the porcine donor heart, and secure with a double-armed 5-0 polypropylene suture (Prolene 5-0). Finally, divide the aorta and the pulmonary trunk and remove the adventitia.
2. For laparotomy, perform a midline incision from the xiphoid process to the pubic symphysis. Cover the bowels by wet, warm gauze and push carefully aside by an abdominal retractor. Now, open the retroperitoneal space inferior to the kidneys and the abdominal aorta and expose the IVC for a length of about 5 cm (see Note 7).
3. Give heparin (400 IU/kg) systemically and clamp the infrarenal abdominal aorta and inferior caval vein with two Cooley cardiovascular clamps.
4. Open the cava vein by an incision, which equals the diameter of the pulmonary trunk of the donor heart. Place the xenograft next to the infrarenal vessels and anastomose the pulmonary artery in an end-to-side technique to the recipients IVC by a continuous polypropylene suture (Prolene 6-0 ACC-1).

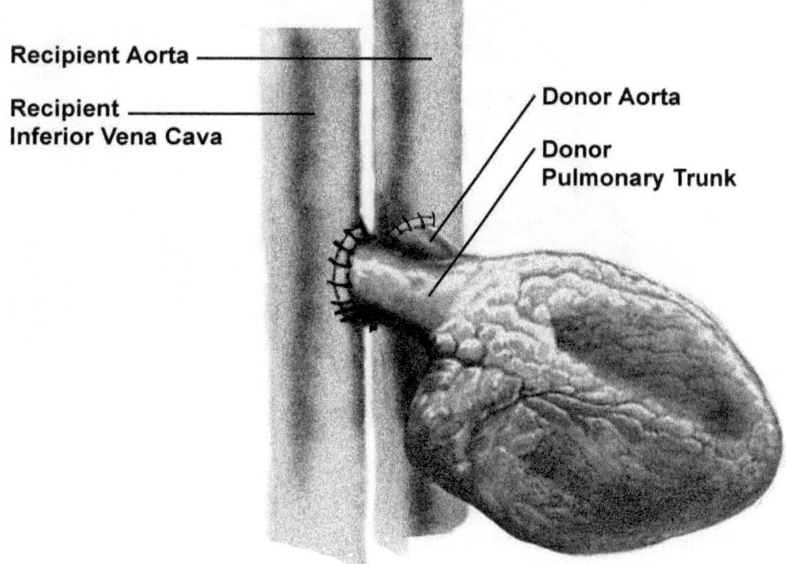


Fig. 1. Heterotopic abdominal heart transplantation: Final in situ position of the two end-to-side anastomoses.

5. Connect the ascending aorta of the donor heart to the abdominal aorta by and end-to-side anastomosis (Prolene 6-0 ACC-1). A drawing of the final in situ situation is shown in Fig. 1.
6. After evacuation of all the air from the xenograft (retrograde filling), open clamps. If there is no spontaneous ventricular rhythm after a few minutes, defibrillate the heart is starting with 10 J. Thereafter, close the abdomen in layers, reduce anesthesia and wean the animal from mechanical ventilation.

### 3.6. Xenogeneic, Orthotopic Heart Transplantation (According to Lower and Shumway)

1. Cool the donor heart at 4°C until the implantation procedure starts. Start the ex vivo preparation for the orthotopic implantation with an incision of the right atrial appendage. Close a possible foramen ovale and incise the left atrium along a line connecting the orifices of the pulmonary veins. Secure the ligatures of the SVC and IVC with polypropylene suture (Prolene 5-0 RB-2). A drawing of the prepared donor heart is shown in Fig. 2a.
2. CPB and posttransplant immunosuppression cause an inevitable loss of red blood cells. Therefore, perform median sternotomy and pericardiotomy very carefully to minimize blood loss and improve postoperative outcome.
3. After systemic heparinization (400 IU/kg), pass tapes around the SVC and IVC and perform cannulations with two separate catheters (Paediatric Cannula made of Polyurethane 16–20°F). Place an arterial cannula (Fem Flex II Femoral Aterial Cannula

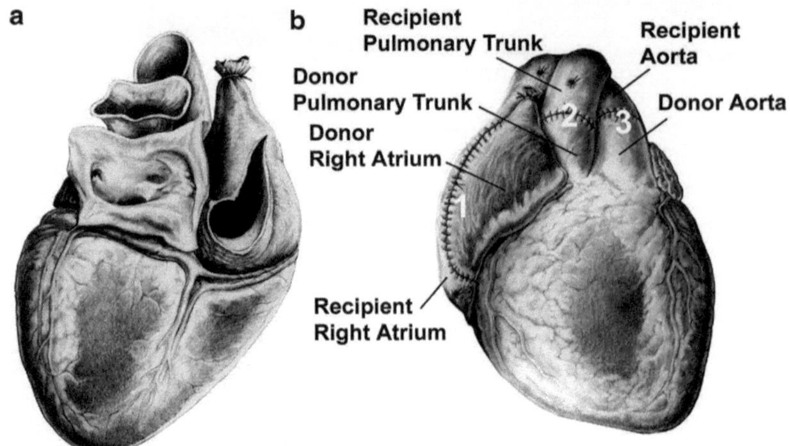


Fig. 2. Orthotopic heart transplantation: Donor heart after final preparation before implantation into the recipient starts (**a**) and the in situ situation after the implantation procedure (**b**) Anastomoses between (1) the right atrium of the donor heart and the atrial remnant of the recipient, (2) the aorta of the donor and the recipient and (3) the pulmonary trunks.

12–16°F) in the ascending aorta as close to the arc as possible. Connect the tubes of the CPB with all cannulas, establish extracorporeal circulation and reduce the body temperature to 32°C.

4. Use HTK solution for cardiac arrest of the recipient's heart, and administer via the aortic root cannula (DLP10016 5°F).
5. Explantation of the recipient's own heart follows occlusion of the caval veins and cross-clamping of the ascending aorta.
6. Start resection at the crista terminalis of the right atrium and continue along the wall of the left atrium. Divide the atrial septum close to the AV-valves, and the ascending aorta and the pulmonary trunk as close as possible to the valvular annulus.
7. Place he donor heart in the recipient's pericardium to the left of the atrial remnant. Perform all anastomoses in continuous technique with a double-armed 6-0 polypropylene suture. Suturing starts at the level of the left upper pulmonary vein. Keep the stitches tight and placed deeply. Thereafter, join the right atrium in a similar fashion of suture.
8. The sutures of the pulmonary artery and the ascending aorta complete the transplantation procedure. A drawing of the final in situ situation is shown in Fig. 2b.
9. Insert an aortic root cannula (DLP10016 5°F) in the ascending aorta of the xenograft to evacuate all the air from the heart and remove the aortic cross-clamp. If necessary, defibrillate the xenograft starting with 10 J. After a minimum reperfusion time of 60 min, wean the animal from the extracorporeal circulation. Follow decannulation of the aortic and the venous lines by antagonization of heparin with protamine sulphate (1 mL/1,000 IU heparin).

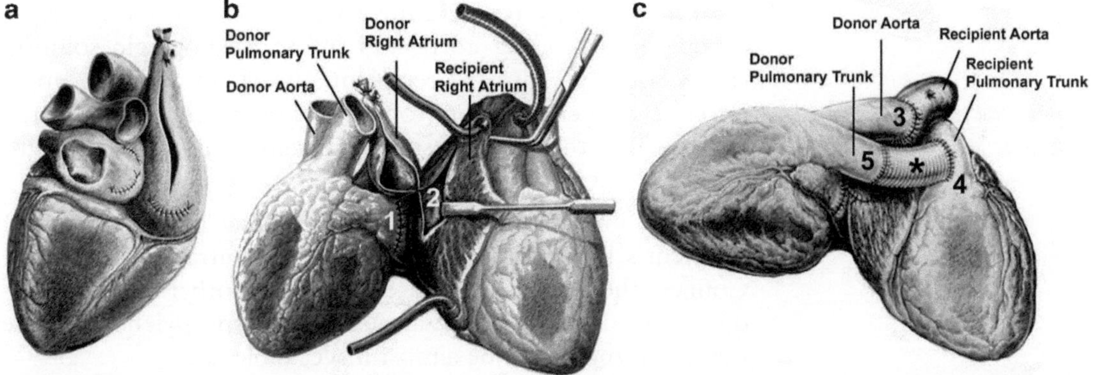


Fig. 3. Heterotopic thoracic heart transplantation: Donor heart before (**a**) both hearts during (**b**) and after the implantation procedure (**c**). Anastomoses between (1) the left atria, (2) the right atria, (3) both aorta, (4) the recipient's pulmonary trunk and the polyester prosthesis (*asterisk*) and (5) the prosthesis and the donor's pulmonary trunk.

10. Close the chest in layers when bleeding has stopped and three thoracic drainages (Argyle 8–16°F) have been placed in the chest. Remove them at the end of surgery when bleeding is minimal.
11. Tapering reduction of inotropic and vasoconstrictor pharmacological support and volume therapy after hemodynamic stabilization of the baboon finally enables a subsequent reduction of anesthesia and weaning of the animal from mechanical ventilation.

### *3.7. Xenogeneic, Heterotopic Thoracic Heart Transplantation (According to Barnard and Losman)*

1. Prepare ex vivo the cardiac xenograft for heterotopic heart transplantation in 4°C Ringer solution. The full length of the SVC is needed; therefore, fix the ligature distal of the azygos vein. Close the three orifices of the right pulmonary veins by a continuous suture (Prolene 5-0) and excise the left ones in order to create one orifice which should have the same size as the orifice of the mitral valve. Finally, create an approximately 5 cm long incision in the right atrium and continue toward the SVC. A drawing of the prepared donor heart is shown in Fig. 3a.
2. After median sternotomy, pericardiotomy and systemic heparinization (400 IU/kg) cannulate the aorta and both SVC and IVC and establish CPB, in the same way as previously described for orthotopic heart transplantation (with the exception that the SVC cannula might be advanced via the right atrial appendage). Create a right-sided pericardial flap by dint of two incisions at the level of the SVC and IVC down to the right phrenic nerve without compressing the phrenic nerve. Laid this pericardial flap over the hilum of the right lung to serve as a cradle for the donor heart.

3. Reduce the body temperature to 32°C during CPB. Cross-clamp the ascending aorta and infuse cardioplegic solution (HTK solution) into the recipient‘s heart. In addition, apply ice-cold Ringer solution topically. Locate the transplanted organ within the right chest and partially compresses the right middle and lower lobe.
4. Start the implantation with a longitudinal incision into the recipient's left atrium, as it is done for mitral valve surgery. Connect the two left atria carefully to each other by a continuous suture line of double-armed 6-0 polypropylene, because this anastomosis will be almost inaccessible later.
5. Afterwards, make an incision in the lower lateral part of the SVC and atrium of the recipient. Start the diamond-shaped anastomosis between the two right atria by a 6-0 double-armed polypropylene suture at the most proximal part of the right atrial incision of the donor and direct towards the mid-portion of a corresponding incision made in the recipient‘s SVC and atrium (see Note 8). Combine the two right atria. A drawing of the implantation procedure is shown in Fig. 3b.
6. Continue the implantation with the aortic end-to-side anastomosis. It is important to leave the donor aorta as short as possible, in order to exert traction in an anterior and superior direction and facilitate a wide opening of the right and left atria. Perform an adequate longitudinal incision into the recipient‘s ascending aorta. It may be made in the most convenient area and should ideally be situated slightly to the right of the midline.
7. Finally, join the two main pulmonary arteries with the help of an adequate interposition of a polyester prosthesis (Gelweave 10 mm), which is gelatin impregnated. Perform first the end-to-side anastomosis on the recipient side, proximal to the bifurcation. Then, complete an end-to-end anastomosis, joining the donor's main pulmonary artery to the vascular graft. A drawing of the final in situ situation is shown in Fig. 3c.
8. Before removing the aortic clamp, evacuate all the air from both hearts. The following procedure is similar to the orthotopic transplantation procedure (as previously described in Subheading 3.6, steps 9–10)
9. Use an implantable telemetry system for postoperative monitoring of the xenograft function and hemodynamic conditions of the recipient (PhysioTel D70-PCTP). The system consists of an implantable transmitter which includes a sensor for temperature and four leads (two ECG leads, one ventricular pressure and one aortic pressure lead). Suture the two ECG leads on the pericardium of the donor heart; insert and secure one pressure lead in the left ventricle of the xenograft and the second one in the ascending aorta of the recipient.

## 4. Notes

1. To reduce the diluting effect of CPB priming and surgical blood loss on hemoglobin concentration, we suggest to use male recipients with a body weight of >20 kg. Recipients with blood group AB enable a blood transfer from baboons with blood groups 0, A, B and AB.
2. The baboons are habituated to the jackets, which are produced in different sizes (from 8 to 25 kg), approximately 4 weeks before transplantation. To ensure optimal safety and mobility for the baboon, the jackets are modified by a professional dress maker.
3. A decrease of blood cell counts is a potential side effect of high dose immunosuppression during rejection therapy; erythropoietin (NeoRecormon, 20 IU/kg) and romiplostim (*n*-plate, 1–10 mg/kg) may be used to increase RBC and platelets by weekly infusion.
4. This custom CPB set was developed by our team (16) based on previous experiments and the comparison of different systems in the past. The HLM set consists of an oxygenator (Hilite 2800, Medos), a hemofilter (model MEHF 05025, Medos), a cardiotomy reservoir (Hilite MC 1640, Medos), a tube set (venous line: 1/4″, 0.6 m; arterial line: 1/4″, 0.6 m; pump line: 1/4″, 1.4 m; vent line $d = 4.8$ mm, 2.0 m), and a reservoir (Hilite MVC 1630, Medos). It enables a reduction of the priming volume to 195 mL. The closed system is powered by a rotation pump (Deltastream DP2, Medos).
5. In the previous experiments we have found that donor-recipient size matching is an important factor to guarantee optimal surgical outcome and minimize complications. To reduce gastrointestinal side effects in the heterotopic abdominal model, donor piglets with a body weight half of the recipients are chosen. Furthermore, donor hearts with the same weight as the recipients own hearts are needed for orthotopic transplantation. However, the ratio of heart weight to body weight for pigs is not the same as for baboons; therefore we use donors that have a descend body weight by about 15%. For the heterotopic thoracic technique, we choose pigs about 2/3 of the recipients body weight.
6. It is important to ensure that all air bubbles are removed within the application line and the cannula. The infusion pressure of the cardioplegia solution should not exceed 150 mmHg).
7. To avoid a lesion of the cisterna chyli with corresponding side effects (e.g., a postoperative lymphatic leakage) abdominal vessels have to be dissected very carefully.

8. According to our own experience, a modification, suggested by Vosloo, might provide an even larger anastomotic area: The (donor) SVC is left open; an incision leads from its orifice into the atrial appendage avoiding, however, the sinus node region.

## References


1. Cooper DK, Keogh AM, Brink J et al (2000) Report of the Xenotransplantation Advisory Committee of the International Society for Heart and Lung Transplantation: the present status of xenotransplantation and its potential role in the treatment of end-stage cardiac and pulmonary diseases. J Heart Lung Transplant 19:1125–1165
2. Schmoeckel M, Bhatti FN, Zaidi A et al (1997) Xenotransplantation of pig organs transgenic for human DAF: an update. Transplant Proc 29:3157–3158
3. Schmoeckel M, Bhatti FN, Zaidi A et al (1998) Orthotopic heart transplantation in a transgenic pig-to-primate model. Transplantation 65:1570–1577
4. Bhatti FN, Schmoeckel M, Zaidi A et al (1998) Three-month survival of HDAFF transgenic pig hearts transplanted into primates. Transplant Proc 31:958
5. Brandl U, Michel S, Erhardt M et al (2005) Administration of GAS914 in an orthotopic pig-to-baboon heart transplantation model. Xenotransplantation 12:134–141
6. Pierson RN, Dorling A, Ayares D et al (2009) Current status of xenotransplantation and prospects for clinical application. Xenotransplantation 16:263–280
7. Phelps CJ, Koike C, Vaught TD et al (2003) Production of alpha 1,3-galactosyltransferase-deficient pigs. Science 299:411–414
8. Kolber-Simonds D, Lai L, Watt SR et al (2004) Production of alpha-1,3-galactosyltransferase null pigs by means of nuclear transfer with fibroblasts bearing loss of heterozygosity mutations. Proc Natl Acad Sci 101:7335–7340
9. Kuwaki K, Tseng YL, Dor FJ et al (2005) Heart transplantation in baboons using alpha1,3-galactosyltransferase gene-knockout pigs as donors: initial experience. Nat Med 11:29–31
10. Yamada K, Yazawa K, Shimizu A et al (2005) Marked prolongation of porcine renal xenograft survival in baboons through the use of alpha1,3-galactosyltransferase gene-knockout donors and the cotransplantation of vascularized thymic tissue. Nat Med 11:32–34
11. Chen G, Qian H, Starzl T et al (2005) Acute rejection is associated with antibodies to non-Gal antigens in baboons using Gal-knockout pig kidneys. Nat Med 11:1295–1298
12. Ezzelarab M, Garcia B, Azimzadeh A et al (2009) The innate immune response and activation of coagulation in alpha1, 3-galactosyltransferase gene-knockout xenograft recipients. Transplantation 87:805–812
13. McGregor CG, Davies WR, Oi K et al (2005) Cardiac xenotransplantation: recent preclinical progress with 3-month median survival. J Thorac Cardiovasc Surg 130:844–851
14. McGregor CG, Byrne GW, Vlasin M et al (2009) Preclinical orthotopic cardiac xenotransplantation. J Heart Lung Transplant 16:S224
15. Bauer A, Postrach J, Thormann M et al (2010) First experience with heterotopic thoracic pig-to-baboon cardiac xenotransplantation. Xenotransplantation 17:243–249
16. Herzog R, Brandl U, Bauer A et al (2008) CPB-set with reduced priming volume for xenogeneic heart transplantation. Kardiotechniker 3:78–82

# Chapter 11

# Xenogeneic Lung Transplantation Models

**Lars Burdorf, Agnes M. Azimzadeh, and Richard N. Pierson III**

**Abstract**

Study of lung xenografts has proven useful to understand the remaining barriers to successful transplantation of other organ xenografts. In this chapter, the history and current status of lung xenotransplantation are briefly reviewed and two different experimental models, the ex vivo porcine-to-human lung perfusion and the in vivo xenogeneic lung transplantation, are presented. We focus on the technical details of these lung xenograft models in sufficient detail, list the needed materials, and mention analysis techniques to allow others to adopt them with minimal learning curve.

**Key words:** Xenotransplantation, Xenogeneic, Pulmonary xenograft, Lung transplantation, Lung perfusion

## 1. Introduction

### 1.1. Perspective

Solid organ xenograft survival and function have improved over the last decade in association with successful generation of genetically modified pigs (1). Development of α1,3-galactosyltransferase knock-out (GalT-KO) pigs has helped to overcome the preformed antibody-mediated hyperacute rejection (2) that porcine organs exhibit when transplanted into nonhuman primates (3). Expression of human complement pathway regulatory proteins (hCPRPs), such as hCD46, hCD55, or CD59, controlled the activation of complement which occurs usually within a very short time with wild-type or GalT-KO organs, and significantly delays the organ injury. Despite these genetic modifications, pig lungs activate and sequester platelets and neutrophils, and activate the coagulation cascade, when perfused ex vivo with human blood or transplanted into baboons. These phenomena illustrate that significant major hurdles remain to be overcome in order to reach clinical application of lung xenotransplantation. Further, we suggest that

Cristina Costa and Rafael Máñez (eds.), *Xenotransplantation: Methods and Protocols,* Methods in Molecular Biology, vol. 885, DOI 10.1007/978-1-61779-845-0_11, 

the processes causing lung xenograft injury may be important to control in order to obtain improved results with other organ xenografts.

### 1.2. Historical and Scientific Background

The first experience with xenogeneic pulmonary organs was reported in 1968 by Bryant et al. (4). In these experiments, an ex vivo apparatus integrating porcine lungs was used to oxygenate human blood, with the idea that they might be used for patients undergoing cardiac surgical procedures. The lungs rapidly exhibited increased PVR and pulmonary edema. Since those early days, many groups have conducted multiple studies in cardiac, kidney, and islet xenotransplantation. In contrast, only a few groups have evaluated pulmonary xenotransplantation (historical groups; Michler, Machiarini, Davis, Fodor, Pierson/Azimzadeh). Whether as a cause or a consequence, lung xenograft outcomes remain relatively poor.

One unique feature of the porcine lung xenograft is the presence of pulmonary intravascular macrophages (PIMs), which we (5) and others (6) have shown elaborate thromboxane and vWF multimers, and are associated with severe constriction of the pulmonary vasculature. In addition, lung xenografts are associated with rapid disappearance of platelets and neutrophils from the circulation even when anti-pig antibody and complement activation are potently inhibited. We believe that both of these phenomena are simply more readily apparent for the lung than for other organs due to the relatively high number of PIMs, and the very large and highly vasoreactive endothelial surface area in the lung. Therefore, although lungs have played a relatively minor role in the field of xenotransplanation to date, in our estimation lung xenograft studies offer a unique, valuable opportunity to contribute to future progress in the field.

### 1.3. Animal Models

To study the outcome of pulmonary xenografts or drugs and methods that might improve the function of those organs, different animal models have been developed.

#### 1.3.1. Rodent Models

The use of rodents as organ donor and recipient offers the possibility to conduct studies in a relatively rapid, high-throughput, and cost-effective fashion. Another advantage lies in the reduced amount of pharmacologic or molecular reagents needed to perform mechanistically informative studies when compared to larger animals, and thus lower costs to perform definitive experiments.

Schroeder et al. (7) published a study in 2003 in which lungs from male wild-type mice were perfused ex vivo with human blood. After removal of the heart–lung block, the main pulmonary artery was cannulated with an intravenous catheter. For the experiment, the lungs were ventilated with room air. During the first 15 min, lungs were flushed with 5% human albumin in saline at a continuous flow rate of 0.2–0.5 mL/min. Thereafter, the albumin solution was replaced by the human blood perfusate at unchanged flow conditions. As a control, lungs were also perfused with autologous blood.

In vivo orthotopic left lung xenotransplantation models from hamster or guinea pig to rat have also been performed (8–11). Orthotopic mouse lung allograft models have recently been developed that can take advantage of the multiple genetic modifications to the mouse, including GalT-KO mice that could be used as recipients for theoretically "discordant" Gal-expressing wild-type lungs (12). To the best of our knowledge this model has not been developed for the study lung xenotransplantation.

However, the relevance of these rodent models to the clinically relevant pig-to-human combination is doubtful, in particular because many approaches (genetic modification to the pig) and reagents (e.g., anti-human or anti-pig monoclonal antibodies) that are developed for clinical use do not cross-react with rodents.

#### *1.3.2. Concordant Large Animal Models*

Fox-to-dog (13) as well as Macaque monkey-to-baboon (14) orthotopic lung transplantations have been performed in the past, using either no immunosuppression or FK506. Another series of concordant transplantations was conducted by Sadeghi et al. (15) who transplanted cynomolgus monkey heart–lung blocks into baboons. Recipient animals received cyclosporine and steroid-based immunosuppression.

#### *1.3.3. Large Animal Models of Discordant Lung Xenotransplantation*

In 1998, a cross-circulation lung perfusion model was described as a method to deplete anti-GalT antibodies from the blood of the "recipient" animal (16). The baboon underwent thoracotomy and anticoagulation, and an "outflow" cannula was placed in the inferior vena cava. Via a roller pump, the blood was perfused through main pulmonary artery of the porcine lung, drained by gravity into a heated reservoir and returned to the baboon via a venous cannula in the right atrial appendage (see Fig. 1).

With the same aim to effect immunoabsorption of anti-GalT antibodies, Macchiarini et al. (17) performed experiments in which the recipient animal (goat) was used to cross-circulate the right lung from a GalT-expressing pig prior to receiving the left lung of the same donor animal in an orthotopic transplantation.

Pigs have been identified as the most feasible source of organs for use in humans. Thus, pig-to-human and pig-to-primate models are now preferred because they closely simulate the intended application in the clinic. Importantly, these models permit evaluation of reagents that interact with pivotal pig proteins or with key human and nonhuman primate molecules. In particular, pig-to-primate orthotopic left lung xenotransplantation and ex vivo perfusion of porcine lungs with fresh human blood are currently the most widely used to study discordant pulmonary xenograft injury.

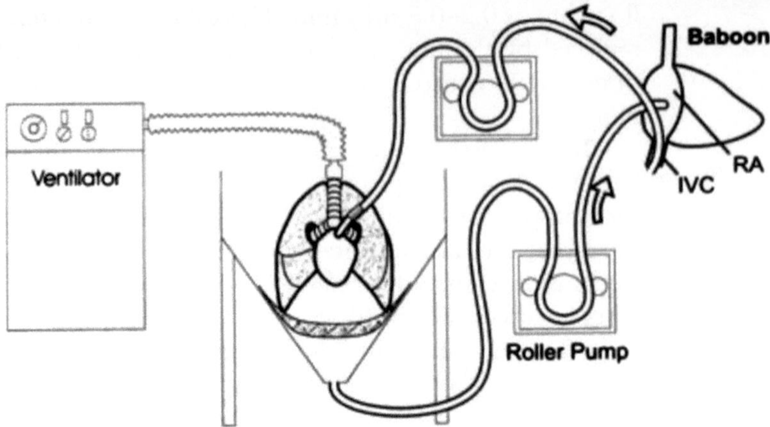


Fig. 1. Diagram of extracorporeal lung perfusion circuit used by Daggett et al. (16). Blood is removed from the baboon inferior vena cava and delivered to the swine main pulmonary artery with a roller pump. Blood is drained into a reservoir and returned to the baboon with a second roller pump.

In the following paragraphs, these two models are reviewed and explained in detail.

## 2. Materials

### 2.1. Drugs and Medications

1. Calcium chloride, 100 mg/mL.
2. Dopamine HCl Inj., 40 mg/mL.
3. Heparin, 1,000 IU/mL.
4. Human albumine 5%, 250 mL.
5. Isoflurane.
6. Ketamine HCl Inj., 100 mg/mL.
7. Noradrenalin (norepinephrine), 1 mg/mL.
8. Perfadex (Xvivo Perfusion AB, Gothenburg, Sweden).
9. Propofol (Abbott Laboratories, North Chicago, IL 60064, USA).
10. Vasopressin Inj., 10 U/0.5 mL.
11. Xylazine (AnaSed Injection), 20 mg/mL.
12. 8.4% Sodium bicarbonate Inj., 1 mEq/mL.
13. 0.9% Sodium chloride.

### 2.2. Ex Vivo Lung Perfusion Circuit and Accessories

1. Glass beaker (ACE Glass Incorporated, Vineland, NJ 08360, USA).
2. Blood reservoir 1,000 mL (ACE Glass Incorporated, Vineland, NJ 08360, USA).

3. Latex tubing 3/8 in.
4. Y-3/8 in. × 3/8 in. × 3/8 in. Blood tubing connector.
5. Straight metal tubing-PA adapter (custom-made).
6. Metal rig.
7. Blood pump (Polyston A/S, Vaerlose, Denmark).
8. Stirrer (Thermolyne Nuova II, Dubuque, ID 52001, USA).
9. Reservoir-water heater (American Pharmaseal Company RK 300, Valencia, CA 91355, USA).
10. Biological atmosphere mixture gas (Airgas, Inc., Salem, NH 03079, USA).
11. Human blood (freshly collected) in blood collection bags containing CPDA-1 (Terumo, Somerset, NJ 08873, USA).
12. Fresh-frozen human plasma (2 bags/lung side).

### 2.3. Monitoring and Anesthesia

1. Isoflurane vaporizer.
2. Endotracheal tube (ETT) and mask.
3. Ventilator.
4. Veterinary anesthesia ventilator.
5. Flowmeter.
6. Pressure monitoring.
7. Pressure transducer.
8. Monitoring line.
9. Syringe pumps.
10. Biological atmosphere mixture gas in tank.
11. 95% Oxygen, 5% carbon dioxide in tank.

### 2.4. Sutures, Catheters, and Disposables

1. Prolene 4-0, 5-0, 6-0, Silk 0 and #1 ties (Ethicon, Somerville, NJ 08876, USA).
2. Vessel loops VL-205 (Key Surgical Inc., Hannover, Germany).
3. Pediatric venous cannula 12-20 Fr (Medtronic Perfusion Systems, Brooklyn Park, MN 55428, USA).
4. Swan–Ganz 7.5 F catheter (Edwards Lifesciences, Irvine, CA 92614, USA).
5. Arrow-flex polyurethane sheath catheter (Arrow, Reading, PA 19605, USA).
6. Angiocatheter 18–22 G (Jelco, Medex Medical Ltd, Rossendale, Lancashire, UK).
7. Tracheal tubes 3.5–6.0 mm.

8. BD Vacutainer (EDTA, Serum, CTAD) (Becton Dickinson, Franklin Lakes, NJ 07417, USA).
9. Ethyl alcohol 190 Proof, 95%.

### 2.5. Equipment and Reagents for Analyses and Assays

1. Computer linked to the detectors (Transonic Systems Inc., Model T206, Ithaca, NY, USA) to record flow and pressure continuously.
2. Automated blood cell analyzer.
3. Flow cytometer and antibodies for P-selectin and CD41 for measuring platelet activation at the cellular level.
4. Asserachrome-βTG enzyme-linked immunosorbent assay (ELISA) from Diagnostica Stago (Parsippany, NJ, USA) to assess platelet activation by beta-thromboglobulin (βTG) level formation.
5. Enzygnost micro F1+2 ELISA (Dade Behring, Marburg, Germany) for measuring thrombin formation.
6. C3a and/or C5a ELISA (Quidel, San Diego, CA, USA).
7. Cytokine Luminex arrays (Luminex Corporation, Austin, TX 78727, USA) for measuring cytokines, chemokines, and other molecules of interest.

## 3. Methods (See Note 1)

### 3.1. Ex Vivo Porcine-to-Human Lung Perfusion Model

Ex vivo xenogeneic perfusion of a porcine lung using human donor blood is the most clinically relevant and thus important model than can be used to assess the impact of interventions on short-term outcomes (within 6–8 h). It closely replicates the kinetics, physiology, and immunohistochemical (IHC) findings seen during in vivo pig lung transplantation in a nonhuman primate. All changes that occur during in vivo hyperacute rejection (HAR), such as increased PVR, loss of transpulmonary blood flow, failure of oxygenation, and tracheal or lung edema, are observed in both models. Only perfusion of pig lungs with whole human blood offers the possibility to perform mechanistic studies in the clinically relevant species combination and to evaluate interaction of formed blood elements across species. Functional parameters as well as blood and tissue samples can be collected at prespecified intervals over a specified time range to study the mechanisms of rejection (see Note 2).

It is possible to perfuse one lung, both lungs en bloc, or to divide the pulmonary bloc into left and right lungs and perfuse the "pair" separately. We primarily utilize the paired model because the impact of modifications to one circuit can be assessed in the context of a contemporaneous contralateral reference ("control") experiment where only the intervention under study differs between the two circuits. This approach allows biologically significant differences to become more readily apparent, especially given biologic variability among pigs, and since uncontrolled variables (such as anti-pig antibody titer) could otherwise exert confounding influences on experimental outcome and complicate study interpretation (see Note 3).

*3.1.1. Donor Organ Procurement*

1. Premedicate the pig with intramuscular ketamine and xylazine and isoflurane by mask (see Note 4).
2. Intubate through a midline surgical tracheostomy, and an ETT of appropriate size secured with multiple heavy ties (see Note 5).
3. Ventilate and anesthetize the pig with inhaled isoflurane through the ETT.
4. Open the chest with a median sternotomy contiguous with the tracheostomy incision.
5. Obtain vascular access peripherally or by cannulating a vessel in the surgical field with IV tubing (see Note 6).
6. After heparinization (see Note 7), introduce a large angiocatheter (>14 G) or an angulated cannula (aortic or venous cannula, 12-20 Fr) into the pulmonary trunk through a Prolene purse-string to introduce the flush solution.
7. Administer premedication (e.g., prostacyclin).
8. Ligate the right superior and inferior venae cavae, and left superior vena cava, and cross-clamp the ascending aorta, before administration of the flush solution (see Note 8).
9. Make an incision in the left atrium, or amputate the left ventricular apex, to allow free drainage of the pulmonary vein effluent.
10. After administration of the flush solution, inflate the lungs, place a clamp on the trachea or endotracheal tube, and explant the lungs en bloc with the heart, removing the trachea, which for convenience can be done, including the tracheostomy site with the ETT in place.
11. If the experimental design calls for perfusion of both lungs at the same time, secure a return cannula into the left ventricle (it can be, but need not be, introduced across the mitral valve into left atrium), with umbilical tape gently but securely tightened around the LV apex.
12. Establish pulmonary artery (PA) perfusion via the angulated cannula in the pulmonary trunk through which administers the flush solution, with care to preserve competence of the pulmonic valve during subsequent perfusion.
13. Alternatively (and more commonly) introduce a flanged metal or plastic cannula through an incision in the right ventricular

outflow tract, through the pulmonary valve, and secure with a heavy tie around the main pulmonary artery distal to the pulmonoplegia site. Cannula sizes and dimensions take into account the size of the vessels to be cannulated (which depend on pig size) and the size of tubing used for the perfusion circuit [typically 1/4 in. (6.4 mm) or 3/8 in. (9.5 mm)].

14. Insert an ETT into the trachea at a convenient cervical location (safely above the high take-off of the right upper lobe "pig" bronchus), and attach securely with umbilical tapes or a strong silk ligature if the tracheostomy tube from the procurement is not transferred to the perfusion apparatus.
15. Put the gently inflated lungs (to avoid barotraumas) with the ETT clamped on ice until the blood for the perfusion has been prepared and the experiment can be started.
16. For the "paired" perfusion system (side-by-side individual right and left lungs), separate surgically the lungs as follows. Because the right main pulmonary artery is relatively short, cannulate it before dividing the left pulmonary artery, securing the right PA cannula with a heavy tie just distal to the main PA bifurcation (see Note 9).
17. Then, divide the left pulmonary artery just past the main PA bifurcation, usually leaving sufficient length to secure a cannula (e.g., stainless steel cannula) with a simple heavy silk suture (see Note 10). The PA cannulas will later be connected to the inflow perfusion tubing.
18. Insert a size-matched balloon-cuffed ETT into the cervical trachea as described above. The trachea must be left with the right lung due to origination of the right upper lobe ("pig bronchus") from the trachea.
19. Divide the left main bronchus flush with the trachea, and close the resulting tracheal defect with a continuous 4-0 or 5-0 Prolene suture.
20. Cannulate the left main bronchus directly, securing the ETT (3.0–6.0 mm) with its tip just proximal to the left main bronchial bifurcation with a 4-0 Prolene suture, taking multiple partial-thickness bites in the outer airway, tieing the suture tightly, and then securing the same ligature over the ET tube (we prefer to use an ET tube with a balloon, to assure secure fixation of the ETT). Thus, inflate the tube balloon gently.
21. To prevent the ETT from being dislodged, extend the bronchial Prolene suture proximally along the ETT, and additionally tie the suture around the ETT just proximal to the proximal end of the balloon.
22. For storage, inflate each lung gently (avoiding barotraumas), clamp the ETT, place the lung in a plastic bag, and cool on water or saline ice.

*3.1.2. Human Blood Preparation*

The preparation of the human blood used for the perfusion of the porcine lungs can be done in various ways. It is essential to use some freshly collected blood from healthy donors in order that the platelet and neutrophil function simulate in vivo biologic phenomena (see Note 11).

1. Use regular clinical blood banking collection bags containing CPDA-1 to collect the fresh blood (~450 mL/1 blood unit). Depending on the size of the pigs and thus the size of the lungs, it needs to be determined how many individual donors are needed (see Note 12).
2. In the "paired" lung model, apportion the blood between the circuits in approximate proportion to relative pig lung weight (60% right and 40% left can be assumed, in our experience).
3. To prevent clotting of the blood in the circuit while reconstituting calcium-dependent physiologic mechanisms (the coagulation and complement pathways are both critically dependent on calcium), add heparin (3 IU/mL blood–plasma mixture) prior to adding calcium chloride (1.6 mg/mL blood–plasma mixture) to neutralize the CPDA chelating agent.
4. Add sodium bicarbonate to obtain a target pH of 7.4.

*3.1.3. Organ Perfusion*

The system described permits perfusion of single lungs or lungs en bloc in a simple, safe, highly reproducible manner (see Fig. 2). The perfusion circuit can of course be modified for other purposes (e.g., to study ischemia reperfusion injury), or for ex vivo lung allograft experiments.

1. Clean the tubing system, and any other system components that will have contact with the blood, rinse extensively with clean water, disinfect (ethyl alcohol 95%), and dry overnight, before blood is introduced into the perfusion circuit.
2. Coat all blood-contacting circuit surfaces with human albumin 5% in 0.9% "normal" saline to minimize blood activation by contact with recycled glass, plastic, and silicon circuit components (see Note 13).
3. Suspend the pig lung in a funnel over the jacketed glass-beaker that will collect the blood issuing from the pulmonary veins.
4. Warm the reservoir beaker with a circulating water pump.
5. Pump blood from the reservoir via the tubing system, and return to the lung via the PA (see Note 14).
6. Connect an appropriately adjusted volume-controlled ventilator to the ETT and ventilate the lung with a custom-ordered gas mixture of 21% oxygen and 5–8% carbon dioxide balanced with nitrogen. This concentration of carbon dioxide maintains $pCO_2$ at 35–45 mmHg, and prevents the profound alkalosis that would otherwise occur in the absence of a significant

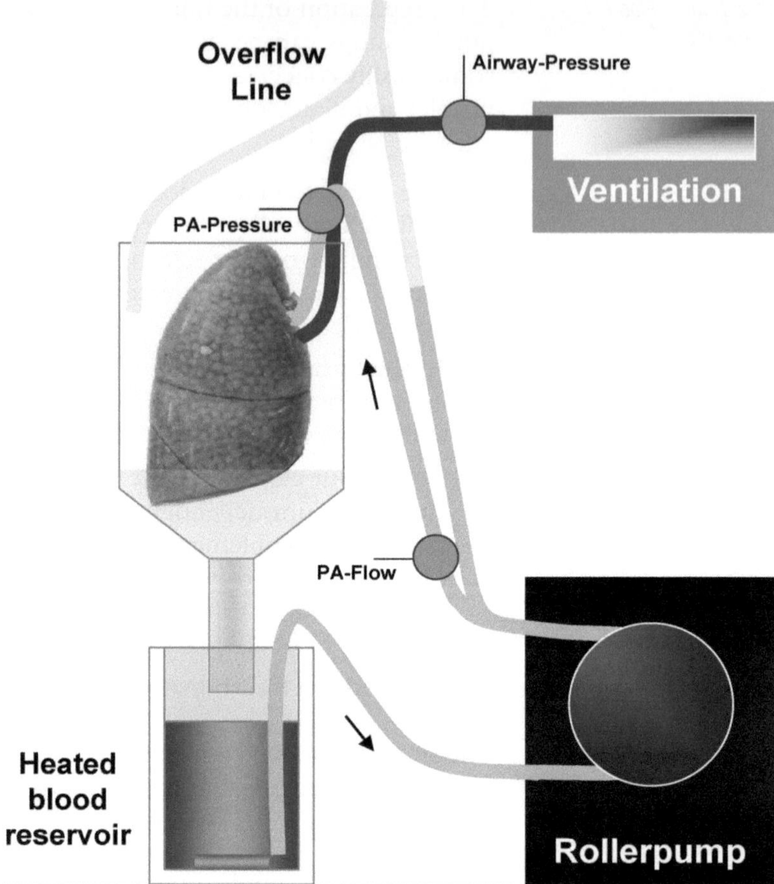


Fig. 2. Diagram of the ex vivo lung perfusion circuit. Blood is pumped via silicon tubing into the pulmonary artery by a roller pump. After the perfusion of the lung, the blood is drained into a heated blood reservoir from which it is recirculated to the lung.

source of tissue metabolism, and thus $CO_2$, in the ex vivo circuit.

7. To evacuate air from the blood-containing tubing, pour the prepared human blood into the reservoir and start the roller pump.
8. De-air the circuit and connect the inflow tubing to the pulmonary artery. The flow setting of the pump should be up to 75 mL/kg body weight (BW)/min (see Note 15).
9. For side-by-side bilateral lung perfusion, devote 60% of blood volume and flow to the right lung, and 40% to the left side, based on the normal proportion of relative lung weight.
10. Ventilate the lungs at 7 mL/kg BW and adjust ventilator tidal volume according to the lung compliance, seeking to achieve peak airway pressure of 12–15 cm $H_2O$ (~8–11 mmHg).

11. To record physiologic lung function parameters, use pressure transducers to record PA and airway pressures, and a flow probe to continuously monitor the blood flow into the lung in the inflow-tubing just proximal to the pulmonary artery. Calculate the PVR by dividing mean PA pressure by blood flow.
12. Collect blood samples directly from the pulmonary vein effluent.
13. Excise lung biopsies behind tissue clamps, leave in place for the duration of the experiment (stapling devices could be used, but add considerable expense).
14. Trisect each lung biopsy and process for later histological, molecular, and immunohistological analyses.
15. Assess lung function by blood gas analyses, measuring $pO_2$, $pCO_2$, and pH over time under consistent circumstances. By intermittently switching from the ventilator gas mixture (21% $O_2$) to 100% oxygen for 3–5 min, measure the capacity of lung oxygenation. If preferred for a specific application (and if the cost of the oxygenator can be afforded), a "deoxygenator" (oxygenator swept with 8% $CO_2$ and 10–14% $O_2$ balanced with nitrogen) can be integrated into the circuit (see Note 16), and oxygenation capacity of the lung assessed by change in step-up on $pO_2$ across the lung over time during ventilation with room air.

### *3.2. In Vivo Porcine Lung Transplantation Model*

As for other xenogeneic organ transplants, the transplantation of a porcine pulmonary graft into a nonhuman primate, represents the gold standard to test the function, observe blood parameter changes, and determine survival of a transplant in a clinically relevant "translational" xenogeneic model. Even though the ex vivo perfusion model facilitates informative mechanistic studies, it will never offer the possibility to discover perhaps unexpected phenomena that may occur during attempted life-support in a living animal, and allow monitoring of perturbations in the physiology of the recipient. Importantly, the in vivo model is not perturbed by the artificial elements of the circuit (blood-tubing and blood–air interactions, for example), or by unappreciated effects of "pooling" multiple human donors.

Although studies in cynomolgus monkeys have been reported (18), most in vivo lung xenotransplant work has been done in baboons, where adult animals are of appropriate size (>10 and preferably >15 kg) to accept left lung grafts from 6- to 10-kg pigs (16, 19, 20) (see Note 17).

#### *3.2.1. Donor Organ Procurement*

1. Perform the anesthesia of the donor pig and surgical procurement of the heart–lung bloc as described in the ex vivo lung perfusion section.
2. For the in vivo lung transplantation, separate the left porcine lung surgically from the heart.

3. Excise the pulmonary veins as an island from the left atrium.
4. Because there is often a large distance between the upper and lower left pulmonary veins, imbricate the intervening left atrial wall to reduce the donor vein cuff size to more closely approximate the smaller baboon pulmonary vein cuff.
5. Divide the PA just proximal to the left main pulmonary artery bifurcation (see Note 18).
6. Finally, clamp the left main bronchus to ensure that the lung stays inflated and is cut close to the bifurcation of the trachea.
7. Store the lung on ice until the implantation of the organ.

*3.2.2. Preparation of the Recipient*

1. Sedate baboon recipients with ketamine (10 mg/kg BW, i.m.).
2. After the intubation, maintain general anesthesia with isoflurane (1–3%, i.h.).
3. For initial intravenous access, place an angiocatheter in a superficial extremity vein.
4. Place an arterial line in the femoral artery for monitoring of hemodynamic parameters (see Note 19).
5. For later positioning of a Swan–Ganz catheter, prepare the femoral vein and cannulate directly with an inducer sheath.
6. To prevent blood clotting at the catheters, give heparin as a bolus (70 IU/kg BW, i.v.) and infuse at 200 IU/h. Flush lines with heparinized saline until use.
7. Open the chest of the animal with a transverse "clamshell" incision at the fourth or fifth intercostal space.
8. Open both, the pericardium and the pleura with longitudinal incisions.
9. Introduce a flow-directed PA balloon catheter into the femoral vein, and manually guide across the tricuspid valve (see Note 20) and out the pulmonary outflow tract to the proximal PA.
10. Position a snared vessel loop around the right pulmonary artery, working between the aorta and SVC. Securing this loop allows reversible selective perfusion of the transplanted left lung, and thus intermittent or continuous rigorous assessment of left lung function.
11. A flow probe (Transonic Systems Inc., Ithaca, NY) on the ascending aorta allows measurement of the cardiac output (see Note 21).

*3.2.3. Explantation of the Left Native Baboon Lung*

1. For the explantation of the native lung, displace the heart from the pericardium and into the right chest, avoiding hemodynamic embarrassment by incising the right inferior pericardium

along the diaphragm. To this end, retract a suture tied on the left atrial appendage toward the right, and protect it with a rubber tube (Rummel snare) from rubbing against the contracting heart. Sutures through the pericardium just anterior to the left hilum, similarly protected with rubber tubing and retracted anteriorly and to the right, aid in exposure with minimal hemodynamic compromise.

2. Working within (particularly helpful!) and outside the pericardium, identify and dissect the left hilar structures.
3. Ligate left upper and lower pulmonary veins centrally and peripherally with heavy suture (Silk 0 or #1) and divide.
4. Use a straight vascular clamp to occlude the left PA as close to the mediastinum as possible, and divide the LPA distally just proximal to the left main PA bifurcation.
5. After dividing the inferior pulmonary ligament, divide the left main bronchus at the hilum and control bronchial vessels with sutures or (for nonsurvival procedures) with cautery.
6. Restore ventilation to the right lung by placing a straight clamp on the left main bronchus (see Note 22).

*3.2.4. Implantation of the Xenogeneic Organ*

1. Place the cooled xenogeneic lung in the left pleura of the recipient baboon.
2. If necessary, tailor the donor or recipient bronchus, pulmonary artery, and vein to enable anastomosis of the structures and accommodate common size discrepancies.
3. First, anastomose donor and recipient bronchus with a continuous suture (4/0 or 5/0 Prolene). It is also possible to use an interrupted 4/0 Prolene suture if intermediate-term survival with bronchial healing is desired.
4. Because the donor lung has intrinsically stiffer than the baboon lung, the clamps on the bronchus can now be removed without significant interference in completing the vascular anastamoses.
5. For the end-to-end anastomosis of the pulmonary artery, use a continuous Prolene 5/0 or 6/0 suture.
6. For the later venting, keep the suture open and not tied and knotted yet (see Note 23).
7. To simplify the connection of the pulmonary veins, free up the recipient pulmonary vein and left atrium from the pericardium posteriorly to allow placement of a clamp proximally on the left atrium.
8. Remove ligatures on the pulmonary veins, preserving maximal recipient cuff diameter, and incise the atrial tissue between the upper and lower recipient vein orifices to obtain one large foramen.

9. Connect the donor and recipient left atrial cuffs with a Prolene 4/0 or 5/0.
10. To restore perfusion to the transplanted lung, open the clamp on the pulmonary veins and assure hemostasis of the venous anastomosis.
11. Open slowly the proximal clamp on the pulmonary artery to evacuate air from the PA.
12. Tie the suture after venting the suture, fully establishing blood flow to the xenograft.

*3.2.5. Longitudinal Graft Assessment*

After transplantation of the xenogeneic organ, it is essential to control the hemodynamic parameters and stabilize the recipient. Inotropic medication is almost always required during graft implantation (see Note 24). In 2007, Nguyen et al. (19) used the described in vivo model to perfuse a porcine lung in a baboon for up to 255 min. At the same laboratory, recent experiments have shown that it is possible to keep the transplanted recipient animal alive under general anesthesia for 24 h (unpublished data). Others have done so for up to 3–5 days (21). During this time, it is possible to collect not only tissue samples from the xenogeneic lung, but also to record ventilation parameters, systemic, pulmonary arterial, and central venous pressures. Flow probes on the ascending aorta and on the left pulmonary artery allow to record cardiac output and flow to the transplanted lung, respectively. By pulling gently on the snared vessel loop on the right pulmonary artery, it is possible to partially or completely occlude this vessel and thereby adjust recipient dependence on the pulmonary xenograft (see Note 25). During this single lung perfusion period, PVR, cardiac output, systemic arterial pressure, and systemic arterial oxygen saturation indicate whether the transplanted left lung is able to support life.

Since rejection of the xenogeneic pulmonary graft usually happens within hours and is often associated with lung edema and oxygenation failure, animals are so far being kept under general anesthesia to avoid stress and discomfort of the animals. If future experiments show an improvement of the graft survival and function, it is in principle feasible to close the chest and recover the animal from the operation.

### 3.3. Analyses and Assays

To compare the function and the outcome of different xenogeneic lung transplantations or lung perfusions, an easy measurable parameter is the time after which the organ is rejected or loses its function. To determine this time point, markers as the loss of blood flow though the lung, an increased vascular resistance, failure of blood oxygenation, occurrence of tracheal edema, or the loss of perfusate due to lung edema should be taken in consideration.

To understand the biology of organs in a xenogeneic surrounding, it is essential to analyze the results concerning functional and hemodynamic, hematologic, immunologic, and immunohistologic changes. The previously described models offer the possibility to have access to all of these parameters by recording pressures and flows and collecting tissue and blood samples in the run of the experiment.

*3.3.1. Functional and Hemodynamic Analyses*

1. Measure the function of the organ by the PVR (see Note 26) and by the ability of the lung to oxygenate the blood. Especially, within the first 30 min of perfusion and also before lung demise, an increase in PVR can be observed. If possible, record those parameters continuously or in short intervals during this time. The best approach is to record flow and pressure continuously on a computer linked to the detectors.
2. Measure the oxygenation of the blood by analyzing the pulmonary vein blood gasses after a short time (3–5 min) on 100% $FiO_2$, and record as the amount of step-up in $pO_2$ relative to pulmonary vein $pO_2$ on 21% $FiO_2$. If a deoxygenator is incorporated in the circuit, this maneuver and calculation becomes unnecessary to measure lung gas transport function intermittently during the experiment.

*3.3.2. Hematologic Analyses*

With the xenogeneic perfusion of an organ, multiple hematological changes can be observed in the perfusing blood.

1. After enabling blood flow through the organ, initially collect blood samples at very short intervals (5–10 min) as sequestration and activation of cells occurs immediately; thereafter predetermined intervals of 30–60 min are usually sufficient to track trends over time, and compare results between groups treated in various ways.
2. For cell counts, enumerate white blood cells, including neutrophils, monocytes, and platelets by standard automated techniques in perfusate/blood samples (see Note 27).
3. For measuring platelet activation at the cellular level, a very useful parameter is the cell surface expression of P-selectin (CD62P) determined by flow cytometry. To measure CD62P expression on platelets, stain blood samples by monoclonal antibodies specific for CD41 (as a marker for platelets) and CD62P (expressed by activated platelets) according to the instructions of the manufacturer. Identify platelets and platelet aggregates by size and by the presence of CD41 staining, and analyze later for expression of CD62P. Express results as the percentage of CD62P-positive cells among CD41-positive cells (see Note 28).
4. To measure platelet activation and coagulation cascade in archived plasma samples, assess platelet activation by βTG level formation using a commercial ELISA (Asserachrome-βTG).

Measure thrombin formation as the production of prothrombin fragments 1+2 (F1+2) generated by the cleavage of prothrombin in thrombin by ELISA (see Note 29).

5. Determine the activation of the complement cascade by measuring C3a or C5a levels in plasma samples collected in EDTA blood tubes and stored in EDTA at −70°C (C3a ELISA; Quidel, San Diego, CA).

*3.3.3. Cytokine Measurements*

As cytokines are signaling molecules, secreted by numerous cells of the immune system, the measurement and level of cytokines can be used to analyze the strength of immune activation and immune response. The analyses may include interleukin 1B (IL-1B), IL-6, IL-8, interferon gamma (IFNγ), TNF inflammatory cytokines, chemokines, soluble CD40L and sCD62P (platelet activation). New high-throughput technologies (Luminex arrays) efficiently measure multiple targets simultaneously in a small volume of plasma (22).

*3.3.4. Histology and Immunohistology*

At various time points throughout the experiments, tissue samples from the xenogeneic organ should be collected to prepare histological slides. A standard staining to visualize cell infiltration of the tissue and vessel thrombosis is the hematoxylin and eosin stain (H&E). Figure 3a shows lung tissue, collected from a GalT-KO. hCD39 transgeneic lung after 4 h of perfusion with human whole blood. Histology shows an intact lung tissue without significant pathological abnormalities. In contrast, Fig. 3b demonstrates findings seen in a hyperacutely rejected organ. This lung tissue from a wild-type pig that was perfused with human blood for only 12 min shows a loss of integrity of alveolar endothelial wall structures, hemorrhage, and vessel thrombosis. Further, IHC staining includes visualization of human monocytes (CD14), neutrophils (myeloperoxidase), T-cells (CD3, CD4, CD8) or B-cells (CD20), total (CD41) and activated (P-selectin) platelets, tissue factor, IgG and IgM, complement pathway products (C4d, C5b-9), vWF (to identify endothelium), and others. Do IHC on snap-frozen lung tissue insufflated with OCT diluted with saline (~1:1) into the airways through a small-gauge angiocatheter (22–24 G) and syringe, to reduce edge artifact.

*3.3.5. PCR/Gene Expression Analyses*

The rejection of an organ does not only affect its function and tissue structure, but also modulates the up- and down-regulation of genes (23). Snap-frozen tissue samples collected at various time points during the experiment can be used to measure gene expression by real-time RT-PCR or cDNA microarray analysis (24). By designing PCR primers and probes that recognize pig or human sequences only, it is possible to monitor changes in gene expression within lung tissue parenchymal cells or within human infiltrating cells.

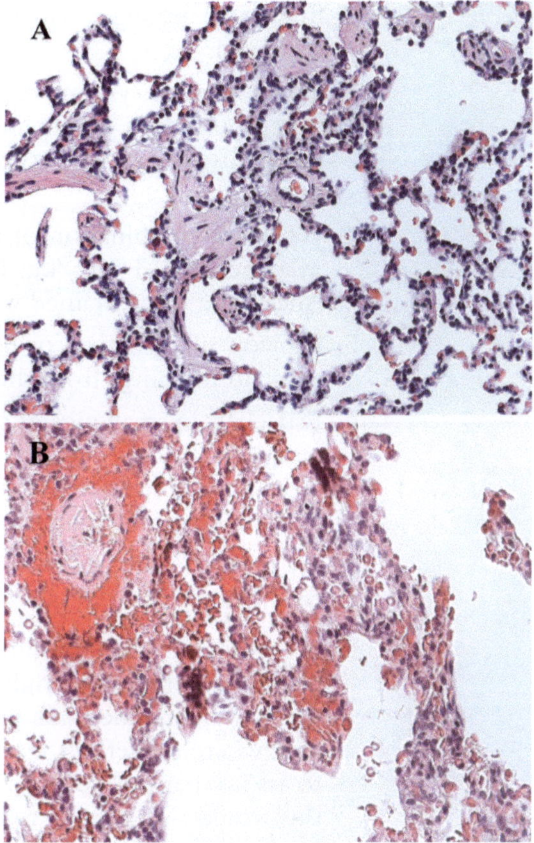


Fig. 3. (**a**) Histological analysis of a GalTKO.hCD39 transgenic pig lung after 4 h of xenogeneic perfusion. Histology shows an intact pulmonary tissue with only very little hemorrhage and cell infiltration. (**b**) Example of a hyperacutely rejected lung. The slide shows a thrombosed vessel and massive hemorrhage in lung tissue from a wild-type pig after 12 min of xenogeneic perfusion.

## 4. Notes

1. All transplant experiments require surgical and anesthesia instruments, equipment, and expertise for the procurement of the organ and, if performed, for the performance and monitoring of the transplant. Involvement of a surgical team trained in transplant and/or cardiothoracic surgical techniques is essential to successfully conduct xenotransplant research.
2. It is advisable to have very short intervals (1–2 min) for the blood and tissue sampling as well as for the recording of the functional parameters at the beginning of the perfusion. Especially during this period, blood (cell sequestration) and

functional changes (increase in PVR) can be frequently observed.

3. The perfusion system (e.g., flow rate and cannula sizes) and other logistics (e.g., the human blood volume) can be adapted to perfuse lungs of pigs with a bodyweight of 2–50 kg or more.
4. We use a combination of ketamine (10–15 mg/kg IM) plus xylazine (0.5–1 mg/kg IM) for the anesthesia induction. Anesthesia is maintained with isoflurane (0.5–4.0% inhaled).
5. Ethicon silk 1-0 ties (or equal) can be used. It is possible to suspend the lungs from those ties.
6. Especially in small pigs, it can be difficult to get IV access on a peripheral vessel (e.g., ear vein). In that case, it is possible to place an angiocatheter in the internal jugular vein through the midline surgical tracheostomy.
7. We give a heparin dose of 500 IU/kg.
8. We use perfadex (50 mL/kg BW).
9. If the right PA is not long enough to insert a cannula to allow perfusion of the upper right lobe, it is instead possible to secure the cannula to the main pulmonary artery and then close the hole where the left PA was dissected with suture. It is essential to assure that the closure suture does not obstruct the vessel as this would significantly reduce the blood flow to the lung and increase the measured PA pressure. We use a Prolene 5-0 double layer mattress/simple continuous suture.
10. As mentioned for the ties that secure the ETT to the airway, we recommend using the ties on the PA catheter/connector as well to hang the lungs. This allows to reduce the weight of the lung on a single anatomic structure and also prevents to vessel (PA) from kinking.
11. Examples in the literature using nondiluted blood (25, 26) or blood diluted with blood-type-matched human plasma (27) for xenogeneic lung perfusions have been conducted. Unpublished work shows that dilution of blood with electrolyte solutions or use of refrigerated or overnight-stored blood significantly attenuates the pace of hyperacute heart or lung rejection.
12. For pigs <20 kg, we utilize two units (two individual donors, each contributing 450 mL) of blood expanded with four 240 mL individual donor "units" of thawed "recovered" or "fresh-frozen" plasma. One fresh human donor and two plasma "units" are usually sufficient for an "en bloc" experiment.
13. While some activation of complement and adhesion of formed blood elements is demonstrable in our circuit, this effect is readily controlled for by measuring these parameters after a

consistent (20 min) interval of perfusate circulation through the circuit. Blood perfused for longer intervals or in the context of iso- or allo-lung perfusion exhibit little additional alteration over relevant study intervals (unpublished observations).

14. Our circuit design incorporates an overflow line that allows the blood to continue to mix and circulate while bypassing the lung when high PVRs occur, to avoid PA pressures above 45 cm $H_2O$ (equivalent to 4.4 kP or 33 mmHg) and avoid stasis in the tubing. This peak PA pressure is chosen to simulate the maximum pressure that we presume a normal right ventricle could overcome for a sustained period in vivo.
15. In our experiments, we have seen that the PA flow can show a wide range of values. Certainly, the genetics of the donor pig (e.g., wild type or GalT-KO pig), but also other parameters (ischemia reperfusion) have influence on the blood flow and contribute to the variability in PVRs in different experiments.
16. The oxygenator adds substantial expense, and is a source of significant blood activation.
17. It is important to size-match donor and recipient for not only physiologic, but also surgical reasons. The main bronchus of the pig has usually a wider lumen than the primate bronchus. To facilitate the bronchus anastomosis, it is thus important to use a donor pig which has a 20–40% lower bodyweight than the recipient baboon.
18. A pig PA 50% larger than the baboon left main PA should be anticipated and managed using standard surgical tailoring approaches.
19. We prefer inserting an arterial line to both of the femoral arteries. This procedure allows us to collect arterial blood without loosing the real-time invasive blood pressure measurement for the time of the drawing which can be crucial in animals with unstable or critical condition.
20. We suggest forwarding the PA balloon catheter after the opening of the chest as the right placement of the catheter can be difficult to achieve without the manual guidance across the tricuspid valve.
21. Due to the movement of the heart, it is often not easy to find a position on the ascending aorta where a stable flow signal can be recorded by the flow probe. We found that ultrasound gel (for nonsterile operations) improves the signal in most cases.
22. Although, only the right lung is ventilated during the implantation of the left lung, it is most likely not necessary to change the settings of the ventilation machine. During the single-lung ventilation, arterial blood gasses should be collected to assure an appropriate oxygenation of the recipient.

23. We typically place a clamp on the donor PA to facilitate alignment, position it adjacent to the recipient LPA, and avoid tension on or stricture of the anastamosis during its performance.
24. In our experiments, we use dopamine (2–20 μg/kg/min), noradrenalin (2–20 μg/kg/min), and vasopressin (0.2–1.0 IU/min). Additionally, we also routinely use inhaled nitric oxide to minimize PVR during graft implantation and subsequent assessment; the utility of this approach has not been rigorously assessed.
25. We recommend to perform this maneuver only after a sufficient reperfusion time and to occlude the vessel only partly at first. As mentioned before, the transplanted lung shows initially a higher vascular resistance and a lower compliance. By completely occluding the right PA shortly after the implantation, this could lead to a cardiac decompensation with acute right heart failure.
26. We estimate the PVR by dividing the pulmonary arterial pressure by the pulmonary flow.
27. Samples are collected in ethylenediaminetetraacetic acid (EDTA) anti-coagulated blood tubes.
28. The blood can be fixed immediately after collection to avoid nonspecific platelet activation during sample processing and to allow simultaneously antibody staining of all time-points at the same time. Additional markers (such as glycoprotein V or activation of GPIIbIIIa receptor) can also be analyzed in a similar way.
29. Importantly, plasma samples used in coagulation studies must be collected in CTAD tubes and processed according to the manufacture's instructions.

## References


1. Pierson RN 3rd, Dorling A, Ayares D et al (2009) Current status of xenotransplantation and prospects for clinical application. Xenotransplantation 16:263–280
2. Schroeder C, Allan JS, Nguyen BN et al (2005) Hyperacute rejection is attenuated in GalT knockout swine lungs perfused ex vivo with human blood. Transplant Proc 37:512–513
3. Hisashi Y, Yamada K, Kuwaki K et al (2008) Rejection of cardiac xenografts transplanted from alpha1,3-galactosyltransferase gene-knockout (GalT-KO) pigs to baboons. Am J Transplant 8:2516–2526
4. Bryant L, Eiseman B, Avery M (1968) Studies of the porcine lung as an oxygenator for human blood. J Thorac Cardiovasc Surg 55:255–263
5. Collins BJ, Blum MG, Parker RE et al (2001) Thromboxane mediates pulmonary hypertension and lung inflammation during hyperacute lung rejection. J Appl Physiol 90:2257–2268
6. Cantu E, Balsara KR, Li B et al (2007) Prolonged function of macrophage, von Willebrand factor-deficient porcine pulmonary xenografts. Am J Transplant 7:66–75
7. Schroeder C, Wu GS, Price E et al (2003) Hyperacute rejection of mouse lung by human blood: characterization of the model and the role of complement. Transplantation 76:755–760
8. Komatsu K, Youm W, Konishi H et al (1996) Prolonged survival of hamster-to-rat pulmonary xenografts by tacrolimus (FK506) and cyclophosphamide. J Heart Lung Transplant 15:722–727
9. Miyata Y, Ohdan H, Yoshioka S et al (1998) Relationship of xenogeneic microchimerism to

graft outcome in hamster-to-rat lung xenotransplantation. J Heart Lung Transplant 17:233–240
10. Nonaka M, Kadokura M, Kataoka D et al (2000) Hyperacute xenorejection of guinea pig-to-rat lung transplantation can be attenuated by blood which has perfused another xenograft. Ann Thorac Cardiovasc Surg 6:146–150
11. De Perrot M, Keshavjee S, Tabata T et al (2002) A simplified model for en bloc double lung xenotransplantation from hamster to rat. J Heart Lung Transplant 21:286–289
12. Yang YG, deGoma E, Ohdan H et al (1998) Tolerization of anti-Galalpha1-3Gal natural antibody-forming B cells by induction of mixed chimerism. J Exp Med 187:1335–42
13. Veith FJ, Richards KU, Hagstrom JWC, Montefusco CM (1981) Intrafamilial lung xenograft from fox to dog. J Thorac Cardiovasc Surg 81:546–552
14. Takeda M, Kawauchi M, Nakajima J, Furuse A (1995) Methorexate in rescue therapy for xenotransplanted lung in primes. Third International Congress of Xenotransplantation, Boston (Abstract po-46)
15. Sadeghi AM, Laks H, Drinkwater DC et al (1991) Heart–lung xenotransplantation in primates. J Heart Lung Transplant 10:442–447
16. Daggett CW, Yeatman M, Lodge AJ et al (1998) Total respiratory support from swine lungs in primate recipients. J Thorac Cardiovasc Surg 115:19–27
17. Macchiarini P, Oriol R, Azimzadeh A et al (1999) Characterization of a pig-to-goat orthotopic lung xenotransplantation model to study beyond hyperacute rejection. J Thorac Cardiovasc Surg 118:805–814
18. Blum M, Chang AC, Collins BJ et al (1997) The effect of nitric oxide and thromboxane blockade on pulmonary vascular resistance in a pig-to-primate lung transplant model. Fourth International Congress for Xenotransplantation, Nantes, France, September
19. Nguyen BN, Azimzadeh AM, Zhang T et al (2007) Life-supporting function of genetically modified swine lungs in baboons. J Thorac Cardiovasc Surg 133:1354–1363
20. Yeatman M, Daggett CW, Parker W et al (1998) Complement-mediated pulmonary xenograft injury—studies in swine-to-primate orthotopic single lung transplant models. Transplantation 65:1084–1093
21. Cantu E, Parker W, Platt JL, Duane Davis R (2004) Pulmonary xenotransplantation: rapidly progressing into the unknown. Am J Transplant 4(Suppl 6):25–35
22. Nguyen BN, Azimzadeh AM, Schroeder C et al (2011) Absence of Gal epitope prolongs survival of swine lungs in an ex vivo model of hyperacute rejection. Xenotransplantation 18:94–107
23. Azimzadeh AM, Dawson H, Nguyen BN et al (2005) aGal antigen modulates the expression of pro-inflammatory genes in hyperacute lung rejection. In: 8th international xenotransplantation congress, Goteborg, Sweden, Sept 2005
24. Knosalla C, Yazawa K, Behdad A et al (2009) Renal and cardiac endothelial heterogeneity impact acute vascular rejection in pig-to-baboon xenotransplantation. Am J Transplant 9:1006–1016
25. Macchiarini P, Mazmanian GM, Oriol R et al (1997) Robert Rieben, Philippe Dartevelle. Ex vivo lung model of pig-to-human hyperacute xenograft rejection. J Thorac Cardiovasc Surg 114:315–325
26. Kim HK, Kim JE, Wi HC et al (2008) Aurintricarboxylic acid inhibits endothelial activation, complement activation, and von Willebrand factor secretion in vitro and attenuates hyperacute rejection in an ex vivo model of pig-to-human pulmonary xenotransplantation. Xenotransplantation 15:246–256
27. Schroeder C, Allan J, Nguyen BN et al (2004) Xenogenic ex vivo perfusion of lungs from Galt K/O pigs: initial results. Transplantation 78(Suppl 2):20

# Chapter 12

# Thymic Transplantation in Pig-to-Nonhuman Primates for the Induction of Tolerance Across Xenogeneic Barriers

Kazuhiko Yamada and Joseph Scalea

## Abstract

With the advent of knockout pigs for α1,3-galactosyltransferease (GalT-KO, which lack a cell-surface antigen to which humans have preformed antibodies), investigators have extended the survival of life-supporting xenorenal grafts. However, despite these increases, nonhuman primates transplanted with GalT-KO renal grafts are susceptible to anti-donor T-cell responses that are strong or stronger than allogeneic responses. In order to prevent rejection, recipients must be subjected to morbidly high levels of immunosuppression. For these reasons, our laboratory has attempted to develop novel methods of xenogeneic tolerance using vascularized porcine thymic grafts in order to reteach the recipient's immune system to accept the xenogeneic organ as self. These strategies, largely developed by Dr. Kazuhiko Yamada, involve the co-transplantation of a vascularized donor thymus with a kidney. This has been successfully done in two ways. The first method involves the preparation of a composite tissue "thymokidney" and the second utilizes the transplantation of an isolated vascularized thymic lobe. Both strategies involve the transplantation of fully vascularized thymic tissue at the time of xenotransplantation, a fact which is crucial for function of the thymic tissue immediately after xenografting and reeducation of recipient T-cells. These strategies have successfully induced tolerance across fully allogeneic models in miniature swine and prolonged graft survival in our pig-to-baboon model of life-supporting xenotransplantation to greater than 80 days with in vitro evidence of donor-specific unresponsiveness. Although it is too early for the development of clinical renal xenotransplantation protocols, this chapter describes the authors' unique experience with one of the most promising preclinical large-animal models of xenotransplantation. Furthermore, understanding the importance and measurement of T-cell responses in xenotransplantation is contingent upon a functional knowledge of these procedures.

**Key words:** Xenotransplantation, Transplantation, Thymus, Thymic transplantation, Thymokidney, T-cell depletion, Miniature swine

## 1. Introduction

*Avoidance of hyperacute rejection (HAR) with GalT-KO pigs.* The critical shortage of human donors, combined with the increasing incidence of end-stage organ disease has prompted physicians to

Cristina Costa and Rafael Máñez (eds.), *Xenotransplantation: Methods and Protocols,* Methods in Molecular Biology, vol. 885, DOI 10.1007/978-1-61779-845-0_12, © Springer Science+Business Media, LLC 2012

search for alternative sources of donor organs (1–3). Miniature swine are generally considered the best match for potential human xenotransplantation partly because of their size, but also because of their breeding patterns and largely known genetic profile (4). Until recently, however, xenotransplantation appeared unfeasible because of HAR to a glycoprotein antigen constitutively expressed on the surface of swine cells to which humans (and old world monkeys) have preformed antibody (5–9). To avoid this problem, in 2002 several groups, including ours, produced knockout pigs which do not express the Gal epitope (10, 11). The Transplantation Biology Research Center (TBRC), in collaboration with Immerge Inc., developed the GalT-KO pigs through targeted disruption of the α1,3-galactosyltransferase (GalT) gene in the nucleus of a pig fibroblast cell line that was isolated from the most inbred subline of miniature swine (11, 12). The GalT-KO animal was thus derived from our most highly inbred line of MGH-miniature swine (13) in order to provide a potentially unlimited supply of genetically near-identical organs that could be used to replace transplants which failed for technical or nonimmunologic reasons, without additional immunosuppression. On confirmatory testing, these homozygous GalT-KO animals expressed no Gal. In fact, they produced natural anti-Gal antibodies (14). Using GalT-KO donors, HAR of xenografts has been successfully avoided (15).

*Xenogeneic T-cell responses.* Due to the intensity of antibody-derived HAR across xenogeneic barriers, early researchers were unable to ascertain the degree of xenogeneic anti-donor T-cell responses to vascularized donor organs. Initially, there was speculation that the xenogeneic T-cell response would be less severe than an allogeneic response because of discordant MHCs. Early investigations into T-cell responses exploited the mouse anti-human or anti-pig models. These studies revealed a defective mouse CD4 T-cell anti-stimulator MHC class II molecule interaction, effectively eliminating T-cell activation via the direct pathway (16, 17) and suggested that if humoral mechanisms were overcome, cell-mediated rejection would be a minor obstacle. However, studies done in the early 1990s in the authors' laboratory and by other investigators showed that the direct pathway of activation did exist in the pig-to-human model (18, 19). We have demonstrated that (1) human T cells responded at least as well to xeno-MHC antigens (Ags) as they did to allo-MHC Ags in MLR. In addition, human-anti-pig T-cell responses appeared to share similar antigen presenting cell (APC) requirements for stimulator (direct pathway) or responder (indirect pathway) and (2) the majority of the primary human-anti-pig xeno-response was directed toward porcine MHC class II Ags and involved interactions with human CD4 accessory molecules. These data indicated that the human-anti-porcine T-cell response was similar in strength and specificity to the allogeneic response (18).

*The importance of inducing tolerance for successful life-supporting xenogeneic renal transplantation.* Efforts to avoid humoral rejection have included the use of transgenic donors expressing complement-inhibitory proteins, such as hDAF (20), CD59 (membrane attack complex inhibitor) (21), and human membrane cofactor protein (hMCP, CD46) (22), as well as the elimination of Gal expression on pig cells. Despite the early success of these efforts to control HAR, both severe delayed xenograft rejection (DXR) and acute cellular xenograft rejection (ACXR) remained obstacles. We have achieved life-supporting hDAF-to-baboon renal survival of up to 29 days using chronic immunosuppression (23). We hoped that GalT-KO pigs would further prolong survival of life-supporting renal grafts. However, to our disappointment, GalT-KO pigs led to minimal improvement in survival (34 days) using immunosuppressive regimens that included ATG and anti-CD154 mAb (15). Zhong and his colleagues reported maximum survivals of only 16 days (24) using GalT-KO kidneys. Even with the use of heavy immunosuppression in these protocols, rejected GalT-KO grafts showed severe cellular and humoral rejection. The glomeruli and peritubular capillaries of these rejected xenografts stained brightly for IgG. Additionally, investigators observed consistent aberrations in the complement cascade.

The development of cellular rejection in this model despite heavy chronic immunosuppression suggested that small numbers of T cells can initiate both cellular and humoral responses. These data are consistent with reports by Korsgren and colleagues which indicated that even very small numbers of T cells were sufficient to initiate rejection of porcine islets by macrophages in T-cell-deficient rodents (25, 26). The level of immunosuppression needed to control these T-cell responses and prolong xenograft survival has been associated with prohibitive morbidity and mortality. Additionally, these regimens have not yielded survival rates that would merit transition to clinical trials. Thus, these results provide a compelling rationale to pursue clinically applicable tolerance strategies that would potentially avoid the high level of immunosuppression employed in current xenotransplantation protocols.

*Yamada's strategy to overcome xenogeneic T-cell responses: vascularized thymic transplantation.* In an attempt to control these very powerful T-cell responses, strategies targeting elements of the immune response have been developed and tested. Since immunosuppression alone is prohibitively morbid, the authors believed a strategy for xenogeneic tolerance induction was imperative for successful xenotransplantation. Transplantation tolerance, which uses the body's own immune system to reteach T cells to suppress the anti-donor response, was first attempted by the author, Yamada, using the simultaneous transplantation of a donor pig thymus and kidney. The thymus is known to play an integral role in self-tolerance and in tolerance across allogeneic barriers. As T-cell education occurs in the thymus, thymic transplantation is thought to lead to tolerance by a central mechanism, much like the education of T cells in a normal developing child.

Thymic tissue, as opposed to vascularized thymic grafts, was first demonstrated to induce xenograft tolerance in a pig-to-mouse transplant model by Sykes et al. Fetal thymic tissue was transplanted directly under the renal capsule in T-cell-depleted thymectomized mice (27). These results showed that murine recipients of porcine thymic tissue developed donor-specific unresponsiveness in mixed lymphocyte reaction (MLR) assays and subsequently accepted skin grafts from the thymic donors (27–29). In order to extend these results to our pig-to-primate model, it was first necessary to demonstrate that thymic transplantation was possible in the pig. Our initial attempts to implant minced thymic tissue as in the mouse model failed, with evidence of graft rejection by days 15–30 (30). We hypothesized that ischemic damage to the transplanted thymus during revascularization led to an active immunologic response, and that thymic tissue should be transplanted as a vascularized organ. Furthermore, during the tenuous revascularization period, unless T-cell depletion was complete, which is very difficult in large animals, thymic tissue was rejected before it had the ability to induce tolerance.

In order to eliminate the revascularization period following allogeneic or xenogeneic transplantation, Yamada et al. developed two procedures for vascularized thymic transplantation in miniature swine. One is the preparation of composite "thymokidneys" in which autologous thymic tissue is allowed to engraft for 1–2 months under the donor's own kidney capsule before allogeneic transplantation (31–33) (see Fig. 1). The other is the transplantation of an isolated vascularized thymic lobe (VTL) graft, or, the so-called Yamada VTL procedure (34) (see Fig. 2). Using these novel techniques, we demonstrated that vascularized thymic tissue induces tolerance and supports thymopoiesis across fully allogeneic barriers in MGH miniature swine (35). In other experiments, when kidneys and nonvascularized thymic tissue from the same donor were transplanted as separate grafts, recipients rejected the nonvascularized allogeneic thymic grafts within 4 weeks and kidney grafts were rejected within the second postoperative month. In contrast, recipients of thymus with simultaneous kidney (thymokidney), heart (heart plus VTL), or kidney following VTL transplant survived long-term with stable graft function and in vitro donor-specific unresponsiveness (32, 33, 35, 36).

Based on these encouraging results, the authors' laboratory has extended this vascularized thymic transplant strategy to our pig-to-baboon model of renal xenotransplantation. The authors' strategy, which utilizes GalT-KO kidney and thymus swine donors,

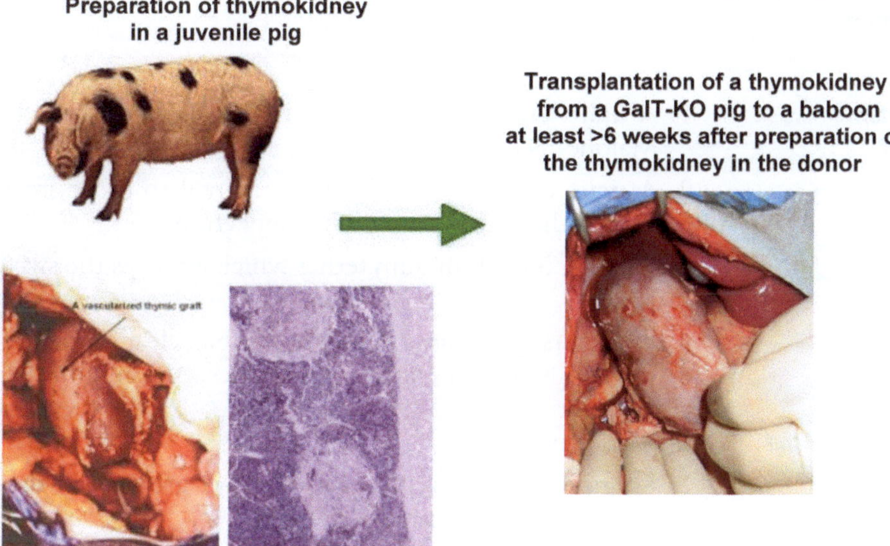


Fig. 1. Schematic diagram of preparation of the thymokidney. HE specimen and a gross thymokidney (*left*) at 12 weeks after preparation in a donor pig (*left*). Gross GalT-KO thymokidney transplanted in a baboon (*right*).

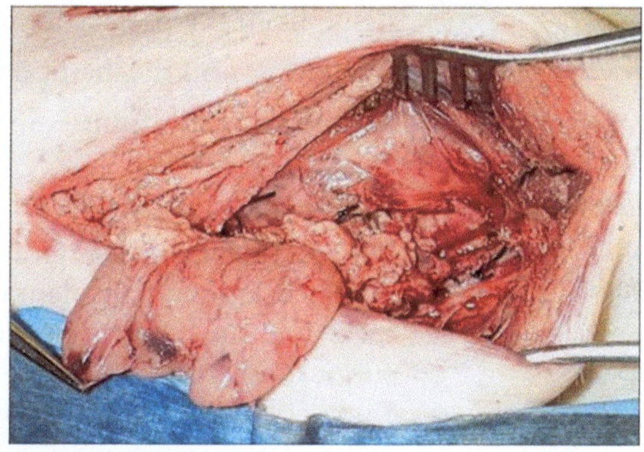

Fig. 2. Surgical picture of isolated VTL transplantation.

has led to prolongation of graft survival to more than 80 days (15). More recently, we have attempted to reduce early postoperative death from infection associated with the induction protocol (37). We successfully eliminated steroids and whole body irradiation from the induction regimen which led to: (1) decreased comorbidity attributable to steroids and irradiation and increased mean survival to greater than 50 days from a mean of 34 days; (2) demonstrable baboon thymopoiesis (reeducated T-cells) in transplanted, vascularized pig thymic grafts; and (3) in vitro evidence of donor-specific tolerance.

## 2. Materials

### *2.1. Catheter Placement*

1. A lightweight spandex/lycra jacket (Lomir Biomedical Inc., Malone, NY).
2. A rigid swivel which allows the animal to move freely in the cage without disturbing line connections.
3. A 1.0–1.5 m aluminum tether which attaches the swivel to the jacket.
4. Catheters, generally 1.5–2.0 m flexible, clear Tygon microbore tubing (Norton Performance Plastics, Akron, Ohio).

### *2.2. Thymectomy and VTL Procurement*

1. Donor pigs.
2. Surgical telescopes (preferably 3.5× or greater) for the primary assistant and the primary surgeon.
3. Surgical draping and surgical instruments including: scalpel, bone saw and/or heavy bone scissors, weitlander retractors, electrocautery, hemo clips, and bone wax as well as suture materials (PDS, prolene, and silk ties).
4. Cotton swab applicators for gentle separation of the thymic tissue from the pericardium.
5. Anesthesia and mechanical ventilator.
6. Euro-collins solution.

### *2.3. Thymokidney, VTL and Kidney Transplantations*

1. Animals: GalT-KO donors and nonhuman primates.
2. Surgical draping and surgical instruments are the similar to clinical pediatric kidney transplantation, including: scalpel, electrocautery, fine-tip/Castro needle drivers, 2.5–4.5×-magnification surgical loupes, 6-0 to 8-0 nonabsorbable surgical suture, Euro-collins solution, balfour retractors, and sterile ice.
3. Anesthesia and mechanical ventilator.
4. Euro-collins solution.

### *2.4. Flow Cytometry to Assess T-Cell and B-Cell Depletion as Well as Host Thymopoiesis in Porcine Thymic Grafts*

1. Anti-human (cross-react to baboons) Ab: CD3 (polyclonal rabbit anti-human CD3, A0452; Dako North America, Inc., Carpinteria, California, USA), CD4 (mouse anti-human CD4 mAb, 1F6; Invitrogen, Carlsbad, California, USA), CD8 (mouse anti-human CD8; BD Biosciences, Franklin Lakes, New Jersey, USA), CD20 (mouse anti-human CD20; BD Biosciences), and CD25 (mouse anti-human, BD Biosciences). All conjugated to a fluorochrome.

2. Anti-pig mAb (available in our research center): Anti-porcine CD1, CD2, CD3, CD4, CD8, CD25, CD45RA, Class I, Class II, PAA (produced internally). All conjugated to fluorochrome.
3. Conjugated anti-mouse IgG1 and anti-mouse IgG2a to be used as controls.
4. Cold FACS media: 1 L Hanks balanced salt solution, 1 g bovine serum albumin (BSA), 1 g sodium azide.
5. FACS tubes.

#### 2.5. Baboon Mixed Lymphocyte Reaction

1. Reagents for peripheral blood monocytes (PBMC) isolation: Hank's balanced salt solution (Life Technologies, Grand Island, NY, USA), Histopaque 1077 (Sigma–Aldrich, St. Louis, MO, USA), ACK lysing buffer (BioWhittaker, Maryland, MD, USA).
2. MLR medium: AIMV media (see Subheading 2 and Note 1).
3. MLR plate: 96-Well, flat-bottom plate (Costar, Cambridge, MA, USA).
4. Cell irradiator (5,000 rads).
5. For thymidine incorporation and cell proliferation: Thymidine –([$^3$H], 2.0 Ci/mmol, aqueous solution, steri-packaged, at 1.0 mCi/mL (37 MBq/mL)) (PerkinElmer, Waltham, Massachusetts, USA).
6. Liquid scintillation (PerkinElmer).
7. Betaplate reader (Pharmacia, Gaithersburg, MD, USA).

#### 2.6. Antibody/Complement-Mediated Cytotoxicity Assay

1. Target pig PBMC.
2. Medium 199 with 2% fetal bovine serum (FBS).
3. Decomplemented serum samples (56°C for 30 min).
4. 7-AAD (Sigma) diluted to 10 μg/mL in FACS media.
5. Rabbit complement (at a predetermined working dilution).
6. 96-Well U-bottom plates and lids, nonsterile.
7. Combitips, combi-pipettor, micropipette tips, multichannel pipettor, Pasteur pipettes, micropipettor, truncated pipette tip, plastic trough.
8. 37°C, $CO_2$ incubator.
9. Centrifuge.
10. FACS tubes.
11. Cold FACS media: 1 L Hanks balanced salt solution, 1 g BSA, 1 g sodium azide.
12. Ice and ice bucket.

## 3. Methods

Though significant academic progress has been made, xenotransplantation has not yet become a clinical reality. Below is a brief introduction to the surgical and immunologic procedures compiled from the most successful pig-to-nonhuman primate renal transplantation protocols (3, 37).

### *3.1. Catheter Placement*

At the authors' institution, kidney xenotransplantation is considered life-supporting, and a bilateral native nephrectomy is typically performed at the time of transplantation. Thus, daily laboratory values are required to monitor the animal's health and graft function. For these reasons, the placement of central venous and arterial catheters or "lines" is required for animal care during the induction, peri-transplantation and posttransplantation periods. Lines provide the caregiver with the ability to both deliver drug therapy and draw blood for diagnostic testing. This is of particular importance in baboons or monkeys because it is difficult to perform a physical examination beyond simple observation. Although central catheters confer a risk of infection, there is greater risk associated with sedation for daily or twice-daily blood draws and drug-administration, especially in potentially fragile post-xenotransplantation recipients.

The number of lines required depends on: (1) whether blood can be drawn off a venous line, (2) the number of drugs administered (continuous and/or intermittent), (3) drug compatibility, and (4) if blood drug-levels are being tested. Typically, three lines (two venous, one arterial) are placed 1 week prior to transplantation. The arterial line is used primarily for blood draws and the two venous lines are used for drug delivery. As is done with human arterial catheters, the line's patency is maintained by keeping it under constant forward-flow pressure and adding low dose heparin to the normal saline carrier. Two venous lines are used in our standard model because: (1) soluble MMF must be placed on a dedicated D5%W carrier line and not mixed with normal saline, and because (2) antibiotic compatibility may vary during the animal's course (see Fig. 3).

1. Generally, place central lines on the left-side.
2. Make a 3-cm transverse incision, one finger-breath above the clavicle, starting from the midline and extending across the SCM using the scalpel.
3. Electrocautery and blunt dissection should be used to transect the platysma and isolate the external jugular vein (EJ), the internal jugular (IJ) and the carotid artery (see Note 2).
4. Tunnel the lines from the animal's back (between shoulder blades) to the open neck wound.

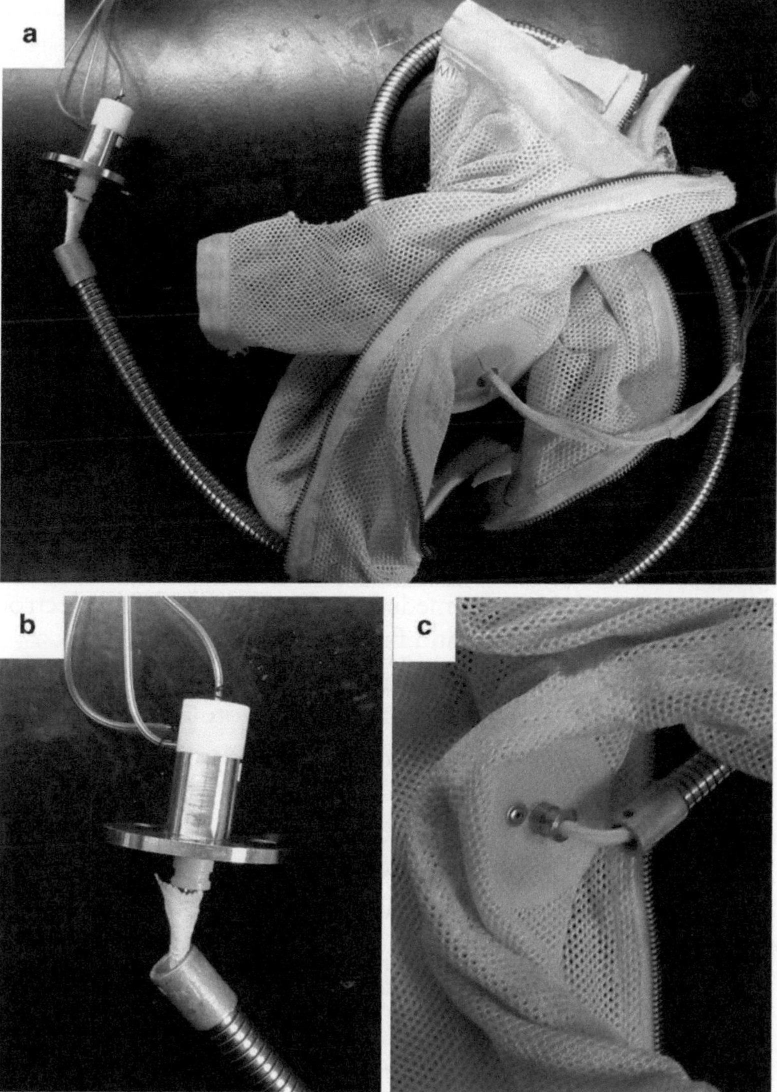


Fig. 3. (**a**) A picture of the line system used for baboon recipients of xenografting procedures. The mesh jacket is rigidly attached to an aluminum tether through which 1–3 catheters are placed. The tethers are 4–6 ft in length. (**b**) An exploded view of the swivel shown in (**a**), which has been unattached from the aluminum tether. As many as three catheters are securely affixed to each of three ports on the swivel. This allows the animal to move within the cage without twisting the lines. (**c**) An exploded view of the connection between the jacket and the tether. In this image, the jacket has been unattached from the tether, showing the cloth tape-wrapped catheters traveling through the jacket and into the tethering system.

5. Dissect the sternocleidomastoid and retract medially or laterally for access to the carotid sheath. The easier approach is to retract laterally.
6. Open the carotid sheath and isolate the carotid, and then tie of distally.
7. Place a small bull-dog clamp on the vessel, 1 cm proximally.

8. Create an arteriotomy using tenotomy scissors and then cannulate the vessel.
9. Release clamps and advance the arterial catheter (see Note 3).
10. A similar procedure is performed for the IJ followed by the EJ (see Note 4).
11. Close the skin with buried subcuticular stitches.
12. Place a protective line-jacket on the animal in a standard fashion (see Note 5 for catheter-related issues).

### 3.2. Thymectomy and VTL Procurement (See Notes 6–9)

#### 3.2.1. Thymectomy

Our previous data in pig-to-mice models have demonstrated that thymectomy is required for the induction of xenogeneic tolerance (27). Three weeks prior to transplantation, and 2 weeks prior to line placement, animals on the tolerance-induction (thymokidney) protocol undergo thymectomy (37).

1. Perform a midline sternotomy incision in the pig and obtain hemostasis of the skin.
2. Expose the median pectoral groove using electrocautery, blunt dissection, and finger fracture.
3. Dissect the manubrium, exposing the jugular notch, and then perform a sternotomy (see Note 7).
4. Dissect and remove thymic gland (see Note 8) using cotton swab applicators.
5. Weigh the excised thymus and take several pieces for histology and immunologic assays.
6. Test for air leak with 5–10 mL of NS. If no leak is observed, go to the next step. If an air leak is observed, repair it.
7. Following meticulous hemostasis, close chest with interrupted 0-PDS sutures.
8. Ask the anesthesiologist to provide a sustained breath to fully expand the lungs. Again, assess for air leak with 5–10 mL of NS. If no leak is observed, the chest closure is completed (see Note 9).

#### 3.2.2. VTL Procurement

The VTL harvest and VTL transplantation is technically challenging. To improve the likelihood of success, the surgeon should have extensive experience with porcine thymectomy (38) as well as vascular surgical techniques.

1. For the procurement of thymic lobe(s) from donors, make a longitudinal cervical incision to expose the porcine cervical thymus (38).

2. Expose the pretracheal muscles and the connective tissue sheath overlying the thymus to provide access to the left and right lobes (34).
3. Carefully dissect the lateral thymus away from the connective tissue and identify the blood supply to the thymic lobe (see Note 10).
4. Ligate and transect the major vessels from/toward thymic lobe(s). The minor accessory vessels can be ligated on the bench.
5. Perfuse excised thymic lobe with Euro-collins.

### 3.3. Thymokidney, VTL, and Kidney Transplantations

#### *3.3.1. Thymokidney Preparation in GalT-KO Pig Donors (Yamada's Thymokidney Procedure)*

The thymokidney model of xenotransplantation is the most successful life-supporting xenograft protocol for large animals (15). In this model, donor-specific T-cell tolerance to the donor graft is induced by co-transplantation of donor thymic tissue. The procedure for the preparation of a composite thymokidney was developed because transplantation of prevascularized thymic tissue helps to prevent revascularization-associated ischemia. For that reason, thymic tissue in the thymokidney model is placed beneath the autologous native renal capsule in the donor (7–12-weeks-old juvenile miniature swine) 4–6 weeks prior to composite transplantation (see Note 11). During this time, it becomes vascularized by the autologous kidney. The prevascularized thymokidney graft is then transplanted as a composite kidney graft using the kidney transplantation procedure described below (31–33) (see Fig. 4a).

1. Excise cervical thymic lobe(s) (see Note 12) through a midline or 1 cm lateral neck incision (38).
2. Mince the excised thymic tissue with metzenbaum scissors or a scalpel on the back table (see Note 13), and then close incision.
3. Expose the kidney using a retroperitoneal flank incision (right and left thymokidney preparation should be performed through two separate flank incisions to prevent abdominal adhesions as both kidneys will be procured in several weeks) and then place thymic tissue beneath the autologous renal capsule (see Note 14).
4. Close the edge of the capsular window using electrocautery or with a single 7-0 prolene suture.
5. Close flank incision in layers with 3-0 or 2-0 vicryl. Nylon can be used for pig skin.

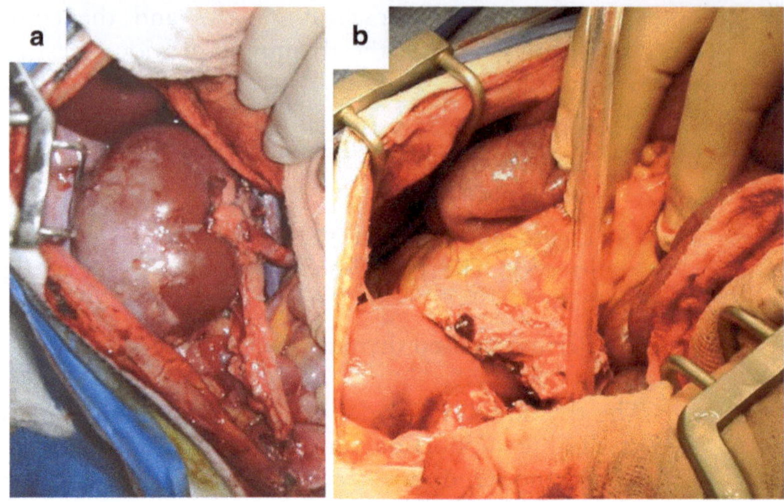


Fig. 4. Surgical pictures of GalT-KO thymokidney (**a**) and simultaneously transplanted GalT-KO VTL and kidney (**b**) in baboons. These pictures were taken 30–60 min after revascularization of xenografts from GalT-KO pigs in baboons. All grafts were well vascularized immediately after reperfusion in baboons and kidney grafts produced urine.

#### *3.3.2. Thymokidney Procurement from GalT-KO Pig Donors*

The thymokidney procurement for pig-to-baboon xenotransplantation is similar to human kidney procurement. However, since the donor kidney contains thymus underneath the kidney capsule, it is necessary to preserve the donor's perinephric fat. This is because local tissues may be adherent resulting from the thymokidney preparation. Careful manipulation technique is required to avoid further inflammation of the fragile thymic tissue vascularized by the kidney.

1. A midline incision from xiphoid to pubis is used to provide access the abdomen (see Note 15).
2. Expose thymokidneys (see above), isolate the renal vein (RV) and artery (RA), transecting each vessel at the level of the aorta (Aa) (see Note 15).
3. Remove thymokidney(s) and flush via the RA on the back table using Euro-collins and immediately place on ice.

#### *3.3.3. GalT-KO Thymokidney Transplantation to Baboons*

Surgical thymokidney transplantation is generally performed using methods similar to those of pediatric kidney transplantation. Although primary kidney transplantation in pig-to-nonhuman primates may be described by other groups, the unique and complex nature of the procedures described in this chapter will help readers perform successful xenogeneic kidney transplantation. While many groups have experimented with xenotransplantation, the procedures developed by Dr. Yamada have resulted in both clinical (minimal ATN) and immunologic (extended graft survival) successes for our group. Because porcine kidneys are predisposed to vasoconstriction, surgeons should minimally manipulate the graft as described below (see Note 16).

1. Make a midline incision on baboon's abdomen from xiphoid to pubis.
2. Gain exposure using a Balfour retractor and retract the animal's bowels to the left, allowing access to the right retroperitoneum.
3. For nephrectomy of the right kidney, focus inferiorly and isolate the host ureter. Ligate the ureter such that 50% of the native length remains attached to the bladder (this is done so that a future uretero-ureteral anastamosis can be performed if a technical complication arises).
4. Skeletonize the host renal vessels to the level of inferior vena cava (IVC) and the aorta (Aa). Then, ligate the RA using 3-0 silk ties followed by ligation of the RV.
5. Place the hemoclips proximally.
6. Left-sided nephrectomy (see Note 17).
7. For kidney transplantation (see Note 16), dissect the Aa and IVC such that the surgeon has sufficient access for the vascular clamp and anastomosis (care is taken to avoid injuring the inferior mesenteric artery).
8. Heparinize the recipient with 70–100 IU/kg at least 2 min prior to beginning the transplant.
9. First, cross clamp recipient IVC with a statinsky (see Note 17). The vein is done first because the RA will overly the IVC following anastamosis (see Note 18).
10. Create venotomy with an 11-blade or Potts' scissors.
11. Typically, place two stay sutures (one superiorly and one inferiorly on the cava) and sew first the back wall, followed by the front wall with 7-0 prolene (8-0 if baboons are <5 kg). The venous anastomosis should be completed under 12 min.
12. Place a bull-dog on the graft RV and gently release the statinsky from the recipient IVC, allowing venous blood to back-fill the RV (see Note 19, important).
13. After clamping the Ao with a statinsky, use an 11-blade and Potts' scissors to create an arteriotomy.
14. Perform arterial anastamosis using 7-0 running prolene (8-0 if baboons are <5 kg). The author highly recommends using a parachute (3 o'clock single knot) technique performed under 15 min as this minimizes manipulation of the graft and ischemia time (see Note 20).
15. Release the aortic statinsky. Obtain hemostasis using interrupted 7-0 prolene sutures as needed, if bleeding is encountered. With this technique, in MGH GalT-KO miniature swine, the kidney becomes pink, warm and produces urine within minutes of revascularization.

16. Perform the neoureterocystostomy using 2 or 3 running 7-0 PDS or prolene sutures. Placement of a 3F or 4F stent is recommended (the stent can be removed after day 30).
17. Cystopexy can be performed if necessary (such as short donor ureter).
18. Close the abdomen in 3–4 layers. Close the peritoneum first, followed by fascia, dermis, and epidermis.
19. Wean off the ventilator and replace the jacket. The baboon (monkey) should be perched within 2 h of extubation. See Note 21 for animal care recommendations.

#### 3.3.4. VT L Transplantation (Yamada's VTL Procedure)

As the VTL harvest, the VTL transplantation is technically challenging and extensive experience with vascular surgical techniques will increase the likelihood of success. The team should practice with a pig-to-pig allogeneic model before applying these techniques to a pig-to-nonhuman primate model. The advantage of the thymokidney is that the prevascularized thymus is transplanted simultaneously with the kidney. However, the VTL procedure can accompany the transplantation of any organ (including the kidney). In contrast to the prevascularized thymokidney, this procedure is more complex as the actual thymic lobe must be anastomosed to the recipient vessels. In theory, a co-transplanted thymic graft should confer tolerance to the recipient regardless of the transplanted organ (i.e., liver, heart, etc.). In an allogeneic model of VTL and simultaneous heart transplantation (39), the recipient became tolerant in a donor-specific manner. Although attempts to prepare thymohearts have been undertaken, these results have not been encouraging. One reason for failure of the thymoheart may relate to the anatomic challenges (limited space) beneath the pericardium. Thymoliver grafts have, as of yet, not been attempted. The VTL transplantation from the GalT-KO pig into baboons is typically completed using end-to-side anastamoses for the thymic artery and vein to the recipient baboon (monkey) Aa or IVC above renal vessel anastomosis for simultaneous VTL and kidney transplantation (see Fig. 4b) or carotid artery and IJ for isolated VTL transplant, respectively (34, 35). This is done in a similar manner to that of kidney transplantation (see above and Subheading 4). Either 7-0 or 8-0 prolene should be used and 3.5× or greater (greater than 4.5× for end-to-end) surgical loupes are recommended. See Note 21 for animal care recommendations.

### 3.4. Flow Cytometry to Assess for T-Cell and B-Cell Depletion as Well as Host Thymopoiesis in Porcine Thymic Grafts

In order to assess T-cell or B-cell levels as well as host T-cell development in the transplanted thymus (thymopoiesis), anti-human monoclonal antibodies (cross-reactive to baboon molecules) are used. Anti-porcine mAb are used to assess pig thymocytes in the thymic grafts after xenotransplantation. Some anti-porcine antibodies may not be commercially available.

1. Resuspend $1 \times 10^6$ PBMC or thymocytes in 100 μL of Hank's balanced salt solution containing 0.1% BSA and 0.05% sodium azide.
2. Incubate for 30 min at 4°C with conjugated antibodies. Use conjugated anti-mouse IgG1 and anti-mouse IgG2a as controls for nonspecific binding.
3. After three washes, analyze cells by flow cytometry using propidium iodide gating to exclude dead cells.

### 3.5. Baboon-Mixed Lymphocyte Reaction

The MLR assesses immunologic responses to the donor (mainly CD4 T-cells responses against donor antigens on MHC class II) as compared to that against third party to determine whether (1) the recipient is immunocompetent and (2) the recipient has developed the donor-specific response. Here, we describe bulk MLR assays using PBMC.

1. Isolate and resuspend responders and stimulators at $4 \times 10^6$ cells/mL in AIMV media (see Note 1).
2. Irradiate stimulator cells with (25 Gy).
3. Plate responder $4 \times 10^5$ cells/well with an equal volume of stimulators in a 96-well, flat-bottom plate. All samples should be tested in triplicate.
4. Incubate in 5% CO, for 5 days at 37°C.
5. Add 1 pCi of thymidine to each well for the final 6 h of incubation.
6. Read thymidine incorporation by liquid scintillation using a Betaplate reader.

### 3.6. Antibody/Complement-Mediated Cytotoxicity Assay

The 7AAD-dye-exclusion FACS procedure is designed for the titration of cytotoxic antibodies to cell surface antigens, using an antibody/complement reaction, followed by a dye exclusion assay and FACS acquisition. A similar procedure is used for SLA genotype testing of swine peripheral blood lymphocytes.

1. Process blood and count PBMC.
2. Prepare cell solutions of $5 \times 10^6$ cells/mL. The dilutions are made in Medium 199 (2% FBS, i.e., 10 mL FBS and 490 mL Medium 199). Keep the diluted suspensions on ice.
3. Identify the test panel and determine the appropriate dilutions for these antibodies. Also, decide which pig PBMC will be used as target cells. Typical setup involves serial dilutions of selected serum samples beginning at 1:2.
4. For sera serial dilutions, add 25 μL of Medium 199/2% FBS to each well of the plate and 25 μL of each serum to the first well of each row, diluting the serum samples across the row using a

multichannel pipette up to 1:8 dilution. Mix the dilutions of the antibody by pipetting up and down five times.

5. You also need to prepare three control wells per target cell containing either media alone, the diluted rabbit complement and both. These are your viability and complement background controls.
6. Add 25 μL of the appropriate cell suspension (pig target cells) to each well, including the complement and medium control wells.
7. Incubate the plate in the $CO_2$ incubator for 15 min, at 37°C.
8. After incubation, add 125 μL of Medium 199/2% FBS to each well in the plate. Centrifuge at 260×*g* for 5 min at 4°C (brake on).
9. Dilute the complement to an appropriate (predetermined) dilution in Medium 199 (1:8, i.e., rabbit complement 0.5 mL and 3.5 mL M 199).
10. Dispense 25 μL of the diluted complement into the appropriate wells.
11. Seal the plate with a plate sealer and shake the plate on the tray shaker for 30 s.
12. Incubate the plate in the $CO_2$ incubator, at 37°C for 30 min.
13. Put the plate on ice.
14. Add 200 μL cold FACS media to all wells containing cells.
15. Transfer the contents of each well to the FACS tubes.
16. Add 10 μL of diluted 7-AAD stain to all tubes.
17. Place tubes either on ice or in refrigerator until the FACS machine is set up for acquisition. 7-AAD should stain cells for approximately 30 min before acquiring.
18. Acquire cells using an FACS machine.

## 4. Notes

1. For the MLR assay, use an AIM-V media which is serum free. This is important because serum-based media are associated with higher levels of background stimulation, especially across a xenogeneic barrier.
2. The EJ should be tagged with a vessel loop. This is done because the vein is very small and frequently experiences vasospasm. To prevent visual obstruction, the EJ catheter is placed after the IJ vein and arterial (A) catheters.
3. The arterial catheter should not be advanced more than 6 cm in a 6 kg baboon.

4. The IJ and EJ are low venous pressure. After distal ligation, a proximal clamp is typically unnecessary prior to venotomy. Because, anatomically, the EJ communicates with the subclavian vein at an acute angle, the EJ line is typically only advanced 1–2 cm. The catheters are tacked to the SCM, deep to the reapproximated platysma. This may help to avoid contamination if a wound infection (rare) occurs.
5. Catheter-related issues: If line infection occurs, they become obsolete, or they become dysfunctional, the lines must be removed. In order to do so, the animal is taken to the OR for line removal through the same incision used for placement, taking care not to injure the lines during dissection, as this would introduce a potentially fatal air embolus, or allow the animal to bleed. If, after removal, additional lines are required, the femoral vessels can be used. In doing so, the femoral artery will need to be tied distally, which is tolerated remarkably well in baboons (40). Both ipsilateral femoral vessels can be used during a single line placement. In some cases, it may be necessary to subcutaneously tunnel a second venous line across the animal's anterior pelvis. Two, and sometimes three, lines can be tunneled safely to the animal's back (at the level of L1). The longitudinal groin incisions, despite not being covered by the jacket are rarely a problem. If needed, all four vessels (two femoral arteries and two femoral veins) can be ligated without causing clinically significant morbidity.
6. Thymectomy and VTL procurement are technically challenging and should be performed by an experienced surgeon.
7. A small finger (typically the fifth digit) is used to develop the plane between the sternum and the mediastinum, caudally. The same is then done from the xiphoid process, rostrally. Once the plane from the manubrium to the xiphoid has been fully developed, electrocautery, a bone saw and heavy scissors are used to perform a sternotomy. A heavy weitlander retractor is used to provide exposure.
8. In juvenile animals, the intact thymus is easily distinguished from fat by its texture and lighter color. Taking care to avoid damaging the pleura and pericardium, the thymus is gently dissected off the heart using cotton-swabbed applicators. Small but major vessels to the thymus are present at the superior aspect of the mediastinal thymus, which should be identified and clipped or ligated with 3-0 or 4-0 silks.
9. Insight following thymectomy: Prior to tying the last two sutures, the anesthesiologist is instructed to provide a sustained breath. If a lung injury is present, air bubbles will rise from the pooled mediastinal saline. Once no more bubbles are encountered, the chest closure is completed and the animal is weaned from the ventilator as quickly as possible, such that poor respiratory drive due

to technical issues, if encountered, can be dealt with prior to leaving the operating room. The chest is closed in three layers and the skin is closed using an absorbable subcuticular stitch.

10. The blood supply to the thymic lobes is highly variable though branches typically arise from the subclavian artery opposite from the internal mammary artery. The thymic veins primarily drain into the native internal jugular vein. At the time of thymic lobe procurement, both the feeding arteries (subclavian or thymic) and the internal jugular vein are taken en bloc. In addition, the rostral thymus may have collateral blood flow from the carotid. In this case, the carotid artery is procured en bloc as well.
11. Fetal porcine thymic tissue has been shown to proliferate under mice renal capsules and subsequently induce tolerance. However, in our experience, when only several pieces of autologous porcine thymic tissue were placed subcapsularly, the tissue did not grow well. Because of poor fetal thymic availability, we chose to use juvenile pigs (7–12 weeks) for thymokidney preparation. In addition, we have observed poor growth of thymi from aged swine (>12 months old) when placed beneath the renal capsule. For this reason, juvenile swine (6–10 kg) are preferred (32).
12. Miniature swine, unlike baboons or humans, have a cervical thymic lobes in addition to the mediastinal thymus. The cervical thymus is ideal for thymokidney preparation or VTL transplantation because it is easily accessed using a longitudinal off-midline cervical incision. A single, fully excised cervical thymic lobe is typically required for the preparation of two thymokidneys. Thus, only half of a thymic lobe is sufficient for a single thymokidney.
13. Upon removal of the thymic tissue, the gland is minced prior to implantation. However, we have observed poor results when the thymus is minced too finely or when the thymus is injected subcapsularly using an angiocatheter. This is likely due to increased cell lysis and inflammatory changes.
14. The kidney is reached using a flank incision and a capsular window is created using an 11-blade or angiocatheter. The tissue must be placed carefully under the renal capsule, taking care not to tear the capsule. The optimal amount of thymus to procure from the juvenile is typically related to the size of the kidney itself. Unpublished data from our laboratory has indicated that when only a few small pieces of thymus were implanted, subsequent growth of the transplanted thymus was insufficient. We have observed improved results when 40–60% of the superior surface of the kidney was covered with thymus (15, 41).

15. If the recipient has a second kidney (or thymokidney) to be used for at a later date, the animal must be recovered. Thus, in transecting the RA, care must be taken not to take much of a Carrel patch. Typically, 1–2 mm of aortic patch is sufficient for the recipient transplant. A small Carrel patch allows the surgeon to perform a primary longitudinal closure of the donor Aa using 6 or 7-0 prolene suture. If the Carrel patch is larger, it will be necessary to close the Aa using a Dacron or Gore-Tex patch. Thereafter, the abdomen is closed in the standard three layer fashion and the animal is recovered.
16. Transplantation should be performed by a skilled surgeon and the vascular anastomosis should be completed within 30 min. If a young surgeon is performing the xenokidney transplant, an attending should be mindful of the anastomosis time and ensure that the graft is minimally manipulated.
17. In our standard model of xenotransplantation, we generally remove both kidneys prior to xenograft transplantation (37). However, in some circumstances, if investigators are expecting an increased risk of graft rejection, the renal graft (pig) may be transplanted before or after the left kidney is removed. In some cases, such as the presence of high levels of preexisting anti-non-Gal antibody, if the transplant is expected early graft rejection, it may be beneficial to leave the contralateral native kidney and ligate the ureter. Removing the ligature will save the life of the animal to follow immunologic assessment even after removal of the rejected graft.
18. It is important to note the blood pressure before and after clamping the IVC as hemodynamics may vary dramatically with clamping and unclamping the IVC. If blood pressure decreases, the distal Aa can be clamped using a bull-dog to improve circulation.
19. This will diagnose bleeding from the venous anastamosis. To avoid narrowing the venous anastamosis, sutures should not be tied prior to unclamping the vein. This will allow the RV to engorge prior to tying the final knot. If the surgeon is concerned about bleeding, sutures can be tied loosely prior to unclamping. The venous anastomosis should be completed quickly (under 15 min), and preferably within 10 min.
20. As porcine kidneys have a tendency to suffer vasoconstriction due to manipulation, the authors recommend an anastamotic approach that minimizes touching the xenograft. To do so, the surgeon should stand on the animal's left side and should be able to perform both the arterial and venous anastomoses quickly and skillfully. A large Carrel patch for the donor vessel is not recommended.

21. Animal care: Following line placement and prior to transplantation, the animals undergo a brief, but stringent induction using T-cell and B-cell depleting agents. These are generally monoclonal or polyclonal antibodies that are commercially available (42). Administration of any cellular depleting antibody should be preempted by flow-cytometric T-and B-cell subset phenotyping. T-cell depleting agents have been tested at multiple doses and at multiple time-points; however, the general goal is to achieve near-complete T-cell depletion (50–150 cells/μL) for the first 2 weeks which protects the transplanted thymic grafts in the induction period (Yamada, K. et al., manuscript in preparation). Because immunosuppression is extensive during the first 2 weeks, recipients are started on antibiotics to avoid opportunistic infection (41). Our standard regimen includes: cefazolin, levaquin, and ganciclovir. The animal's WBC typically falls below 1,000 cells/μL for the first week. While the animal is leukopenic, platelet counts are a good surrogate for WBC concerning infection (43). Clinically, we have observed that platelets fall in the setting of bacteremia, despite a very low WBC, in a potentially normal appearing animal. Like humans, immunosuppressed baboons may display no signs of sickness while immunosuppressed, despite active infection. For this reason, any clinically concerning signs (decreased appetite, activity, etc.) should prompt blood cultures and empiric widening of antibiotic coverage (41). Typically, coverage is increased to vancomycin and cefazolin is stopped. If there is concern for gram negative infection, piperacillin tazobactam can be started. In these situations, it is best to consult an infectious diseases expert (41).

## References


1. Cooper DKC, Ye Y, Rolf LL Jr, Zuhdi N (1991) In: Cooper DKC, Kemp E, Reemtsma K, White DJG (eds) Xenotransplantation: the transplantation of organs and tissues between species. Heidelberg, Springer
2. Cooper DK, Gollackner B, Sachs DH (2002) Will the pig solve the transplantation backlog? Annu Rev Med 53:133–147
3. Yamada K, Griesemer A, Okumi M (2008) Pigs as xenogeneic donors. Transplant Rev 19:164–177
4. Sachs DH, Galli C (2009) Genetic manipulation in pigs. Curr Opin Organ Transplant 14:148–153
5. Cooper DK, Good AH, Koren E et al (1993) Identification of alpha-galactosyl and other carbohydrate epitopes that are bound by human anti-pig antibodies: relevance to discordant xenografting in man. Transpl Immunol 1:198–205
6. Galili U, Clark MR, Shohet SB, Buehler J, Macher BA (1987) Evolutionary relationship between the natural anti-Gal antibody and the Gal alpha 1–3Gal epitope in primates. Proc Natl Acad Sci USA 84:1369–1373
7. Galili U (1993) Evolution and pathophysiology of the human natural anti-.alpha.-galactosyl IgG (anti-Gal) antibody. Springer Semin Immunopathol 15:155–171
8. Galili U, Gregory CR, Morris RE (1996) New World monkeys as a primate model for xenografts in the absence of anti-Gal. Transplant Proc 28:567–568
9. Galili U, Shohet SB, Kobrin E, Stults CL, Macher BA (1988) Man, apes, and old world monkeys differ from other mammals in the

expression of alpha-galactosyl epitopes on nucleated cells. J Biol Chem 263: 17755–17762

10. Phelps CJ, Koike C, Vaught TD et al (2003) Production of alpha 1,3-galactosyltransferase-deficient pigs. Science 299:411–414
11. Kolber-Simonds D, Lai L, Watt SR et al (2004) Production of alpha-1,3-galactosyltransferase null pigs by means of nuclear transfer with fibroblasts bearing loss of heterozygosity mutations. Proc Natl Acad Sci USA 101:7335–7340
12. Lai L, Kolber-Simonds D, Park K et al (2002) Production of alpha-1,3-galactosyltransferase knockout pigs by nuclear transfer cloning. Science 295:1089–1092
13. Mezrich JD, Haller GW, Arn JS et al (2003) Histocompatible miniature swine: an inbred large-animal model. Transplantation 75:904–907
14. Dor FJ, Tseng YL, Cheng J et al (2004) Alpha1,3-galactosyltransferase gene-knockout miniature swine produce natural cytotoxic anti-Gal antibodies. Transplantation 78:15–20
15. Yamada K, Yazawa K, Shimizu A et al (2005) Marked prolongation of porcine renal xenograft survival in baboons through the use of alpha1,3-galactosyltransferase gene-knockout donors and the cotransplantation of vascularized thymic tissue. Nat Med 11:32–34
16. Moses RD, Pierson RN, Winn HJ, Auchincloss H Jr (1990) Xenogeneic proliferation and lymphokine production are dependent on CD4+ helper T cells and self antigen-presenting cells in the mouse. J Exp Med 172:567–575
17. Deschamps JY, Roux FA, Sai P, Gouin E (2005) History of xenotransplantation. Xenotransplantation 12:91–109
18. Yamada K, Sachs DH, DerSimonian H (1995) Human anti-porcine xenogeneic T cell response. Evidence for allelic specificity of mixed leukocyte reaction and for both direct and indirect pathways of recognition. J Immunol 155:5249–5256
19. Murray AG, Khodadoust MM, Pober JS, Bothwell AL (1994) Porcine aortic endothelial cells activate human T cells: direct presentation of MHC antigens and costimulation by ligands for human CD2 and CD28. Immunity 1:57–63
20. Cozzi E, White DJ (1995) The generation of transgenic pigs as potential organ donors for humans. Nat Med 1:964–966
21. Mollnes TE, Fiane AE (2003) Perspectives on complement in xenotransplantation. Mol Immunol 40:135–143
22. Adams DH, Kadner A, Chen RH, Farivar RS (2001) Human membrane cofactor protein (MCP, CD 46) protects transgenic pig hearts from hyperacute rejection in primates. Xenotransplantation 8:36–40
23. Buhler L, Yamada K, Kitamura H et al (2001) Pig kidney transplantation in baboons: anti-Gal (alpha)1-3Gal IgM alone is associated with acute humoral xenograft rejection and disseminated intravascular coagulation. Transplantation 72:1743–1752
24. Chen G, Qian H, Starzl T et al (2005) Acute rejection is associated with antibodies to non-Gal antigens in baboons using Gal-knockout pig kidneys. Nat Med 11:1295–1298
25. Sandberg JO, Benda B, Lycke N, Korsgren O et al (1997) Xenograft rejection of porcine islet-like cell clusters in normal, interferon-gamma, and interferon-gamma receptor deficient mice. Transplantation 63:1446–1452
26. Benda B, Karlsson-Parra A, Ridderstad A, Korsgren O (1996) Xenograft rejection of porcine islet-like cell clusters in immunoglobulin- or Fc-receptor gamma-deficient mice. Transplantation 62:1207–1211
27. Zhao Y, Swenson K, Sergio JJ et al (1996) Skin graft tolerance across a discordant xenogeneic barrier. Nat Med 2:1211–1216
28. Rodriguez-Barbosa JI, ZhaoY BR et al (2001) Enhanced CD4 reconstitution by grafting neonatal porcine tissue in alternative locations is associated with donor-specific tolerance and suppression of preexisting xenoreactive T cells. Transplantation 72:1223–1231
29. Nikolic B, Gardner JP, Scadden DT et al (1999) Normal development in porcine thymus grafts and specific tolerance of human T cells to porcine donor MHC. J Immunol 162:3402–3407
30. Haller GW, Esnaola N, Yamada K et al (1999) Thymic transplantation across an MHC class I barrier in swine. J Immunol 163:3785–3792
31. Yamada K, Shimizu A, Ierino FL et al (1999) Thymic transplantation in miniature swine. I. Development and function of the "thymokidney". Transplantation 68:1684–1692
32. Yamada K, Shimizu A, Utsugi R et al (2000) Thymic transplantation in miniature swine. II. Induction of tolerance by transplantation of composite thymokidneys to thymectomized recipients. J Immunol 164:3079–3086
33. Yamada K, Vagefi PA, Utsugi R et al (2003) Thymic transplantation in miniature swine: III. Induction of tolerance by transplantation of composite thymokidneys across fully major histocompatibility complex-mismatched barriers. Transplantation 76:530–536

34. LaMattina JC, Kumagai N, Barth RN et al (2002) Vascularized thymic lobe transplantation in miniature swine: I. Vascularized thymic lobe allografts support thymopoiesis. Transplantation 73:826–831
35. Kamano C, Vagefi PA, Kumagai N et al (2004) Vascularized thymic lobe transplantation in miniature swine: thymopoiesis and tolerance induction across fully MHC-mismatched barriers. Proc Natl Acad Sci USA 101:3827–3832
36. Nobori S, Shimizu A, Okumi M et al (2006) Thymic rejuvenation and the induction of tolerance by adult thymic grafts. Proc Natl Acad Sci USA 103:19081–19086
37. Griesemer AD, Hirakata A, Shimizu A et al (2009) Results of gal-knockout porcine thymokidney xenografts. Am J Transplant 9:2669–78
38. Yamada K, Gianello PR, Ierino FL et al (1997) Role of the thymus in transplantation tolerance in miniature swine. I. Requirement of the thymus for rapid and stable induction of tolerance to class I-mismatched renal allografts. J Exp Med 186:497–506
39. Nobori S, Samelson-Jones E, Shimizu A et al (2006) Long-term acceptance of fully allogeneic cardiac grafts by cotransplantation of vascularized thymus in miniature swine. Transplantation 81:26–35
40. Sengupta D, Harper M, Jennett B (1974) Effect of carotid ligation on cerebral blood flow in baboons. J Neurol Neurosurg Psychiatry 37:578–584
41. Fishman JA, Patience C (2004) Xenotransplantation: infectious risk revisited. Am J Transplant 4:1383–1390
42. Preville X, Flacher M, LeMauff B et al (2001) Mechanisms involved in antithymocyte globulin immunosuppressive activity in a nonhuman primate model. Transplantation 71:460–468
43. Gollackner B, Mueller NJ, Houser S et al (2003) Porcine cytomegalovirus and coagulopathy in pig-to-primate xenotransplantation. Transplantation 75:1841–1847

# Chapter 13

# Isolation of Porcine Pancreatic Islets for Xenotransplantation

Karin Ulrichs, Sibylle Eber, Bianca Schneiker, Sabine Gahn, Armin Strauß, Vasiliy Moskalenko, and Irina Chodnevskaja

## Abstract

This chapter deals with a technique for isolating intact islets of Langerhans from the pig pancreas based on our experience performing approximately 750 isolations. The procedure we describe involves identification of an optimal donor pancreas, purification and in vitro culture of islets, diabetes induction in recipients, and transplantation of islets and their immunomodulation. Besides the sophistication of the technical equipment employed, the major factors influencing the isolation outcome are the pig breed, the number and morphology of the islets in the donor pancreas, the quality of the collagenase/neutral protease, and the skill of the team members.

**Key words:** Pig breed, Pancreatic islet isolation, Collagenase, Neutral protease, Porcine islet of Langerhans, In vitro culture, Diabetes induction, Islet transplantation, Microencapsulation, Immunomodulation

## 1. Introduction

The species *Sus scrofa* (pig) is favored for isolating intact islets of Langerhans from the pancreas. Among the reasons for this are the following: pig insulin is well tolerated by diabetic patients and differs from human insulin in only one amino acid; pigs breed well and are easy to keep and their gastrointestinal tract resembles that of humans. Compared to isolating human islets, however, it can present greater difficulties. One difficulty is posed by the variation in the number, size, and morphology of islets within the pancreas among the different pig breeds, a phenomenon that has been reported by us and many other groups. The reasons for these

Cristina Costa and Rafael Máñez (eds.), *Xenotransplantation: Methods and Protocols,* Methods in Molecular Biology, vol. 885, DOI 10.1007/978-1-61779-845-0_13, 

variations remain unknown and are currently being investigated with the support of federal and local breeding institutions. If pigs with good-quality islets are available, isolation of their islets is easy as we describe in detail below. This is particularly the case since high-quality enzymes can be obtained that disintegrate the pancreatic tissue but preserve the islets. This chapter focuses on the process for selecting the best possible pancreas for islet isolation, and describes the purification and in vitro culture of the islets. Only briefly discussed here are diabetes induction, in vitro testing of islet function, transplantation into diabetic animals, and microencapsulation to immunomodulate the xenogeneic immune response in the recipient, since these subjects are extensively covered elsewhere in the literature.

## 2. Materials

### *2.1. Media, Solutions, and Supplements (with Modifications for Use in Experiments) in Chronological Order of Their Application*

1. Hanks' balanced salt solution (HBSS, 500 mL) to be used as manufactured to transport organs.
2. Fresh dithizone (DTZ) solution: Add 9 mL of HBSS to 10 mg DTZ (Sigma-Aldrich, Taufkirchen, Germany) dissolved in 1 mL dimethylsulfoxide (DMSO, 1-mL aliquot stored at −20°C until use). The solution is then passed through a 0.2-µm filter. This solution must always be prepared freshly. Prepare as described to discriminate between islets and acinar tissue (1).
3. Double-concentrated University-of-Wisconsin (UW-2×) stock solution: Dissolve 250.78 g lactobionic acid (Fluka, Steinheim, Germany) in 2,187.5 mL of distilled water and adjust pH to 7.0 with 5 M KOH. Add 47.6 g $KH_2PO_4$, 8.6 g $MgSO_4$, 124.78 g raffinose (Sigma-Aldrich), 472.5 mg allopurinol (Sigma-Aldrich), 6.4 g glutathione reduced (Sigma-Aldrich), 9.3 g adenosine (Sigma-Aldrich), and 350 g HES 200.000/0.5 (Fresenius Kabi, Bad Homburg, Germany), vortex until the solution is clear, adjust to pH 7.4 with 5 M NaOH, fill up to 3.5 L with distilled water, pass solution through a 0.2-µm filter, and store in 1-L glass flasks at 4°C until needed.
4. Single concentrated UW solution (UW-1×): Dissolve 135 mg allopurinol in 500 mL of distilled water (5 min sonic bath), fill up to 1 L with distilled water, transfer solution to a 2-L sterile glass flask, add 1 L UW-2×, and adjust pH (7.4–7.5) as above and density ($\delta = 1.048$) with density meter DMA 35N (Anton Paar, Graz, Austria). Store at 4°C until needed.
5. Enzymatic digestion solution: Prepare 0.5 mM trypsin inhibitor pefabloc SC (Roche Diagnostics, Mannheim, Germany), 2.5 mM $CaCl_2$ (to stabilize collagenase activity) in 200 mL

UW-1× solution; subsequently add 4 PZ units of NB8 collagenase (Nordmark, Uetersen, Germany) and 0.65–0.7 DMC units of neutral protease (Nordmark) per gram of pancreas. Vortex gently (no foam!) until all ingredients are dissolved. Prepare solution immediately before use, keeping it at room temperature.

6. $KH_2PO_4$ buffer stock solution (1 M): Dissolve 136.09 mg $KH_2PO_4$ in 1 L of distilled water and pass through a 0.2-μm filter. Store at room temperature.
7. HBSS solution with 13% fetal calf serum (FCS) (HBSS-13%, to stop enzymatic digestion): Add 65 mL of pretested heat-inactivated FCS (Cell Concepts, Umkirch, Germany) and 12.5 mL of $KH_2PO_4$ buffer stock solution (25 mM) to 500 mL of HBSS and adjust pH to 7.4 with 5 M NaOH. Store on ice until needed.
8. HBSS solution with 10% FCS (HBSS-10%, to wash, digest, and purify islets): Prepare solution as just described, but use 50 mL of heat-inactivated FCS instead of 65 mL. Store on ice until needed.
9. Nicotinamide solution (to restore islet viability and function after isolation): Dissolve 3.1 g nicotinamide (Sigma-Aldrich) in 15 mL of UW-1× solution, pass through a 0.2-μm filter, and store at 4°C until needed.
10. HAM's F12 islet culture medium (for in vitro culture of isolated islets): Add 50 mL of heat-inactivated pretested FCS (10%), 5 mL of penicillin/streptomycin (1%), 5 mL of amphotericine B (1%), 5 mL of glutamine (1%), and 3.1 g nicotinamide (50 mM) to 500 mL of HAM's F12 medium (Cell Concepts).
11. Fluorescein diacetate (FDA) solution (to test viability of islets): Dissolve 1 mg FDA (Sigma-Aldrich) in 1 mL of acetone. Prepare freshly and store in the dark.
12. Propidium iodide (PI) solution (to mark dead cells): Dissolve 5 mg PI in 10 mL of phosphate-buffered saline (PBS). Store in aliquots of 500 μL at 4°C.
13. Wash solution (WS, for islet purification gradient) (460 mL): Mix 230 mL of UW-2× and 230 mL of OptiPrep solution (Progren, Heidelberg, Germany), transfer to sterile flasks, and adjust optical density to $\delta = 1.206$.
14. Low-density solution (LDS) for islet purification gradient (309 mL): Mix 85 mL of WS and 224 mL of UW-1×, transfer to sterile flasks, and adjust optical density to $\delta = 1.090$.
15. The beta cell toxin streptozotocin (STZ, Sigma-Aldrich) for inducing diabetes in rats and mice.

16. Citrate buffer to dissolve STZ: A 1 M stem solution C is prepared from solution A and solution B as follows: To prepare 10 mL of solution A, add 2.1 g citric acid monohydrate to 0.9% NaCl solution (pH 1.7). To prepare 10 mL of solution B, add 2.94 g trisodium citrate dihydrate to 0.9% NaCl solution (pH 7.8). To prepare a 1 M stem solution C, mix 6.3 mL of solution A with 8.2 mL of solution B (pH 4.5). Dilute stem solution C with 0.9% NaCl solution 1:10 to produce a 0.1 M solution C.

### 2.2. Harvesting Donor Organs at Slaughterhouse or Farm

1. Two 10-L polystyrene transportation boxes filled with chipped ice.
2. Four 50-mL Falcon tubes filled with 20 mL HBSS transportation medium and 100 mL HBSS as reserve.
3. Sterile examination gloves, face masks, protective clothing, and rubber boots.
4. Sterile surgical tools for organ preparation, including small and large scissors and forceps (Aesculap, Tuttlingen, Germany), disposable scalpels, and cotton wool swabs.
5. Four sterile stainless steel kidney trays (AKULA Medizintechnik, Lauf, Germany) for on-site organ preparation.
6. Eight sterile plastic bags (2 L, four bag-in-bag), each double bag filled with 300 mL sterile HBSS, and metal clips to close bags.
7. Waterproof permanent markers and documentation sheets.
8. Four 45-cm-long 18 G Cavafix® Certo® 255 catheters (Braun, Melsungen, Germany).
9. Surgical suture material.
10. Clean plastic transportation basins (50 L) to collect the visceral organ package in toto.
11. Large sterile cotton sheets to cover preparation area or desk.
12. 70% Ethanol for on-site disinfection.

### 2.3. Microscopic Prescreening for a Suitable Organ

1. Dewar vessel with 2 L liquid nitrogen (−196°C) and safety attire and equipment to protect from splashing nitrogen.
2. Standard laboratory light microscope with ×10 and 20 objectives.
3. Cryostat Leica CM 3050S (Leica, Wetzlar, Germany) with microtome stainless steel blades S35 (Feather Safety Razor Co., Ltd., Medical Division, Hartenstein, Wuerzburg, Germany).
4. Standard glass slides and glass cover slides for histology (Histobond™, Marienfeld, Lauda-Koenigshofen, Germany).

5. Cotton wool swabs, disposable scalpels, small scissors and forceps to prepare tissue blocks, large forceps to transfer cryotubes to nitrogen in Dewar vessel.
6. Tissue Tek deep freeze medium (Sakura Finetek Europe B.V., Zoeterwoude, the Netherlands).
7. Cryotubes (2 mL, Brand, Wertheim, Germany).
8. Ten milliliter of freshly prepared DTZ solution.

### 2.4. Islet Isolation

1. C. Ricordi's stainless steel chamber (500 mL for a 100-g pancreas) with lid and triple lock, medium inlet, medium outlet, sample outlet, temperature and pH controls, four stainless steel marbles (2 cm in diameter), sieve with 500-μm mesh size, and O-ring (all items from Sauer Feinmechanik, Wuerzburg, Germany).
2. Sterile hood HS 18/2, approx. 185 cm wide (Heraeus Instruments, Hanau, Germany). The sterile hood should be wide enough to allow two team members to work side by side.
3. Scales EW 600-2 M (Kern, Albstadt, Germany) max. 600 g; fine scales ALJ 220-4 (Kern & Sohn, Balingen, Germany) max. 220 g.
4. Vortex MR 3001 (Heidolph, Schwabach, Germany).
5. Microscope Axiovert 25 CLF (Carl Zeiss, Jena, Germany).
6. pH meter 315i (WTW, Weilheim, Germany).
7. Water bath E200 (Lauda, Lauda-Koenigshofen).
8. Centrifuge Rotana 46 R (Hettich, Tuttlingen, Germany) with four plastic centrifuge tubes (600 mL).
9. Peristaltic pump ISM 404B with pump head model 7015-20 and Masterflex precision tubing (IsmatecSA, Glattbrugg-Zuerich, Switzerland).
10. Digital temperature control ama-digit ad 15th (Amarell, Koethen, Germany).
11. Laboratory stopwatch, sockets, and clamps.
12. Dimroth condenser, about 50 cm long (Hartenstein).
13. Stainless steel kidney trays, surgical scissors and forceps, small stainless steel clips, pipette set of range 10–1,000 μL, 50-mL Perfusor® syringe (Braun), Omnifix-F 1-mL Tuberculin syringes with injection needles (Braun).
14. Fresh DTZ solution.
15. Standard 24-well plastic plates.
16. Masterflex silicon tubes, 60–70-cm-long pieces with 0.5 cm inner diameter (Hartenstein; the tube ends are covered with

aluminum to keep them sterile); plastic T-junction stopcocks to connect silicon tubes (Hartenstein).

17. A 500-mL glass flask with HBSS-13% stopping medium.
18. A 500-mL glass flask with HBSS-10% washing medium.
19. Two 500-mL glass flasks to collect digest.
20. Sterile plastic ware, e.g., pipette tips, Falcon tubes, Petri dishes.
21. Disinfectant and cotton swabs.
22. Plastic basin with chipped ice.

### 2.5. Islet Purification

1. Sterile hood (see above).
2. COBE 2991 cell processor model 1 without cooling system (COBE BCT, Inc., Lakewood, CO, USA), 2991™ blood cell processing set (Gambro BCT, Inc., Lakewood, CO, USA).
3. Peristaltic pump ISM 404B with pump head model 7015-20 and Masterflex precision tubing.
4. 50-mL Plastic tubes, four glass beakers (250 mL).
5. Centrifuge Rotana 46 R.
6. Standard 24-well plastic plates.
7. Nuaire™ US auto flow water-jacketed incubator (IBS Integra Biosciences GmbH, Fernwald, Germany; adjusted to 22–24°C low temperature and 5% $CO_2$ in air, humidified).
8. Fresh DTZ solution.
9. Axiovert 25 microscope.
10. UW-1× solution, WS, and LDS stored on ice.
11. Pipettes, sterile glass, plastic ware.

### 2.6. Viability Test

1. Darkroom.
2. Standard glass and cover slides for histology.
3. Fluorescence microscope BX50 with ×10, 20, and 40 magnification and digital camera ColorView12 (Olympus, Hamburg, Germany).
4. Fresh FDA, PI, and DTZ solutions (see above).
5. IS mounting medium (Dianova, Hamburg, Germany).
6. Pipette set and cotton swabs.

### 2.7. Determination of Islet Purity and Islet Yield

1. Standard 24-well plastic plate.
2. Fresh DTZ solution.
3. Axiovert 25 microscope with ×50 magnification and a measuring eye piece (1 cm with 100 graduation lines, Carl Zeiss).
4. Manual cell counter (Hartenstein).
5. Pipette set and cotton swabs.

#### 2.8. In Vitro Culture of Islets

1. Sterile hood.
2. Nuaire™ US auto flow water-jacketed incubator.
3. Axiovert 25 microscope.
4. Standard culture plastic flasks (250 mL).
5. HAM's F12 islet culture medium (see above).
6. Fresh FDA, PI, and DTZ solutions.
7. Pipette set, sterile glass, and plastic ware.

#### 2.9. In Vitro Function of Islets

1. Sterile hood.
2. Nuaire™ US auto flow water-jacketed incubator.
3. EASIA or ELISA kit for insulin determinations (see Note 1).
4. Pipette set, sterile glass, and plastic ware.
5. ELISA reader (Thermo Max Microplate Reader, MWG Biotech, Ebersberg, Germany) and computer software MikroWin Version 3.0 (Mikrotek Laborsysteme GmbH, Overath, Germany).

#### 2.10. Diabetes Induction and Islet Transplantation

1. A license to perform animal experiments.
2. Facilities to keep small and large animals, including intensive care units.
3. Fully equipped operation theaters for small and large animals.
4. Specially trained surgeons.
5. Tools for macro- and microsurgery.
6. Inbred rat or mouse strains (Harlan Laboratories, AN Venray, the Netherlands) and/or Goettingen minipigs (Ellegaard, Dalmose, Denmark).
7. STZ dissolved in "citrate buffer" to induce diabetes. To induce diabetes in rodents, dissolve STZ in this 0.1 M solution C and inject it intraperitoneally within 10 min. Give 200 mg/kg STZ for mice and 35 mg/kg for rats, adjusting injection volume precisely to the body weight (take into account recommended injection volumes for mice and rats).
8. Appropriate suture materials (Braun).
9. Light microscope (Wild M650, Heerbrugg, Switzerland, with ×6, 10, 16, 25, and 40 magnification) to magnify operation field.
10. Special medication for anesthesia, analgesia, hyper- and hypoglycemia.
11. Three blood glucose test systems: (a) Accu-Chek Sensor with Sensor Comfort Pro test sticks (Roche Diagnostics) for rats and pigs requiring 4-μL blood samples, (b) Ascensia Elite with test sticks (Bayer Diagnostics Europe Ltd., Dublin, Ireland) for mice requiring 2-μL blood samples, (c) CGMS® (Medtronic MiniMed, Northridge, USA) for pigs to measure blood glucose real time.

## 3. Methods

### *3.1. Media, Solutions, and Supplements*

1. Most media can be prepared well before the day of isolation (day 0).
2. Total media volumes should be calculated for a number of subsequent isolations. Exceptions are the enzymatic digestion solution containing collagenase and neutral protease, DTZ solution to discriminate between endocrine and exocrine cells, and FDA medium and STZ in citrate buffer to induce diabetes. These solutions must be always prepared freshly.
3. Solutions and media should not be stored for more than 1 month.
4. Cleanliness and sterility of all devices, solvents, and supplements are essential.
5. Double-check each step carefully, as the time between media preparation and islet transplantation may span many weeks.

### *3.2. Harvesting Donor Organs at Slaughterhouse or Farm*

Two team members are needed for this procedure as one member collects the organ packages and assists the second member with preparation and storage of the pancreata. Remember to take extra materials in case of unforeseen events if the slaughterhouse or farm is far from the laboratory.

1. Cover preparation area or desk at the slaughterhouse/farm with large sterile cotton sheets and use 70% ethanol for on-site disinfection.
2. Collect at least four donor pancreata following the procedure described below to identify the most suitable organ by light microscopy prior to isolation. The best donors may be retired female breeder pigs of the local landrace (1–2 years old; 150–300 kg body weight) (2). Microscopic histological prescreening of four potential donor organs prior to isolation saves disappointment, money, time, and man power. If you wish, you can collect more or fewer organs.
3. Transfer abdominal organs in toto from a brain-dead healthy pig to the clean transportation basin. To avoid bacterial and fungal contamination of the pancreas (and the isolated islets), make sure that all organs are intact, with no leakage from stomach and bowels.
4. Identify the pancreas and prepare it locally under the half-sterile conditions (see Note 2). To this end, touch and squeeze the organ as little as possible; remove fat, lymph nodes, and connective tissue with scissors and forceps while the organ is still warm; and avoid any cuts of the pancreas capsule. Carefully inspect the organ after preparation and exclude organs of

unusual anatomy and color, e.g., grey, dark red, or speckled organs or organs that contain excessive fat or a hematoma.

5. Cut the splenic lobe off the pancreas (it contains body and tail of ~100 g) and transfer it into a stainless steel kidney tray on ice containing 100 mL HBSS.
6. Cut off two tissue blocks of 1 $cm^3$ each from the pancreas body for histology and transfer them to a Falcon tube with ice-cold HBSS.
7. Identify the main pancreatic duct holding the organ upright and cannulate the duct with the Cavafix® catheter while the organ is still warm. Be gentle and do not use force when cannulating approximately 2/3 of the duct. The organ preparation and cannulation must be quick to avoid unnecessary (warm) ischemic time (cannulation of the pancreatic duct can also be performed later in the laboratory).
8. Fix catheter with suture material at the duct entrance to avoid its slipping during transportation.
9. Store pancreata in ice-cold HBSS inside the double plastic bags. Close bags with metal clips, put them in the polystyrene transportation box, and cover them with ice.
10. Make sure that each Falcon tube and double bag is correctly labeled to avoid mixing tissue samples and donor organs.
11. Transport donor organs immediately to the isolation laboratory (see Note 3). Warm and cold ischemic time should be as short as possible.

#### 3.3. Microscopic Prescreening for a Suitable Organ

1. Swab the tissue blocks from the Falcon tubes with cotton wool, transfer to cryotubes containing the cryoprotection medium Tissue Tek, and then quick-freeze in liquid nitrogen.
2. Immediately, with the cryostat at –20°C, cut 10-μm-thick tissue sections from four frozen tissue blocks representing the four donor pancreata.
3. Place sections on glass slides, immediately stain with 100 μL fresh DTZ solution, and inspect under the microscope at ×10 and 20 magnification. Islets of Langerhans stain red within seconds, whereas acinar tissue remains unstained. The number, size, and morphology of islets in the DTZ-stained tissue sections correlate with the number, size, and morphology of islets in situ and after isolation. Only pancreata containing sufficient numbers of large islets (200–250 μm in diameter), showing nice morphology and evenly distributed strong red staining, should be considered as donor organs for subsequent isolation (3). A sufficient number on a 1-$cm^2$ pancreatic tissue section would be a minimum of five islets. No red staining or only slight red staining (always compared to a positive control)

indicates a lack of insulin vesicles in the islets, which may be due to trauma, stress, or inadequate conditions of feeding and/or keeping of the donor animals. Such pancreata should be excluded as donor organs, as it is not yet clear if the function of such poorly stained islets can be restored.

### 3.4. Islet Isolation

The basic isolation methodology is described in great detail and schematically outlined by its developer, Ricordi (4, 5). The methodology is used with great success by many groups around the globe. Fully automated, it can be used by beginners to avoid unnecessary costs. Details of the automated method for pancreatic islet separation, as designed by C. Ricordi, can be viewed on the Internet (6). The automation may save time and labor, but has no particular influence on final isolation results.

#### 3.4.1. Assembly of Equipment on Day −1

1. Assemble all isolation equipment on day −1 under the sterile hood, ensuring that all parts are clean, dry, and sterile. Have enough isolation equipment parts in reserve in case of unforeseen events during isolation as isolations can be saved if sterile, clean spare parts are ready at hand. The isolation chamber can be made of stainless steel, transparent plastic, or glass. A plastic or glass chamber allows viewing of the tissue disintegration, but we use the stainless steel chamber with great success. The size of the Ricordi chamber must correlate with the size of the donor organ. For a 100-g pancreas, we use a 500-mL chamber.
2. Set up the connections, without media, in preparation for day 0. The Masterflex silicon tubes connect the glass flask containing the recirculation medium (enzymatic digestion solution) to the peristaltic pump, water bath, Dimroth condenser, chamber inlet, and chamber outlet, which is then connected to the glass flask containing the recirculation medium. Silicon tubes should be as short as possible to avoid unnecessary use of expensive enzymatic digestion solution. In the elution phase, silicon tubes are reconnected so that the stopping medium and then the washing medium elute the digest from the Ricordi chamber into the collection flask.
3. After assembly, double-check all parts, cover isolation equipment with sterile cotton sheets, and close the hood. All procedures concerning the handling of the equipment during isolation should be well learned in advance, as the viability of islets also depends on the total time between transfer of the pancreas pieces to the chamber and the moment the digest with the islets is put on ice prior to purification.

#### 3.4.2. Islet Isolation on Day 0

1. First, turn on the water bath, as a digestion temperature of 37°C is required in the chamber when digestion begins.
2. Fill 2/3 of the chamber with UW-1× solution and start pumping the medium through the system to preheat the chamber.

3. Transfer the previously selected donor pancreas (see above) into an empty sterile kidney tray on ice inside the hood, and remove residual fat (see Note 4), connective tissue, and lymph nodes with small scissors and forceps. Do not damage the pancreas capsule to avoid leakage.
4. Weigh the organ.
5. Fill the 50-mL Perfusor® syringe with the enzymatic digestion solution, connect it to the catheter in the pancreatic duct, and inject the solution carefully, putting very little pressure on the syringe, via the duct into the organ. All parts of the 100-g pancreas must expand well as organ parts that do not expand will not digest and will not release islets. Use small stainless steel clips to fix capsule leakages. A 100-g pancreas takes as much as 200 mL of enzymatic digestion solution to be sufficiently expanded (blown up) (see Note 5).
6. Cut the organ into 3–4 pieces, put them into the chamber together with the four marbles, fill the chamber with the rest of the enzymatic digestion solution, put the sieve and O-ring into position, and close and lock the lid carefully.
7. Begin the recirculation phase by starting the peristaltic pump (85 rpm: flow rate 75 mL/min at inner tubing diameter of 0.5 cm). This step is $t_0$ of digestion. During digestion, shake the chamber by hand (10 hubs/10 s, 5 s pause, 10 hubs/10 s, etc.) to guarantee good mixture of its contents. An automatic shaker may be preferable, but shaking by hand leads to equally good results. Normal digestion time ranges from 15 to 25 min, depending on the activity and the amount of collagenase/neutral protease used, and the time to achieve the temperature optimum inside the chamber (range 35–37°C). If necessary, use prewarmed UW-1× solution to extend the recirculating medium volumes in order to remove air bubbles.
8. When the enzymatic digestion solution becomes cloudy (circulating tissue particles), take a 300 μL tissue/digest sample from the chamber every minute with a sterile Omnifix 1 mL Tuberculin syringe and transfer it to the first well of the 24-well plastic plate containing 300 μL fresh DTZ staining solution per well.
9. Quickly inspect the digestion status under the Axiovert 25 microscope (see Note 6).
10. Carefully check and document the temperature and pH value inside the chamber at various digestion times, as the pH tends to decrease during digestion from pH 7.45 (starting point) to about pH 7.3 (elution point). If it drops below pH 7.2—for reasons that are not yet well understood—the final islet preparation will have very poor viability. Thus, pH 7.2 is our cutoff point. This steep drop in pH during digestion is never seen with organs that are explanted in the operation theatre.

Therefore, we think a prolonged warm ischemic time during organ harvest leads to this otherwise unexplained drop of pH. We tested a number of different buffers and found that the $KH_2PO_4$ buffer gave the best results.

11. As soon as the sample shows 1–2 well-isolated normal-sized red-stained islets without an exocrine rim, halt recirculation of the medium and elute the digest from the chamber with ice-cold HBSS-13% (stop solution) (see Note 7). Then, collect the digest in the centrifuge tubes and wash three times with HBSS-10% at 250×*g* in the centrifuge (slow acceleration, 3 min, 4°C, no brake).
12. Resuspend gently and pool the sediments before transferring to a 500-mL glass flask. Then, add the UW-1× solution up to a volume of 200 mL. Three samples of 100 μL each are transferred to the 300 μL DTZ-containing wells in the 24-well plastic plate to count islets within the digest prior to purification.
13. At this point, add 15 mL nicotinamide solution and 50 mL FCS to the pancreas digest plus UW-1× solution to bring the digest volume to a final 500 mL.
14. The pancreas digest is then put on ice for 1 h.
15. The undigested pancreatic tissue is collected from the Ricordi chamber and weighed (see Note 8).

#### 3.5. Islet Purification

1. Place the COBE cell processor close to sterile hood and prepare it for islet purification. To this end, insert the 2991™ blood cell processing set very accurately (folds in the plastic material will interfere with the developing gradient), arrange and fix colored tubing according to processor function, and connect tubing of processing set with tubing of peristaltic pump (inside sterile hood). If you use a COBE cell processor that has no built-in cooling system, keep all solutions and gradients on ice.
2. Centrifuge at 200×*g* the 500 mL digest that was rested for 1 h, discard the supernatant, and resuspend the digest carefully with 200 mL of ice-cold UW-1× solution and 120 mL of WS.
3. Calibrate the COBE cell processor as follows: 1,500 rpm, supernatant flow rate at 100 mL/min, minimal mixing time 60 s, supernatant volume 600 mL, valve selection V 1.
4. Pump the cold digest suspension into the processor at 150 rpm (peristaltic pump) and start centrifugation. Then, pump 96 mL of ice-cold LDS at 45 rpm into processor, followed by 120 mL of ice-cold UW-1× solution, also at 45 rpm. Centrifuge digest for 1–2 min to allow gradient build. The tissue separation inside the processing set can be viewed macroscopically.
5. Operating the processor buttons manually, discard the first 50 mL of the gradient, elute the subsequent gradient, and collect

samples of 30 mL each in seven carefully numbered 50-mL Falcon tubes (nos. 1, 2, …, 7). Keep tubes with contents on ice.

6. Transfer three 100-μL samples from each tube to a new 24-well plastic plate containing fresh DTZ solution in 7×3 carefully numbered wells (nos. 1.1, 1.2, 1.3, …, 7.3).
7. Select tubes containing the most pure islets (usually nos. 3, 4, and 5). These 3 islet fractions are then pooled and resuspended in 100 mL of ice-cold UW-1× solution. Then, three 100-μL samples are transferred to DTZ-containing wells to count the purified islets.
8. Transfer the main islet suspension to two Falcon tubes and centrifuge (200×*g*, slow acceleration, 3 min, 4°C, no brake). Then, discard the supernatant and resuspend the sediment with 32 mL of HAM's F12 islet-culture medium (with supplements).
9. Transfer 30 mL of islet suspension to a 250-mL culture plastic flask and put into the water-jacket incubator for low-temperature (24°C) in vitro culture. The final 2-mL islet suspension is used for subsequent testing of viability and purity, and for quantification of the islet yield.

### *3.6. Viability Test*

The viability tests are performed in the darkroom (see Note 9).

1. Add 1 μL of FDA solution to 50 μL of islet suspension (the reaction time of this mixture is 15 min at 37°C in the dark).
2. Then, add 10 μL of PI solution to the suspension, transfer 100 μL of this mixture to a glass slide, add 25 μL of mounting medium to stabilize the fluorescence, and cover the mixture with a cover slide.
3. Remove superfluous solution with a cotton swab.
4. Immediately analyze islets under the fluorescence microscope at 488 nm wavelength with a 530-nm filter at ×10, 20, or 40 magnification. Document the result with the ColorView12 digital camera. Viable islets show a bright green color, whereas dead cells display a red nucleus. Count viable and dead islets and indicate as percentage of the total cell number (see Note 10).

### *3.7. Determination of Islet Purity and Islet Yield*

1. Transfer 100 μL of the final islet suspension to each of the five wells of a 24-well plastic plate, each well containing 100 μL of fresh DTZ solution.
2. To determine islet purity, inspect and measure the red-stained islets and unstained residual acinar tissue fragments (preferably by two team members independently) under the Axiovert 25 microscope at ×50 magnification using the measuring eyepiece. Then, calculate mathematically the percentage of endocrine tissue in relation to the total amount of tissue. This calculation is performed for each well.

**Table 1**
**The number of islet equivalents (IEQ = an islet with a diameter of 150 μm) in an islet preparation is estimated by measuring the islet diameters in a representative sample observed under the light microscope and calculating their volumes mathematically. The islet volumes are converted with the help of the conversion factor**

| Islet diameter, range (μm) | Mean volume ($\mu m^3$) | Conversion factor (counted islet number × conversion factor) |
|---|---|---|
| 50–99 | 294.525 | 0.16 |
| 100–149 | 1,145.373 | 0.66 |
| 150–199 | 2,977.968 | 1.7 |
| 200–249 | 6,185.010 | 3.5 |
| 250–299 | 11,159.198 | 6.3 |
| 300–349 | 18,293.231 | 10.4 |
| ≥350 | 27,979.808 | 15.8 |

This table was first published by Bretzel et al. (9) and is reproduced here with kind permission of Springer Science and Business Media

3. To determine the islet yield, measure the size of each single islet in the five wells with the eyepiece and grade it according to Table 1. Depending on the type of pancreas selected for isolation, smaller or bigger islets may predominate. To harmonize differing results among various research groups, count islets of each size group (see Table 1) according to an imaginary islet 150 μm in diameter. Such an ideal islet is called an "islet equivalent" or IEQ (see Note 11).

### *3.8. In Vitro Culture of Islets*

1. Inspect in vitro-cultured islets every day under the light microscope at sufficient magnification to identify possible bacterial and fungal contamination. Do not mistake the smaller fragments of dying islet cells (e.g., as a consequence of mechanical stress during isolation) as bacterial contamination; again, sterility of all materials is essential (see Note 12). Islets should not aggregate or disintegrate.
2. Perform FDA/PI viability testing according to research aims. DTZ staining is not necessary to inspect islets.
3. Change half of the HAM's F12 culture medium (with supplements) every second day to put islets under minimal stress. Use sedimentation instead of centrifugation to separate islets and medium. We have also cultured porcine islets in the (very expensive) human islet cell-specific hCELL medium—exactly as produced by the manufacturer (hCELL technologies Inc.,

Reno, USA)—with very good success for up to 6 weeks (islet viability and function in vitro were excellent).

### 3.9. Testing In Vitro Function of Islets

In vitro assessment of the function of isolated islets of various species is a routine procedure in all laboratories that deal with islet isolation/transplantation on a regular basis. The physiology of insulin secretion in standard assays for testing the dynamic insulin response upon a glucose challenge is extensively described by Luzi et al. (7) and is not repeated here in detail. Such assays are easy to perform, either by static incubation or perifusion of the islets. Assays for the rat (8) and the human system (9) are basically also applicable to porcine islets, although the islet stimulus "glucose" must be dissolved in culture medium specific for porcine islets.

1. Set up assay with pig islets as desired (see Note 13).
2. Collect pig insulin-containing culture supernatants at predetermined time points and store at –20°C before being used in the commercial EASIA/ELISA kit exactly as described in the manual.
3. After conducting the immunodetection, results are analyzed at two wavelengths (450 and 490 nm) against a reference filter (650 nm) with the ELISA reader.
4. Determine final insulin levels following the standard curve and show in μU/mL.

### 3.10. Diabetes Induction and Islet Transplantation

In principle, the technique of transplanting pig islets into specific recipients, e.g., diabetic mice or rats, does not differ from transplanting rodent or human islets into these diabetic recipients. However, whether it is biologically sensible to transplant them into an STZ diabetic recipient or an animal whose diabetes was induced by pancreatectomy depends on the particular research problem. Mice and rats usually become diabetic within 2–3 days after intraperitoneal STZ injection (in case they do not, the full dosage of STZ is injected a second time) (see Note 14). As single beta cells can survive the STZ treatment, residual insulin production may be observed. Pancreatectomy requires the skills of an experienced surgeon or microsurgeon and results in complete elimination of the insulin-producing cells. It is, thus, the technique of choice for diabetes induction when the commercially available insulin detection assay does not discriminate between recipient and donor insulin. Our own group recently described an STZ and a pancreatectomy diabetes model in miniature pigs (10). Details of both methods and their pros and cons are presented in the publication.

Appropriate in vivo testing of isolated pig islets requires their transfer into small diabetic animals, e.g., mice and rats, or, to mimic the clinical situation, into much larger diabetic animals, e.g., diabetic pigs or nonhuman primates. Like in vitro assays, in vivo assays are a standard procedure in most if not all "islet laboratories" around the world for testing the function of isolated pig islets in living animals. All materials and microsurgical techniques for transplanting xenogeneic islets into diabetic rodents are identical to those used in syngeneic or allogeneic transplantation. They are described at length in the book *Experimental Transplantation Models in Small Animals* (11). Porcine islets can also be used in these models; however, since these islets are xenogeneic to the recipient, specific modulation of the recipient's immune system is required after transplantation (see below). Many "classical" immunosuppressive drugs are applied with great success in small animal transplantation. Regarding larger animal models, transplantation of porcine islets into nonhuman primates is usually restricted to those few institutions performing biomedical research in rhesus monkeys or baboons. Transplantation of porcine islets into nonhuman primates was recently reviewed by Hering and Walawalkar (12). Details on the equipment, surgical procedures, immunosuppressive regimes, etc. can be found in this review and in many original publications cited there. In addition, an increasing number of universities provide the specific facilities to perform islet transplantation in diabetic pigs. Pig-to-pig transplantation is allogeneic by nature but can be used to test the in vivo function of islets, different transplantation sites, and techniques and to prepare all the logistics step by step for future clinical application. In this setting, blood glucose concentrations are monitored real time with the CGMS® as previously described by our group (10) or conventionally with the Ascensia Contour® blood glucose monitoring system using capillary blood harvested from the pig ear. In each case, it is advisable to train the pig for compliance well in advance of the blood glucose assays. To test the functional capacity of the transplanted islets, it is necessary to perform the intravenous glucose tolerance test and the hyperglycemic clamp; both were successfully performed in normal, diabetic, and transplanted pigs as described (10).

There is an ongoing debate as to which is the best transplantation site: for small animals, the portal system of the liver, the subrenal capsule, the subcutis, the spleen, the omentum, or the peritoneal cavity (11); for large animals and humans, the portal system of the liver (13), the omentum, or the peritoneal cavity (14). Microencapsulated islets, particularly double capsules, may be too large in diameter for the portal system of the liver. If grafted to the omentum, this site can be removed in case of an emergency, e.g., an infection or non-function. It may be impossible to remove every single islet or microcapsule from the pig's peritoneal cavity.

### 3.11. Immunomodulation of Islets

This section contains only theoretical consideration, as there is a wide range of concepts that have to be dealt with regarding this important subject. The strong immunological barrier between donor islets and the recipient's immune system in the xenogeneic pig-to-human or pig-to-nonhuman primate situations requires manipulation of the donor islets' immunogenicity prior to transplantation, as well as of the recipient's immune system after transplantation. The book *Xenotransplantation* (15) has the subtitle *The Transplantation of Organs and Tissues between Species*, which gives an excellent overview of this type of transplantation and its specific immunology. There are multiple successful approaches to reduce the immunogenicity of isolated islets, as well as regimes to induce immunosuppression; these are extensively described in the book *Pancreatic Islet Cell Transplantation* (16). Our group favors microencapsulation of the isolated pig islets with highly purified alginates. There are already a few companies worldwide providing products and services in this fascinating field (17) (see Note 15). Knowledge and training are also provided by *The Bioencapsulation Research Group* (18). The different methods of encapsulation, e.g., micro- and macroencapsulation, and encapsulation materials are splendidly discussed in the book *Cell Encapsulation Technology and Therapeutics* (19). In addition, the promise and progress of this fascinating concept are critically discussed in an excellent review paper (20). As the field of bioencapsulation rapidly develops and expands, each group has to find its own type of encapsulation material and technique tailored to its particular research aims.

## 4. Notes

1. Always ensure that the antibodies for commercial insulin react in a species-specific manner and that the C-peptide ELISAs cross-react appropriately. Sophisticated antibodies and detection reagents are readily available for mouse, rat, and human experimentation, but rarely for pig; thus, cross-reaction should be tested in advance. Read the company's instructions carefully, but do not rely on them—do your own testing.
2. Sufficient test runs should be performed at the slaughterhouse or farm to learn sterile explantation, preparation, and storage to perfection before organs are shipped to the laboratory for isolation. A visceral surgeon with knowledge of pancreas anatomy/morphology may be helpful in this learning phase.
3. Warm ischemic time is detrimental to the viability of isolated islets—autolysis of the pancreas begins immediately in the brain-dead animal. Optimal organ harvesting is essential for successful islet isolation. In fact, warm and cold ischemic time of the donor organ can be completely avoided if it is explanted with expertise in the operation theatre of the same hospital, where the islets will be isolated.

4. Never use a pancreas containing fat that cannot be removed as fat dissolves during digestion at 37°C and clogs sieve pores and digestion equipment.
5. Prepare the enzymatic digestion solution *after* the pancreas is prepared, well cannulated, and weighed. The expensive enzymatic digestion solution would be wasted if cannulation of the pancreatic duct fails, which can happen due to unusual duct anatomy. Moreover, pancreas tissue that does not distend when the enzymatic digestion solution is injected via the pancreatic duct will not release islets! To proceed with such tissue is a waste of time, labor, and money.
6. Should the DTZ solution fail to stain the islets, do not panic. Isolated islets can be easily identified under the light microscope because they look like beige or brown potatoes; acinar tissue is grey/dark grey.
7. Determining the optimal digestion time point, i.e., the time point when islets are optimally digested in the recirculation phase, requires great experience (and intuition!). The optimal digestion time varies with each type of collagenase, each batch of the same collagenase, collagenase concentration, and temperature inside the chamber. We prefer collagenase NB8 for porcine islet isolation as collagenase NB1 produces much lower islet yields.
8. If you want to perform two isolations within 24–48 h, have a complete second isolation set ready for use. In case of unforeseen events, do not panic, stay calm, and rethink every step; always try to rescue the isolation by being inventive and using unconventional measures.
9. A parallel DTZ-stained control may help to discriminate between islets and rare small lymph nodes; an inexperienced eye may mistake small lymph nodes for islets that also stain green with FDA/PI. Photographic documentation should be carried out first thing as FDA/PI staining is very light sensitive and the preparation with viable cells cannot be preserved.
10. It is best if the number of dead cells is determined independently by two team members. The viability of "slaughterhouse islets" that have suffered up to a 20-min warm ischemic time—due to the slaughtering process—ranges from 80 to 95%, whereas islets isolated from organs explanted in the operation theatre without warm ischemic time have 95–98% viability.
11. Thus, an islet 300 μm in diameter equals 2 IEQ and an islet 0.75 μm diameter equals 0.5 IEQ. A very good donor organ may have a final islet yield of 400,000 IEQ, a yield of 5,714 IEQ/g of organ (100 g of the pancreas were put into the chamber, 30 g remained undigested, thus 70 g of digested pancreas released the 400,000 IEQ). Using a much smaller

minipig pancreas for isolation may result in similar IEQ/g of organ, but in much smaller total IEQ numbers.

12. Sterility and appropriate staff training are the highest priority in islet in vitro culture. It is extremely expensive and very frustrating to have to repeat a culture experiment with isolated islets because of neglect or lack of concentration.
13. Islet numbers and the total medium volume per individual test have to be adapted to the total number of islets available for such assays. In order to save sufficient islets for transplantation, rat and pig assays may require different islet numbers and need to be designed accordingly.
14. The sensitivity of inbred mouse and rat strains to STZ differs greatly; therefore, dosage, timing (single or repeated injections), and diabetes status should be predetermined in each single strain. Do not rely too much on specifications given in publications, as the same inbred rat or mouse strain/substrain may react differently according to where and when it was generated.
15. Cooperation with appropriate physicists, biotechnologists, material researchers, and companies that produce encapsulation devices can be extremely helpful.

## Acknowledgments

The authors want to thank Mr. Alois Reichert, Mr. Peter Heuler, and Mr. Jens Weinberger for their excellent technical assistance during organ procurement, and Mrs. Lorraine Stevenson-Knebel for critically reading the manuscript and her untiring assistance with the paperwork. Furthermore, the authors want to thank Professor Dr. Dr. h.c. Arnulf Thiede, former director of the Surgical Clinic I, for his never-ending theoretical and practical support of our work.

### References

1. Latif ZA, Noel J, Alejandro R (1988) A simple method of staining fresh and cultured islets. Transplantation 45:827–830
2. Krickhahn M, Meyer T, Bühler C, Thiede A, Ulrichs K (2001) Highly efficient isolation of porcine islets of Langerhans for xenotransplantation: numbers, purity, yield and in vitro function. Ann Transplant 6:48–54
3. Krickhahn M, Bühler C, Meyer T, Thiede A, Ulrichs K (2002) The morphology of islets within the porcine donor pancreas determines the isolation result: successful isolation of pancreatic islets can now be achieved. Cell Transplant 11:827–838
4. Ricordi C, Socci C, Davalli AM et al (1990) Isolation of the elusive pig islet. Surgery 107:688–694
5. Ricordi C, Rastellini C (1995) Automated method for pancreatic islet separation. In: Ricordi C (ed) Methods in cell transplantation. RG Landes, Austin, pp 433–438
6. http://www.biorep.com/products-diabetes.aspx

7. Luzi L, Secci A, Pozza G (1992) Metabolic assessment of posttransplant islet function in humans. Methodological considerations and possible pitfalls: a lesson from pancreas transplantation. In: Ricordi C (ed) Pancreatic islet cell transplantation. RG Landes, Austin, pp 361–382
8. Lacy PE, Finke EH, Conant S, Naber S (1976) Long-term perfusion of isolated rat islets in vitro. Diabetes 25:484–493
9. Bretzel RG, Hering BJ, Federlin KF (1995) Assessment of adult islet preparations. In: Ricordi C (ed) Methods in cell transplantation. RG Landes, Austin, pp 455–463
10. Strauss A, Moskalenko V, Tiurbe C, Chodnevskaja I, Germer CT, Ulrichs K (2010) Goettingen minipigs (GMP): comparison of two different models for inducing diabetes. Diabetol Metab Syndr. 2012, 4:7
11. Leow CK (1995) Pancreatic islet transplantation in rodents. In: Green MK, Mandel TE (eds) Experimental transplantation models in small animals. Harwood Academic, Chur, pp 85–105
12. Hering BJ, Walawalkar N (2009) Pig-to-nonhuman primate islet xenotransplantation. Transpl Immunol 21:81–86
13. Bühler L, Deng S, O'Neil J et al (2002) Adult porcine islet transplantation in baboons treated with conventional immunosuppression or a non-myeloablative regimen and CD154 blockade. Xenotransplantation 9:3–13
14. Calafiore R, Basta G, Luca G et al (2006) Microencapsulated pancreatic islet allografts into nonimmunosuppressed patients with type 1 diabetes. Diab Care 29:137–138
15. Cooper DKC, Kemp E, Platt JL, White DJG (1997) Xenotransplantation. Springer, Berlin
16. Ricordi C (1992) Pancreatic islet cell transplantation. RG Landes Company, Austin
17. http://www.lctglobal.com
18. http://impascience.eu/bioencapsulation
19. Kühtreiber WM, Lanza RP, Chick WL (1999) Cell encapsulation technology and therapeutics. Birkhäuser, Boston
20. Orive G, Hernández RM, Gascón AR et al (2003) Cell encapsulation: promise and progress. Nat Med 9:104–107

# Chapter 14

# Pig Neural Cells Derived from Foetal Mesencephalon as Cell Source for Intracerebral Xenotransplantation

**Xavier Lévêque, Véronique Nerrière-Daguin, Isabelle Neveu, and Philippe Naveilhan**

## Abstract

Intracerebral cell transplantation offers the possibility of replacing lost neurons in case of neurodegenerative disorders. To date, the best functional recovery for Parkinson's patients has been obtained using neuroblasts derived from human foetal mesencephalon, but the ethical and practical problems relative to the use of human foetal tissue lead to consideration of alternative sources of cells. In this regard, porcine neuroblasts appear as a valuable source as these cells are available in large quantity and programmed to extend long neurites as human neurons. However, the potential use of pig neural cells in the clinical setting depends on efficient and safe immunosuppression. So, most experimental work in this domain aims at developing immunosuppressive treatments specifically adapted to the central nervous system. In such perspective, transplantation of porcine mesencephalic neuroblasts into the striatum of the adult rat brain is of great interest. Indeed, rejection of intracerebral xenografts has been quite well described in rats, and graft survival can be easily monitored in a rat model of Parkinson's disease. In the present chapter, we describe the methods for isolating neuroblasts from foetal porcine mesencephalon as well as the technique of intracerebral transplantation in adult immunocompetent rats.

**Key words:** Xenotransplantation, Neural cells, Pig neuroblasts, Immune response, Neurodegenerative disease, Parkinson's disease, Intracerebral transplantation, Restorative strategy

## 1. Introduction

Neurodegenerative diseases are characterized by the death of specific neuronal populations in the brain. Indeed, Parkinson's disease is due to the loss of nigral dopaminergic neurons that causes a depletion of dopamine in their target area, the striatum (caudate nucleus/putamen). Pharmacological treatment with L-DOPA compensates partially for the deficit in dopamine but, in most

Cristina Costa and Rafael Máñez (eds.), *Xenotransplantation: Methods and Protocols*, Methods in Molecular Biology, vol. 885, DOI 10.1007/978-1-61779-845-0_14, 

instances, improvement of the patient is temporary. Interesting results have been also observed following deep brain stimulation, but only a restricted number of Parkinson's patients with particular symptomatology can benefit efficiently from this therapeutic neuromodulation. Alternative and complementary therapeutic approaches are, therefore, currently investigated, including the transplantation of foetal neurons that allow the replacement of lost cells. Use of animal models of Parkinson's disease demonstrated that intrastriatal transplantation of foetal dopaminergic neurons isolated from the ventral mesencephalon (VM) has interesting restorative potentials, as such grafted neurons differentiate in situ, emit axons, form functional synapses with the host striatal neurons, and promote significant motor recovery (1, 2). Functional improvements have also been observed in Parkinson's patients who received an intrastriatal, bilateral graft of human foetal VM cells (3–5), but the clinical assays have been temporarily stopped because of dyskinesias (6). Experimental studies are currently performed to improve the restorative strategy (7, 8), but alternative sources of cells are needed due to the ethical and practical problems associated with the use of human foetal neural cells. In this perspective, porcine neuroblasts have sparked a lot of interest as these cells are available in large quantity and emit long neurites like human neuroblasts. Furthermore, the presence of blood–brain barrier, lack of conventional lymphatic drainage, and poor expression of major histocompatibility antigens in the central nervous system are great advantages for long-term survival of xenografts with minimal immunosuppression. Experimental assays demonstrated that porcine neuroblasts isolated for the VM of porcine foetus display a strong potential to differentiate and extend long neurites in a xenogeneic host, such as the rat brain (9–12). Furthermore, axons tend to reconstitute the normal cytoarchitecture and restore the adequate connections between different areas of the host brain (13). Accordingly, motor recovery has been observed after the transplantation of porcine mesencephalic neuroblasts into the striatum of the 6-hydroxydopamine rat model of Parkinson's disease (10).

Clinical trials are being performed on the basis of these encouraging results (14). Porcine mesencephalic tissue treated with anti-MHC1 antibodies was grafted into the striatum of Parkinson's patients immunosuppressed with cyclosporine A. No transmission of pig retroviruses was detected and post-mortem analysis of one patient revealed the presence of porcine dopaminergic neurons and axonal projection in the host striatum 7 months after graft (14). Some patients exhibited clinical improvement, but the general absence of increase in fluorodopa uptake (15, 16) suggests a low survival of dopaminergic neurons possibly due to the host immune response. Indeed, an alloimmunization to donor antigens and immune rejection following foetal neural grafts to the brain in patients with Huntington's disease have been recently reported (17), indicating that an immune response to donor cells occurs in the brain, even in allotransplantation. These observations underline the need for an immunosuppressive strategy adapted to the central nervous system. That is the reason why we have been studying the immune mechanisms that lead to the rejection of porcine neuroblasts following their transplantation into the striatum of immunocompetent adult rats (18, 19). The results suggest the activation of dendritic cells and microglial/macrophagic cells which in turn promotes activation and recruitment of T lymphocytes at the graft site. Analysis of the deep cervical lymph nodes shows an alteration of the T-cell repertoire (20). On the other hand, accumulation of pro-inflammatory molecules such as IL-1α and TNF-α, together with IFN-γ, IL-2, and RANTES expression, indicates an infiltration of the graft by Th1 cells (21). This observation, together with the fact that rat nigral xenografts survive in the brain of MHC class II- but not class I-deficient mice (22), suggests that xenograft rejection in the brain is mainly due to cellular immune responses with a major role of the indirect pathway mediated by host MHCII and Th1 T cells. Treatment with cyclosporine A or minocycline (23) delays, but does not prevent, xenograft rejection. So, complementary or alternative immunosuppressive strategies have to be found. In this perspective, a transgenic pig expressing hCTLA4-Ig under the neuron-specific enolase (NSE) promoter has been created (24). We expect that local expression of the immunosuppressive molecule, following intracerebral transplantation of hCTLA4-Ig-expressing neurons derived from the VM of transgenic foetal pigs, will promote or contribute to the long-term survival of xenogenic cells in the host brain. This hypothesis is currently being tested.

In this chapter, we describe the preparation of VM porcine neuroblasts and detail the protocol of cell transplantation into the striatum of adult rats.

## 2. Materials

### *2.1. Dissection of the Porcine Foetal Mesencephalon*

1. Domestic large white porcine embryos at 28 days of gestation (G28) (INRA, Nouzilly, France).
2. Low-magnification stereoscopic binocular dissection microscope.
3. Sterilized fine forceps: Ref. 9983 or 9980 of Moria (Antony, France).
4. Sterilized fine scissors: Ref. 8023A or 9850 of Moria.
5. Sterile 10-cm cell culture or bacteria dishes.

### 2.2. Preparation of the Neuroblasts from Foetal Porcine Ventral Mesencephalon

1. Hanks' balanced salt solution (HBSS) with phenol red supplemented with 100 U/mL penicillin and 0.1 mg/mL streptomycin.
2. Basal medium: Dulbecco's modified Eagle's medium (DMEM)/ Ham's F12 1:1 supplemented with 33 mM D-glucose, 5 mM HEPES (pH 7.2), 100 U/mL penicillin and 0.1 mg/mL streptomycin, and 2 mM L-glutamine.
3. Complete medium: Basal medium supplemented with 10% heat-inactivated foetal calf serum (FCS) (see Note 1).
4. Defined medium: Basal medium supplemented with N2 supplement.
5. Phosphate-buffered saline (PBS): Prepare 10× stock with 1.37 M NaCl, 27 mM KCl, 100 mM $Na_2HPO_4$, and 18 mM $KH_2PO_4$ (adjust to pH 7.4 with HCl if necessary). Prepare working solution by diluting one part with nine parts of water and autoclave before storage at room temperature.
6. Trypsin TPCK treated from bovine pancreas is dissolved in PBS at 25 mg/mL. Store in aliquots at –20°C and then use for tissue dissociation as required.
7. Deoxyribonuclease I from bovine pancreas (DNase I) is dissolved in HBSS at 10 mg/mL. Store in aliquots at –20°C and then use for tissue dissociation as required.
8. Cell counting chamber (Bürker or Malassez).
9. Sterile 70-μm filter.
10. Sterile cell culture dishes.
11. Sterile disposable conical centrifuge tubes.
12. Vital dye: Eosin 0.15%.
13. Paraformaldehyde (PAF): 4% solution is prepared from PAF 32% (1:8 dilution in 1× PBS) and stored at 4°C.
14. Axioskop 2 plus microscope and digital camera.

### 2.3. Transplantation of Porcine Neuroblasts into the Rat Striatum

1. Lewis 1A adult rats (Janver, Le Genest-Saint-Isle, France).
2. Neuroblasts isolated from porcine foetal mesencephalons.
3. Stereotaxic frame.
4. Ten-microlitre Hamilton syringe.
5. Automated microinjector (Model KDS 310, Phymep, Paris, France).
6. A drill (Foredom C094369, Phymep).
7. Sewing needle.

### 2.4. Perfusion of the Transplanted Rat

1. Anaesthetic: Mix of 20 mL Ketamine (50 mg/mL) and 5 mL Rompun 2%.

2. PBS at 4°C.
3. 4% PAF at 4°C.
4. Peristaltic pump (Cole Palmer Masterflex Pump Drive easy-load model 7518-10) and 2.5-mm tubing.
5. Blunted cannula (Masterflex, 16 gauge).
6. Cryoprotectant solution: 15 and 30% sucrose in 1× PBS.

### *2.5. Coronal Sectioning of the Frozen Brain*

1. Cryostat with a drive mechanism for moving the head.
2. Isopentane.
3. Freezing mounting medium.
4. Gelatinized slides: Dissolve 5 g of gelatine in 800 mL of distilled water at 40°C. Add 4 g of $KCr(SO_4)2{\cdot}12H_2O$ and complete to 1 L with distilled water. Dip slides into the liquid gelatine for 1 h and let dry the slides at room temperature. Cover and store at 4°C.

### *2.6. Immuno-histochemistry*

1. 3% $H_2O_2$ in PBS.
2. Blocking/permeabilizing solution: 4% bovine serum albumin (BSA), 10% donkey normal serum (DNS), and 0.1% Triton in 1× PBS.
3. Primary antibodies: Polyclonal antibody directed against the tyrosine hydroxylase (TH, 1/500, Peel Freeze) to label catecholaminergic neurons; monoclonal antibody directed against the 70 kDa porcine neurofilament subunit (NF70, 1/1000, University of Paris VII) to label porcine neurons.
4. Secondary antibodies: Anti-mouse IgG or anti-rabbit IgG.
5. Avidin-biotinylated enzyme complex (ABC) kit.
6. Diaminobenzidine (DAB) or very intense purple (VIP) staining kit.
7. Ethanol and xylene.
8. Round coverslips (diameter 20 mm).
9. Eukitt quick-hardening mounting medium.

## 3. Methods

### *3.1. Dissection of the Porcine Foetal Mesencephalon*

1. Perform hysterectomy on pregnant sow, 28 days after artificial insemination (G28).
2. Collect the foetuses in HBSS.
3. Perform the dissection on a bench or in a horizontal laminar flow hood, as part of the dissection requires a binocular dissecting microscope.

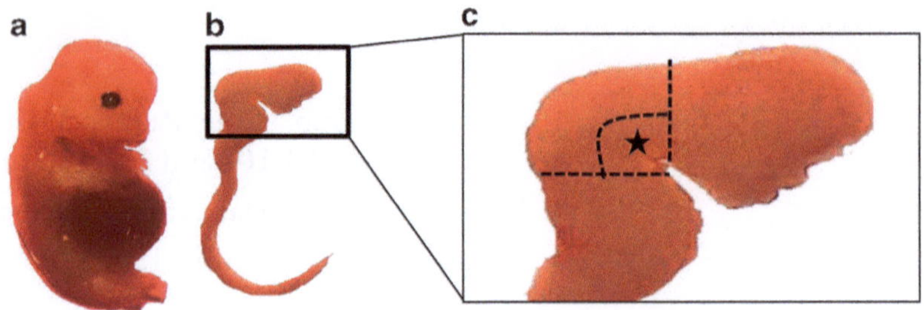


Fig. 1. Porcine ventral mesencephalon dissection. (**a**) G28 porcine embryo. (**b**) Central nervous system. (**c**) *Asterisk*: Ventral mesencephalon.

4. To this end, separate the foetuses from uteri and transfer to 10-cm Petri dishes.
5. Cut the heads with sharp scissors and transfer them into 10-cm Petri dishes filled with 10 mL of HBSS.
6. Remove partly the skull with fine forceps to expose the brain.
7. Remove the meninges with fine forceps.
8. Isolate the mesencephalon (Fig. 1) with fine scissors.
9. Remove the dorsal part of the mesencephalon and transfer the ventral part into a 15-mL conical tube filled with 2 mL of ice-cold HBSS (see Notes 2–4).

### 3.2. Preparation of Neuroblasts from Foetal Porcine Ventral Mesencephalon

Perform the tissue dissociation under sterile conditions in a vertical laminar flow hood as follows:

1. Transfer the VMs into an empty 10-cm Petri dish to cut them into small pieces with a scalpel blade.
2. Transfer minced foetal tissues into a 50-mL conical tube and resuspend in a total volume of 5 mL of HBSS.
3. Add trypsin (final concentration: 0.5 mg/mL) and incubate for 15 min in a 37°C water bath.
4. Add 10 mL of complete medium to inhibit the enzymatic reaction. Let the tube settle for 5 min.
5. Add DNase I (final concentration: 0.1 mg/mL). Incubate for 10 min in a 37°C water bath (see Note 5).
6. Dissociate tissue pieces using a wet 5-mL pipette. Pipet up and down approximatively ten times very carefully and slowly, avoiding air bubbles.
7. Filter the preparation using a sterile 70-μm filter.
8. Centrifuge at 115 ×*g* for 10 min at 20°C.
9. Remove the supernatant carefully.
10. Resuspend the pellet by hitting the tube gently against a hard surface and complete up to 2 mL with HBSS/DNase 0.05%.

11. Count viable cells using a vital dye, such as eosin 0.15%.
12. Centrifuge at 115 ×*g* for 10 min at 20°C.
13. Remove carefully the supernatant and resuspend the cells in HBSS/DNase 0.05% to obtain an amount of 200,000 cells/μL.

***3.3. Transplantation of Porcine Neuroblasts into the Rat Striatum***

1. Anaesthetize Lewis 1A adult rats by intramuscular injection of Rompun/Ketamine (1.6 mL/kg).
2. Place animals into a stereotaxic frame. The incisor bar is set at 3.3 mm above the interaural line.
3. Do an incision in the scalp. Position the needle on the bregma to use it as a reference point (zero). Calculate the injection point coordinates according to a rat brain atlas (25). For intrastriatal transplantation, the injection point is at the following coordinates: anterior +0.7 mm, lateral ±2.8 mm, ventral—6 and 5.6 mm, and incisor bar—3.3 mm.
4. Perform a craniotomy over the injection site using a drill. Remove the bone flap to expose the dura mater.
5. At each injection site, deliver 1 μL of cell suspension with a 10-μL Hamilton syringe (0.8 μL/min) mounted on an automated microinjector.
6. Withdraw gently the syringe 4 min after injection.
7. Replace the bone flap on the skull and carefully sew the skin (see Note 6).

***3.4. Perfusion of the Transplanted Rat***

1. Anaesthetize the rat by intramuscular injection of Rompun/Ketamine (2 mL/kg).
2. Fix the rat on its back on the top of an elevated board and place the support in a large basin for collecting effluents (see Note 7).
3. Rapidly perform an incision along the thorax to expose the heart, insert the blunted cannula into the heart apex (left ventricle), make a small incision into the right cardiac auricle for the blood to escape, and switch on the peristaltic pump to perfuse the rats with 100 mL of physiological serum (0.9% NaCl).
4. Fix the animal by perfusing the rats with 250 mL of cold 4% PAF.
5. Control so that the neck is very rigid.
6. Cut the head and carefully remove the fixed brain.
7. Postfix the brain by immersing the brain for 1 h in cold PAF.
8. Replace PAF by a cryoprotection solution (15% sucrose in PBS) and immerse the brain until its complete submersion (1 or 2 days).

9. Achieve cryoprotection by soaking the brain in 30% sucrose and again, wait for its complete submersion at the bottom of the tube.
10. Freeze the brain by dipping it in a solution of isopentane previously cooled down at –40°C on dry ice.
11. Wrap the brain in aluminium foil and transfer it into a 50-mL Falcon tube for long-term storage at –80°C.

### *3.5. Coronal Sectioning of the Frozen Brain*

1. Set the temperature of the cryostat at –25°C.
2. Place the frozen brain in the cryostat chamber for at least 30 min to bring it to –25°C.
3. Cut coronally the caudal part of the brain with a blade and mount the brain on the specimen disc with freezing mounting medium.
4. Place the specimen disc in the specimen head, orient the brain to get symmetric section of the left and right striata, and adjust the plane of the specimen.
5. Section the brain into 16-μm slices and collect serial sections onto gelatinized slides.
6. Store the slides at –80°C until required.

### *3.6. Immunohistochemistry*

1. Thaw the brain slices at room temperature, and wash them three times in PBS.
2. Incubate the slides for 10 min in 3% $H_2O_2$ to block endogenous peroxidase.
3. Wash three times with PBS.
4. Saturate the non-specific sites by immersing the slides for 45 min in the blocking/permeabilizing solution.
5. After removal of the blocking/permeabilizing solution, add anti-NF70 monoclonal antibody to label the porcine neurons or anti-TH polyclonal antibody to stain the catecholaminergic neurons.
6. After an overnight incubation at 4°C, wash three times with PBS, 5 min each time.
7. Add anti-rabbit or anti-mouse IgG diluted in 1× PBS.
8. After a 2-h incubation at room temperature, wash three times with PBS and amplify the signal using the ABC kit.
9. After a 1-h incubation at room temperature, wash three times with PBS and reveal the signal using as substrate the DAB kit or the VIP kit according to the manufacturer's instructions.
10. Stop the reaction by incubating the slides in distilled water for 5 min.

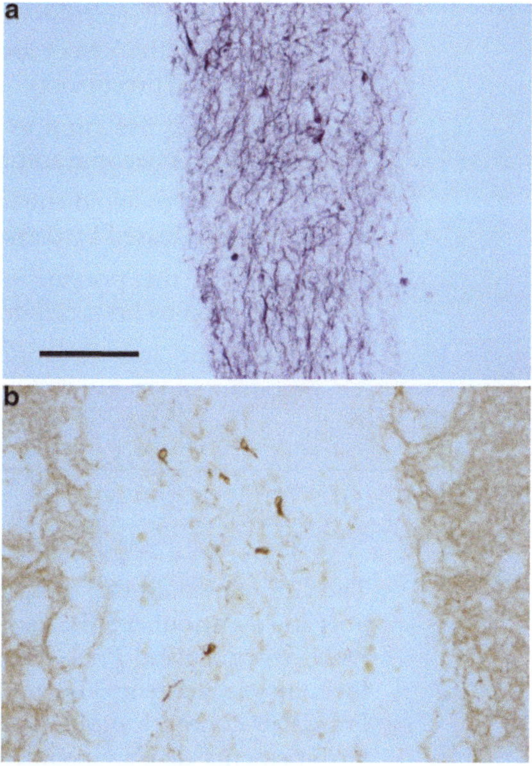


Fig. 2. Immunohistochemistry performed on rat brain 28 days post transplantation with NF70 (**a**) and tyrosine hydroxylase (**b**) antibodies. Scale bar: 100 μm.

11. Dehydrate slices through ethanol series (70, 90, and 95% ethanol, 2 min each) and immerse the slides in xylene for at least 10 min.
12. Fix coverslips on the top of brain slices using the Eukitt quick-hardening mounting medium.
13. Analyze the slides using a phase-contrast microscope.
14. Acquire micrographs using a digital camera (AxioCam HRC, Zeiss) driven by AxioVision Release 4.2 software (Fig. 2).

## 4. Notes

1. Before using the serum, test it to control its innocuity on porcine neural cells.
2. Monitor carefully the dissection of the VM to limit the number of serotonergic neurons in the graft cell preparation as the serotonergic neurons present in the pontine raphe region exacerbate L-DOPA-induced dyskinesia.

3. The time of cell rejection may vary since the graft cell preparation derived from the VM of foetal pig is heterogeneous (mix of glial and neuronal precursors).
4. Perform the dissection as quickly as possible (within 2 h) as foetal tissues become soft and sticky over time.
5. Add DNase to avoid the formation of viscous medium due to the DNA released by damaged cells.
6. Transplant the porcine neuroblasts within 3 h after tissue dissociation to limit cell death.
7. Perfuse the rats in a fume hood to avoid inhaling PAF vapours.

## Acknowledgments

The authors would like to thank Dr. P. Brachet for his steady support and contribution. We also gratefully acknowledge Dr. Vanhove, Dr. I. Anegon, and Pr. J-P Soulillou for their helpful advices and their encouragement. The work was supported by the "Association Française contre les Myopathies" (AFM), the "Fédération des Groupements de Parkinsoniens", and Centaure. X. Lévêque was supported by fellowships from CECAP and Progreffe.

### References

1. Isacson O, Deacon T (1997) Neural transplantation studies reveal the brain's capacity for continuous reconstruction. Trends Neurosci 20:477–482
2. Herman JP, Abrous ND (1994) Dopaminergic neural grafts after fifteen years: results and perpectives. Prog Neurobiol 44:1–35
3. Kordower JH, Freeman TB, Snow B et al (1995) Neuropathological evidence of graft survival and striatal reinnervation after the transplantation of fetal mesencephalic tissue in a patient with Parkinson's disease. N Engl J Med 332:1118–1124
4. Lindvall O, Sawle G, Widner et al (1994) Evidence for long-term survival and function of dopaminergic grafts in progressive Parkinson's disease. Ann Neurol 35:172–180
5. Piccini P, Brooks DJ, Bjorklund A et al (1999) Dopamine release from nigral transplants visualized in vivo in a Parkinson's patient. Nat Neurosci 2:1137–1140
6. Hagell P, Piccini P, Bjorklund A et al (2002) Dyskinesias following neural transplantation in Parkinson's disease. Nat Neurosci 5:627–628
7. Carlsson T, Carta M, Winkler C, Bjorklund A, Kirik D (2007) Serotonin neuron transplants exacerbate L-DOPA-induced dyskinesias in a rat model of Parkinson's disease. J Neurosci 27:8011–8022
8. Olanow CW, Gracies JM, Goetz CG et al (2009) Clinical pattern and risk factors for dyskinesias following fetal nigral transplantation in Parkinson's disease: a double blind video-based analysis. Mov Disord 24:336–343
9. Freeman TB, Wojak JC, Brandeis L et al (1988) Cross-species intracerebral grafting of embryonic swine dopaminergic neurons. Prog Brain Res 78:473–477
10. Galpern WR, Burns LH, Deacon TW, Dinsmore J, Isacson O (1996) Xenotransplantation of porcine fetal ventral mesencephalon in a rat model of Parkinson's disease: functional recovery and graft morphology. Exp Neurol 140:1–13
11. Deacon TW, Pakzaban P, Burns LH, Dinsmore J, Isacson O (1994) Cytoarchitectonic development, axon-glia relationships, and long distance axon growth of porcine striatal xenografts in rats. Exp Neurol 130:151–167

12. Isacson O, Deacon TW, Pakzaban P et al (1995) Transplanted xenogeneic neural cells in neurodegenerative disease models exhibit remarkable axonal target specificity and distinct growth patterns of glial and axonal fibres. Nat Med 1:1189–1194
13. Isacson O, Deacon TW (1996) Specific axon guidance factors persist in the adult brain as demonstrated by pig neuroblasts transplanted to the rat. Neuroscience 75:827–837
14. Deacon T, Schumacher J, Dinsmore J et al (1997) Histological evidence of fetal pig neural cell survival after transplantation into a patient with Parkinson's disease. Nat Med 3: 350–353
15. Fink JS, Schumacher JM, Ellias SL et al (2000) Porcine xenografts in Parkinson's disease and Huntington's disease patients: preliminary results. Cell Transplant 9:273–278
16. Schumacher JM, Ellias SA, Palmer EP et al (2000) Transplantation of embryonic porcine mesencephalic tissue in patients with PD. Neurology 54:1042–1050
17. Krystkowiak P, Gaura V, Labalette et al (2007) Alloimmunisation to donor antigens and immune rejection following foetal neural grafts to the brain in patients with Huntington's disease. PLoS One 2:e166
18. Remy S, Canova C, Daguin-Nerriere et al (2001) Different mechanisms mediate the rejection of porcine neurons and endothelial cells transplanted into the rat brain. Xenotransplantation 8:136–148
19. Michel DC, Nerriere-Daguin V, Josien R et al (2006) Dendritic cell recruitment following xenografting of pig fetal mesencephalic cells into the rat brain. Exp Neurol 202:76–84
20. Melchior B, Nerriere-Daguin V, Degauque N et al (2005) Compartmentalization of TCR repertoire alteration during rejection of an intrabrain xenograft. Exp Neurol 192:373–383
21. Melchior B, Remy S, Nerriere-Daguin V et al (2002) Temporal analysis of cytokine gene expression during infiltration of porcine neuronal grafts implanted into the rat brain. J Neurosci Res 68:284–292
22. Duan WM, Westerman MA, Wong G, Low WC (2002) Rat nigral xenografts survive in the brain of MHC class II-, but not class I-deficient mice. Neuroscience 115:495–504
23. Michel-Monigadon D, Nerriere-Daguin V, Leveque X et al (2010) Minocycline promotes long-term survival of neuronal transplant in the brain by inhibiting late microglial activation and T-cell recruitment. Transplantation 89:816–823
24. Martin C, Plat M, Nerriere-Daguin V et al (2005) Transgenic expression of CTLA4-Ig by fetal pig neurons for xenotransplantation. Transgenic Res 14:373–384
25. Paxinos G, Watson C (1986) The rat brain in stereotaxic coordinates. Academic Press Inc, London

# Chapter 15

## Hepatocyte Xenotransplantation

**Katia R.F. Lima-Quaresma, Andre Gustavo Bonavita, Matheus Kafuri Cytrangulo, Marcelo Alves Pinto, and Luiz Anastácio Alves**


### Abstract

Xenotransplantation of hepatocytes is a future promise to treat liver diseases when there is a formal indication for transplantation. In this chapter, we describe techniques for hepatocyte xenotransplantation. The process was divided into three main steps: hepatocyte isolation, transplantation, and identification of donor cells in the recipient. Tips for each procedure are described at the notes section at the end of this chapter.


**Key words:** Hepatocyte, Xenotransplantation, Protocols, Liver failure, Cell therapy

## 1. Introduction

Cell transplantation has been proposed as a promising method to support patients of several diseases, offering some advantages over whole organ transplantation. Many investigators are carrying out studies with different cell types, such as stem cells, adult allogeneic cells, or xenogeneic cells which may be a bridge for patients on the waiting list for an organ transplant (1–3).

Concerning liver diseases, hepatocyte transplantation has been suggested as an option instead of whole liver transplant, being a less invasive procedure and capable of being performed repeatedly (4–6). However, human hepatocytes are difficult to be obtained and have limited sources (4). Therefore, other cellular sources as adult cells from other species have been suggested. Pig hepatocytes appear to be the best cells for human treatment due to the similarity of metabolism between these species (7). Some studies have shown that the rejection problem could be minimized by different ways, such as the use of immunossupressant therapy, cellular encapsulation, and selection of transgenic pigs with less antigens or

Cristina Costa and Rafael Máñez (eds.), *Xenotransplantation: Methods and Protocols,* Methods in Molecular Biology, vol. 885, DOI 10.1007/978-1-61779-845-0_15, 

inhibitors of complement system (5). The use of cells, tissues, and organs from other species for transplantation, called xenotransplantation, has been pointed as a potential solution to overcome the problem of the severe shortage of transplant donors for liver diseases, such as acute liver failure and other liver-based inherited metabolic disorders.

Hepatocyte xenotransplantation from swine offers additional advantages compared with organ transplantation since xenogeneic cells might not be susceptible to certain human viruses such as HIV, as well as they offer a way for "gene delivery" that could overcome some current hurdles of gene therapy. Thus, genetic changes could be carried out at the level of egg/embryo genome (8).

The method described here aims to obtain fresh hepatocytes from pig liver tissue modified by standard collagenase perfusion methodology described by Berry et al. (9) and Seglen (10). Studies involving hepatocyte xenotransplantation are normally conducted in animal models. Here, we present a protocol for xenotransplantation of swine liver cells into mice. In order to verify if the liver function recovers thanks to cells successfully implanted, we describe a simple method for hepatocyte localization in transplanted animals. In addition, swine hepatocytes can be readily available in enough number to change the clinical status of the patient on a short-term basis.

## 2. Materials

### 2.1. Hepatocyte Isolation

1. HEPES/EDTA solution (S1): 50 mM HEPES, 20 mM EDTA in pure water, and pH 7.4.
2. Cell culture medium (S2): Cells are maintained in William's medium solution supplemented with 10% fetal bovine serum (FBS), 10 μg/mL transferrin, 10 U/mL insulin, 50 ng/mL epidermal growth factor, 100 U/mL penicillin, 100 μg/mL streptomycin, 40 μg/mL dexamethasone, 1.3 μg/mL hydrocortisone, 20 ng/mL hepatocyte (liver) growth factor, and 7 μg/mL acid linoleic.
3. Digesting solution (S3): This solution is prepared on the day of the experiment and contains 75 mg/mL collagenase and 1 mM $CaCl_2$ in HEPES buffer solution. Filter with 0.22-pore membrane. For details, see Note 1.
4. Phosphate-buffered saline (PBS).
5. Percoll® gradient (S4): To prepare the continuous Percoll® gradient, we first dilute it to a density of 1.076 g/mL as follows: 9 mL of Percoll® in 1 mL of PBS (10×) and filter with a 0.22-pore membrane. After this step, 3 mL of this solution

are transferred into a 15-mL conical tube and 2 mL of William's medium are added. The solution is then kept at room temperature until use.

6. Plate coating solution (S5): Collagen stock solution from rat tail at 1 μg/μL in 0.1 M acetic acid. Dilute collagen in 1× PBS to obtain 0.05 μg/μL ready for use. Store at 4°C.
7. Trypan blue and Neubauer chamber.
8. Equipment: Perfusion pump, surgical scissors, 70–100-μm cell strainer, light microscope, centrifuge, and $CO_2$ cell culture incubator.

### 2.2. Cell Labeling and Transplantation of Pig Hepatocytes

1. Animals: Balb/c mice weighing 30 g housed in cages with free access to food and water.
2. Anesthesia: Ketamine hydrochloride and xylazine hydrochloride diluted in sterile saline for a dose of 100 and 10 mg/kg, respectively (see Note 5).
3. DiI (1,1′-dioctadecyl-3,3,3′,3′-tetramethylindocarbocyanine perchlorate) at final concentration of 5 μM (see Note 6 for details).
4. A 25-gauge needle connected to a 1-mL insulin syringe to perform cell transplantation.
5. Scalpel, small microvascular clamps, needle holder, and 5–0 suture.

### 2.3. Tissue Analysis

1. Formalin (10%).
2. Tissue Tech (OCT).
3. Hematoxylin and eosin.
4. Equipment: Cryostat, fluorescence and/or confocal microscope, processors for tissue paraffin blocks, and light microspcope.

## 3. Methods

### 3.1. Hepatocyte Isolation

1. After cannulation of a major liver blood vessel seen in the liver fragment, perfuse the tissues with a HEPES/EDTA S1 solution at 37°C using a perfusion pump set to a flow rate of 10 mL/min for a duration of 20 min or until the tissue becomes pale. The time and flow vary according to the liver sample size.
2. Prepare on the day of the experiment the tissue digestion solution (S2) containing collagenase (see Note 1).
3. Digest tissue with the S2 solution at 37°C at a perfusion flow rate of 10 mL/min during 10 min (see Note 2).
4. At the end of this process, perfuse tissue again with solution S1 at 37°C under the same conditions described for the first step.

5. Disrupt and mince the liver capsule using a pair of scissors to release hepatocytes which are filtrated using a cell strainer of 70–100 μm (see Note 3).
6. Centrifuge the cell suspension at 50×*g*, 4°C, for 1 min. Suspend the cell pellet in 1 mL of William's medium solution (S3) (see Note 4).
7. Prepare the continuous Percoll® gradient as described in the Subheading 2 and use it to separate the hepatocytes from other cells and debris. To this end, add 1 mL of cell suspension into the Percoll gradient and centrifuge for 2 min.
8. Harvest the cellular band from the gradient and wash the obtained cells with S2 solution.
9. To estimate cell number and viability, use the standard Trypan blue exclusion technique.
10. For culture, place cells in a 25-cm$^2$ culture flask coated with rat collagen (0.05 μg/μL).

#### 3.2. Cell Labeling and Transplantation of Pig Hepatocytes

1. Anesthetize the animals by an association dosage of ketamine hydrochloride and xylazine hydrochloride diluted in sterile saline for a dose of 100 and 10 mg/kg, respectively (see Note 5).
2. Label the freshly isolated hepatocytes with fluorescent lipophilic dye DiI at 5 μM (see Note 6 for details).
3. Wash the cells twice using 20 mL of S2 solution to remove any excess of fluorescent dye.
4. Before cell transplantation, leave the cells in the $CO_2$ incubator for 30 min to prevent the diffusion of dye to the injected tissue. After this step, centrifuge the cells at 190×*g* and resuspend the pellet in 1 mL.
5. Perform an incision of 2–3 cm on the abdominal wall close to the spleen using a scalpel. Clamp the splenic circulation and inject the pig hepatocytes suspension (1.5–2×10$^6$ cells) labeled with Dil in a volume of 0.3 mL into the spleen tissue. Keep the occlusion for 2–3 min before releasing spleen circulation.
6. Close the abdominal wall in layers and allow the animal to recover.

#### 3.3. Tissue Analysis

1. Sacrifice the transplanted animals using a $CO_2$ chamber. The time point for euthanasia will depend on the experimental design.
2. After opening the peritoneal cavity, remove the spleen and liver and fix fragments of these organs in 10% formalin and/or include in Tissue Tech (OCT).
3. Cut the fragments included in OCT with a cryostat and analyze the material by fluorescence microscopy or confocal microscopy. Xenotransplanted hepatocytes appear labeled in red (if DiI were used).

4. Process the fragments fixed in formalin 10% for histological analysis, stain with hematoxylin and eosin, and analyze by light microscopy.

## 4. Notes

1. Before conducting the experiment, it is suggested to test the enzymatic activity of the collagenase solution. The test can be attained by means of zymography of collagen degradation or other methods.
2. The concentration of collagenase could vary between batch and the tissue to be digested (if you intend to use another species). For a better digestion result, one should perform a dose concentration curve of the enzyme.
3. The cell strainer used in hepatocyte isolation proceedings can be substituted by sterile gauze.
4. If it is desirable to maintain the cells in culture for more days, it is interesting to test the best concentration of FBS since there are differences between manufacturers.
5. The ketamine anesthesia can be substituted by inhalation of isoflurane at a concentration ranging from 0.5 to 5%.
6. There are several fluorescent cell-tracking dyes on the market. We suggest using the lower concentration that guarantees the hepatocyte labeling. This will avoid overspending and false labeling of other cells. Some of these dyes can pass to unlabelled cells in vitro through unspecific mechanisms (11).

### References


1. Aleem KA, Parveen N, Habeeb MA et al (2006) Journey from hepatocyte transplantation to hepatic stem cells: a novel treatment strategy for liver diseases. Indian J Med Res 123:601–614
2. Habibullah CM, Syed IH, Qamar A et al (1992) Human fetal hepatocyte transplantation in patients with fulminant hepatic failure. Transplantation 58:951–952
3. Galvão FH, de Andrade Júnior DR, de Andrade DR et al (2006) Hepatocyte transplantation: state of the art. Hepatol Res 36:237–247
4. Fox IJ, Chowdhury JR (2004) Hepatocyte transplantation. Am J Transplant 4:7–13
5. Bonavita AG, Quaresma K, Cotta-de-Almeida V et al (2010) Hepatocyte xenotransplantation for treating liver disease. Xenotransplantation 17:181–187
6. Alves LA, Bonavita A, Quaresma K et al (2010) New strategies for acute liver failure: focus on xenotransplantation therapy. Cell Med B Cell Transplant 1:47–54
7. Soucek P, Zuber R, Anzenbacherová E et al (2001) Minipig cytochrome P450 3A, 2A and 2C enzymes have similar properties to human analogs. BMC Pharmacol 1:11
8. Kanazawa A, Platt JL (2000) Prospects for xenotransplantation of the liver. Semin Liver Dis 20:511–522
9. Berry MN, Friend DS (1969) High-yield preparation of isolated rat liver parenchymal cells: a biochemical and fine structural study. J Cell Biol 43:506–520
10. Seglen PO (1976) Preparation of isolated rat liver cells. Meth Cell Biol 13:29–83
11. Fonseca PC, Nihei OK, Savino W et al (2006) Flow cytometry analysis of gap junction-mediated cell-cell communication: advantages and pitfalls. Cytometry A 69:487–493

# Chapter 16

## In Vitro Repair Model of Focal Articular Cartilage Defects in Humans

**Díaz Prado SM, Fuentes-Boquete IM, and Blanco FJ**

### Abstract

Articular cartilage lesions, which do not affect the integrity of subchondral bone, are not able to be repaired spontaneously, thus inducing cartilage degeneration and developing an arthrosic process. To avoid the need for prosthetic replacement, different cell treatments were developed with the aim of generating a repaired tissue with structure, biochemistry composition, and functional behavior equal or similar to those of natural articular cartilage.

The following protocols describe the methods for harvesting articular cartilage explants both from pig and human specimens and isolating and culturing pig chondrocytes. Moreover, the methodology for an in vitro model of xenoimplant of pig chondrocytes in focal defects of human articular cartilage is described.

**Key words:** Articular cartilage, Xenoimplant, Chondrocytes, Cell therapy, Chondral defects, Cartilage explants

## 1. Introduction

The in vitro repair model of focal articular cartilage defects in humans has some advantages compared with experimental animal models or autologous chondrocyte implantation (ACI) in patients. Because of anatomical and physiological differences between experimental animals and humans, there is no animal model of articular cartilage repair that accurately represents the human diarthrodial joint (1). Moreover, the histological evaluation of in vivo implants of autologous chondrocytes requires the collection of repair tissue biopsies, which is an added damage to the joint.

The in vitro models allow not only analyzing the influence in the process of chondrogenesis of different growth factors, such as insulin-like growth factor 1 (IGF-1), members of the transforming growth factor (TGF) superfamily (such as the TGFβ), bone morphogenetic

Cristina Costa and Rafael Máñez (eds.), *Xenotransplantation: Methods and Protocols,* Methods in Molecular Biology, vol. 885, DOI 10.1007/978-1-61779-845-0_16, © Springer Science+Business Media, LLC 2012

proteins (BMPs), growth differentiation factor 5 (GDF5), fibroblast growth factor (FGF), platelet-derived growth factor (PDGF), and cartilage-derived morphogenetic proteins (CDMPs), but also determining the influence of exhibition time to these factors.

The quality of the repaired tissue in the in vitro model is limited by the source cell, the absence of biochemical and biomechanical stimuli, and the time of cultivation. In the in vitro model, the cell source is limited to pig chondrocytes. Instead, in the in vivo model, various types of cells as native cartilage chondrocytes, implanted chondrocytes or stem cells, and stem cells from periosteum, perichondrium, and bone marrow may participate in repairing (2). On the other hand, the absence of biomechanical stimuli may cause fibrillation of the articular surface, decreased thickness of articular cartilage, proteoglycan loss, increased water content, and alteration of the collagen network (3). Furthermore, the biochemical signals from diarthrodial joint tissues, such as synovial membrane, and from the osteochondral explants are absent in our in vitro model. With regard to cultivation time, it is well known that repair tissue acquires more features of the articular cartilage with increasing cultivation time (4).

Xenotransplantation of chondrocytes is a therapeutic alternative to the ACI and the allotransplantation of chondrocytes. This procedure may resolve problems like the need of an additional surgical intervention to obtain cartilage explants, adding to the articular cartilage damage that increases the osteoarthritic process; the reduction of cell proliferation and chondrogenic capacity with aging; cell-culture long times (3–6 weeks) to obtain a sufficient number of cells for implantation; and decreased survival and proliferation of chondrocytes caused by the storage of cartilage or chondrocytes by cryopreservation (5).

The immune barrier is an important objection to the use of xenotransplantation and allotransplantation of chondrocytes. Isolated chondrocytes result in immunogenic reaction, but alloimplantation of chondrocytes encapsulated in their extracellular matrix (6) or embedded in collagen gel or agarose (2, 7) resulted in few or no rejection reactions. Notably, xenotransplantation in vivo of cultured pig chondrocytes into rabbit chondral defects closed with periosteal membrane showed no signs of infiltration by immune cells (8).

Here, we describe the model of xenoimplant in vitro of pig chondrocytes in focal defects of human articular cartilage.

## 2. Materials

### 2.1. Harvest of Pig Articular Cartilage Explants

1. The rear pig legs used must be harvested in aseptic conditions following the "Ratification instrument of the European agreement about protection of the vertebral animals used for experimental purposes" (see Note 1).
2. Sterile stainless-steel disposable scalpels (no. 22 or equivalent).
3. 70% Ethanol.

4. Sterile 60-mL specimen container with attached screw cap.
5. Dissecting forceps.
6. Laminar air flow cabinet.

#### 2.2. Isolation of Pig Chondrocytes

1. Materials defined in Subheading 2.1.
2. Dulbecco's modified Eagle's medium (DMEM).
3. Fetal bovine serum (FBS).
4. Penicillin/streptomycin (p/s): 5,000 U/mL penicillin and 5,000 μg/mL streptomycin.
5. Trypsin–EDTA solution (10×): 5 g/L porcine trypsin, 2 g/L EDTA.4Na in 0.9% NaCl.
6. Trypsin–EDTA solution (1×) (v/v) in DMEM with 1% (v/v) p/s without FBS. This solution is sterilized by passing through a 0.22-μm pore size filter.
7. Clostridial collagenase type IV.
8. Nylon mesh filter (size: 33 mm; pore size: 0.22 μm).
9. 100-mm Sterile plastic Petri dishes.
10. 0.4% Trypan blue solution. Store in dark bottle.
11. Basic medium: DMEM supplemented with 1% (v/v) p/s.
12. Extraction medium: Basic medium supplemented with 5% (v/v) FBS and 2 mg/mL clostridial collagenase type IV. This extraction medium is sterilized by passing through a 0.22-μm pore size filter.
13. Expansion medium: Basic medium supplemented with 10% (v/v) FBS.
14. Physiologic saline solution with 1% p/s.
15. Hemocytometer.
16. Sterile 50-mL centrifuge tubes.
17. Sterile Pasteur pipettes.
18. Sterile syringe.
19. Graduated pipettes and micropipettes.
20. Centrifuge.
21. Orbital shaker at 37°C.
22. Incubator at 37°C and humidified 5% $CO_2$ atmosphere.
23. Inverted optical microscope.
24. Sterile 20-mL conical-bottom tubes.

#### 2.3. Culture of Pig Chondrocytes

1. Materials defined in Subheading 2.2.
2. Trypsin–EDTA solution (2×) (v/v) in saline serum with 1% (v/v) p/s. This solution is sterilized by passing through a 0.22-μm pore size filter.
3. 162-$cm^2$ Culture flasks.

#### 2.4. Producing Chondral Defects and Xenoimplants

1. Human articular cartilage, with no history of joint disease, obtained in aseptic conditions from the femoral condyles and tibial plate or hip joint of human donors within 48 h of post-mortem (see Note 2).
2. Dental drill (2-mm diameter) (Gebr. Brasseler Gmbh & Co. KQ, Lemgo, Germany).
3. Microrrotor (EWL K9, KAVO, Biberach, Germany) (rotational speed: 1,000 rpm).
4. Cellular suspension: $2 \times 10^5$ pig chondrocytes resuspended in 5 mL of complete medium (10% FBS).
5. 96-Well plates.
6. 6-mm Dermal punch (Stiefel Laboratories, Milan, Italy).
7. Physiologic saline solution.
8. Sterile needle.
9. Sterile stainless-steel disposable scalpels (no. 22 or equivalent).
10. 100-mm Sterile plastic Petri dishes.
11. Complete culture medium: DMEM supplemented with 10% FBS and 1% p/s.
12. Plate centrifuge.
13. Incubator at 37°C and humidified atmosphere of 5% $CO_2$/95% air.
14. Laminar air flow cabinet.

## 3. Methods

#### 3.1. Harvest of Pig Articular Cartilage Explants

The following procedure describes a widely used method for excising articular cartilage from freshly obtained pig knee joint (see Note 1). Analogous dissection procedures are used for freshly obtained bovine and rabbit shoulder, knee, and hip joints. In humans, cartilage is similarly excised from hip and knee specimens (see Note 2). Cartilage should be dissected within 48 h after donation and frozen cartilage cannot be used since it does not yield viable cells (9). Carry out all procedures in a laminar flow cabinet.

1. Wash the skin surface with water to remove dirt and then clean with ethanol. Spray tissue with 70% ethanol before incision.
2. Make a large longitudinal incision in the skin with a disposable scalpel. Continue the incision at each end in the transverse direction until femoral quadriceps tendon and the anterior joint capsule are exposed. Make the whole process taking care not to pierce the joint capsule.
3. Make a transverse incision in the femoral quadriceps tendon and the anterior joint capsule with a sterile stainless-steel disposable

scalpel (no. 22 or equivalent). Continue the incision at each end in the longitudinal direction until the joint is exposed. Carefully cut out the cruciate ligaments and open the joint fully. Make the whole process taking care not to cut the cartilage.

4. Aseptically remove cartilage slices, not including mineralized cartilage or subchondral bone, from pig diartrodial joints using sterile stainless-steel disposable scalpels (no. 22 or equivalent).
5. Transfer the cartilage slices to a sterile 60-mL specimen container with attached screw cap containing basic medium.

#### *3.2. Isolation of Pig Chondrocytes*

Carry out all procedures in a laminar flow cabinet.

1. Mince pig cartilage slices with a sterile stainless-steel disposable scalpel (no. 22 or equivalent).
2. Treat cartilage slivers for 10 min at 37°C with 1× (v/v) trypsin in basic medium, under agitation.
3. Remove supernatant.
4. Add a digestion buffer containing extraction medium.
5. Incubate from 2 to 8 h at 37°C in an orbital shaker (until complete digestion).
6. Centrifuge the digest for 10 min at 250×*g* at room temperature.
7. Remove supernatant.
8. According to pellet volume, resuspend the pig chondrocytes by pipetting up and down in expansion medium. The precise volume is not critical (~2 mL of medium per hoof).
9. Determine the cell number using a hemocytometer.

#### *3.3. Culture of Pig Chondrocytes*

Carry out all procedures into laminar flow cabinet.

1. Seed freshly isolated chondrocytes in a 162-$cm^2$ culture flask in expansion medium.
2. Expand cells in an incubator at 37°C in a humidified atmosphere of 5% $CO_2$.
3. Change the expansion medium every 3–4 days, until cells are almost confluent.
4. After first confluence (14–21 days of culture), split the chondrocyte culture. For this purpose, remove expansion medium. Wash cell monolayers twice with physiologic saline solution and dissociate them by treatment at 37°C for 2–3 min with a trypsin–EDTA solution (2×) (v/v) in saline serum with 1% (v/v) p/s. Block enzyme activity by adding extraction medium.
5. Transfer cell suspension to a 50-mL centrifuge tube and wash carefully the culture flask in order to remove all the cells. Centrifuge chondrocytes at 300×*g* for 10 min.

6. Remove supernatant; the cell pellet obtained will be known as passage 0 or S0 (subculture zero).
7. Resuspend cells in expansion medium and seed on culture flasks. The number of viable cells (adjusted by trypan blue exclusion) should be routinely >95%.
8. Expand again chondrocytes until almost confluence, as described above, and repeat the disassociation procedures to obtain S1 chondrocytes.
9. Split cells following this protocol to obtain consecutive passage chondrocytes (Sx).

### 3.4. Producing Chondral Defects and Xenoimplants

Carry out all procedures in a laminar flow cabinet.

#### 3.4.1. Harvest of Human Articular Cartilage Explants

1. Spray the human cartilage tissue obtained from the hospital with 70% ethanol before incision.
2. Aseptically slice cartilage in full thickness, not including mineralized cartilage or subchondral bone, from human diartrodial joints using new sterile stainless-steel disposable scalpels (no. 22 or equivalent) to take cartilage slivers. Make the cut following the curve of the joint surface to obtain long, uniformly shaped slices.
3. Place explants in a physiologic saline solution.
4. Transfer the human cartilage slices into a sterile 60-mL specimen container with attached screw cap.
5. Store at 4°C until drilling the focal chondral defects.

#### 3.4.2. Seeding of Pig Chondrocytes in Focal Defects of Human Articular Cartilage

Carry out all procedures in a laminar flow cabinet (Fig. 1).

1. Transfer the human cartilage slices to a 100-mm sterile, plastic Petri dish.
2. Orient the slice of cartilage, placing the surface layer up.
3. Manually immobilize the cartilage slice with the help of a sterile needle.
4. Drill chondral defects of 2-mm diameter and 1-mm deep on cartilage explants beginning over the superficial layer using a dental drill rotating at 1,000 rpm.
5. Using a sterile dermal punch, cut away cartilage discs of 6-mm diameter including centrally the focal defect.
6. Transfer discs of cartilage to 96-well plate, placing each disc into a well.
7. Resuspend $2 \times 10^5$ pig chondrocytes in 5 μL of expansion medium.
8. Deposit carefully over each focal defect the $2 \times 10^5$ pig chondrocyte suspension using a 10-μL micropipette. Control samples comprise discs without cell implantation.

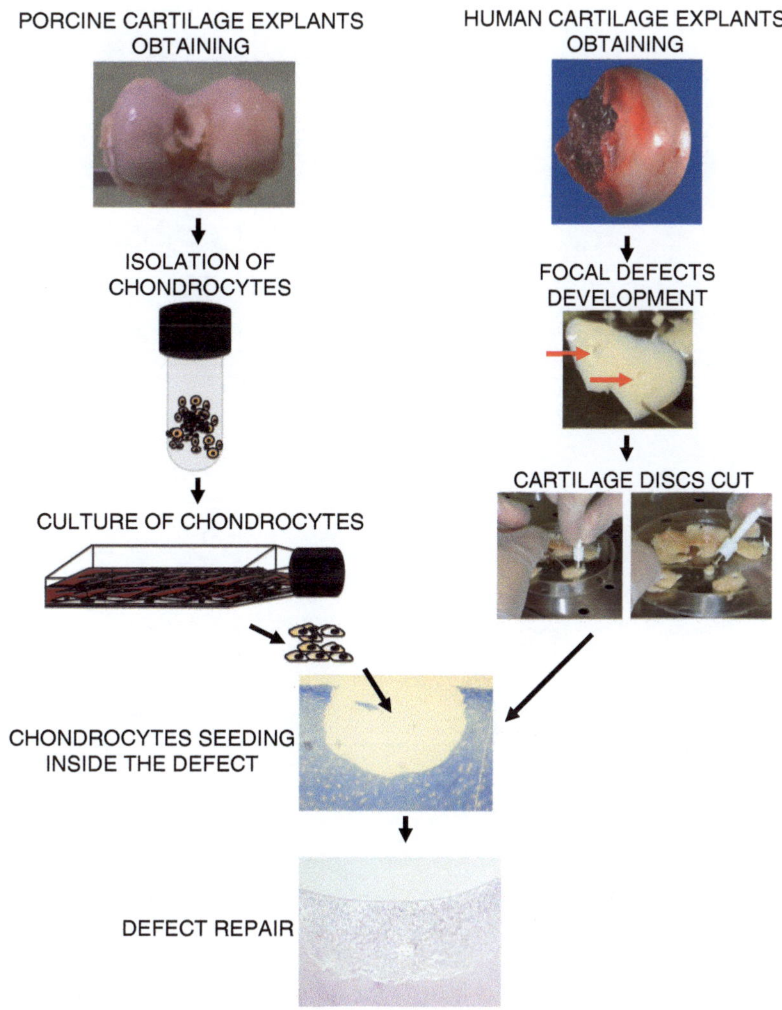


Fig. 1. Schematic representation showing the development of the in vitro repair model of focal articular cartilage defects in humans.

9. Incubate the 96-well plate with the discs, containing the cells inside the defect, for 10 min in an incubator at 37°C in a humidified atmosphere of 5% $CO_2$ (10).
10. Seal with parafilm the 96-well plate containing the implants and transfer to a centrifuge.
11. Centrifuge at $300 \times g$ for 8 min to promote the adhesion of chondrocytes to the injured area of cartilage.
12. Add 150 μL of expansion medium into each well of the plate.
13. Incubate at 37°C in a humidified atmosphere of 5% $CO_2$ for 4, 8, 12, or 16 weeks.
14. Refeed the cultures with fresh expansion medium every 3–4 days.

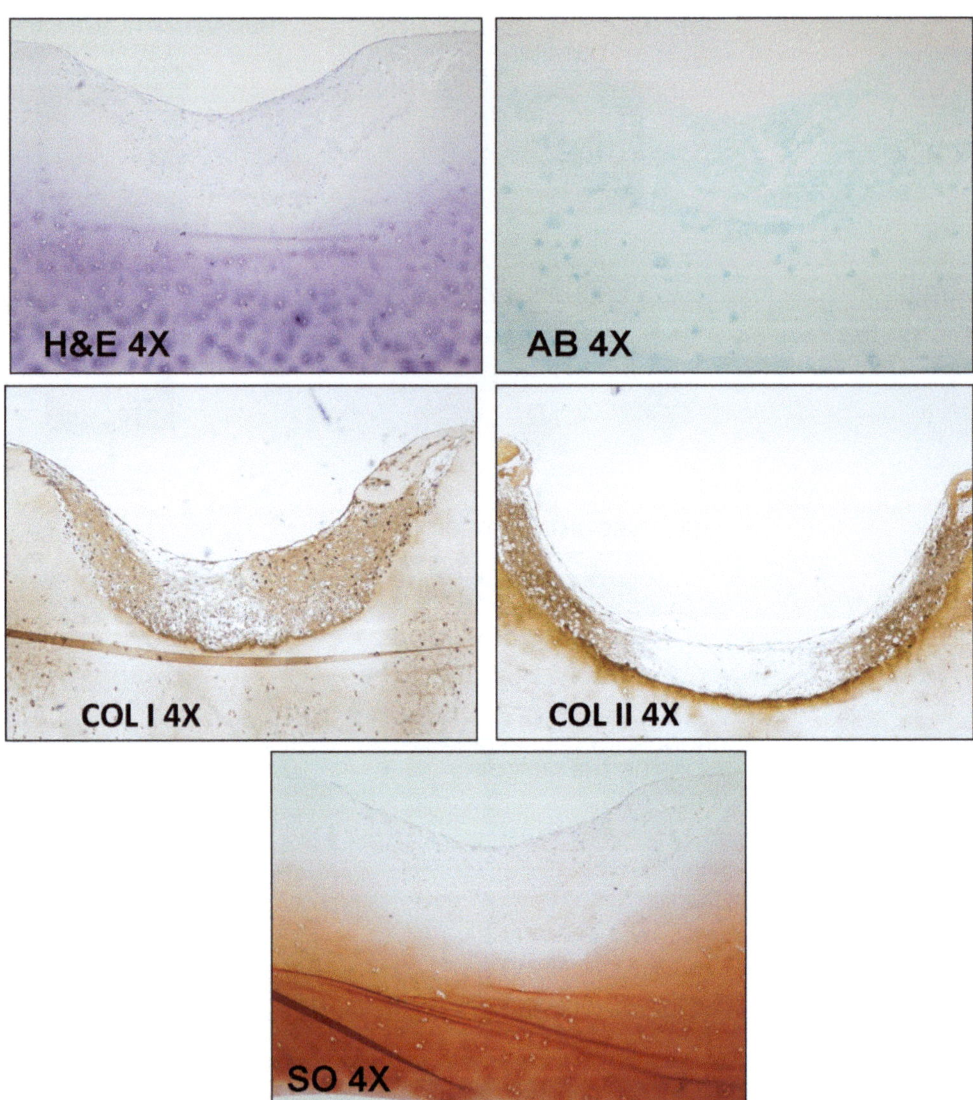


Fig. 2. Histology of the in vitro implantation of pig chondrocytes in human articular cartilage defects after 8 weeks in culture. Hematoxylin–eosin (H&E), alcian blue (AB), type I and II collagens (COL I and COL II), and safranin O (SO) staining.

### 3.5. Assessment of Tissue Repair

In order to determine the presence of characteristic components of hyaline cartilage in the repaired tissue, we perform histochemistry (hematoxylin–eosin, alcian blue, toluidine blue, Masson´s trichrome, and safranin O) and immunohistochemistry (type I and II collagen, chondroitin-4-sulfate, aggrecan, CD46, and integrin β1) stains (Fig. 2). For histological evaluation of repaired tissue, we use an image analysis program named AnalysisD (Olympus, Hamburg, Germany) (see Note 3). For assessing the degree of tissue repair of cartilage defects, we use a semiquantitative histological counting system. In particular, the following parameters are studied for this purpose: tissue morphology, defect

**Table 1**
**Semiquantitative histological counting system for the assessment of repaired tissue of cartilage defects**

| Parameters |
|---|
| Tissue morphology<br>Similar to hyaline cartilage: 2; fibrous tissue: 1 |
| Filling defect<br>75–100%: 3; 50–75%: 2; 25–50%: 1; 0–25%: 0 |
| Regularity of the surface<br>Regular, smooth: 2; irregular: 1 |
| Cellularity<br>High: 2; normal: 1; low: 0 |
| Integration<br>Both edges of the implant well integrated with the native cartilage: 2; one edge of the implant well integrated: 1; no edge of the implant well integrated: 0 |
| Cell morphology<br>Similar to chondrocyte: 1; similar to fibroblast: 0 |
| Matrix staining with safranin O<br>Intense: 3; normal: 2; moderate: 1; without staining: 0 |

filling, regularity of the surface, cellularity, integration, cell morphology, and matrix staining with safranin O. The percentage of repair in each section is calculated by the formula (RTA/CDA) × 100, in which RTA is the repaired tissue area and CDA is the chondral defect area (Table 1).

## 4. Notes

1. Animals must be hosted in an experimental facility with surgery unit (having an official registry number of "Establishments of reproduction, deliveries and users of experimental animals"). All procedures must be carried out according to the "Ratification instrument of the European agreement about protection of the vertebral animals used for experimental purposes" (11).
2. The human study must be approved by the local ethics committee and informed consent must be obtained from each patient. The development of this methodology must be made in accordance with the fundamental principles established in the "Declaration of Helsinki," the "European convention on human rights and biomedicine," the "UNESCO universal declaration on the human genome and human rights," as well as

requirements of country's legislation in the field of biomedical research, protection of personal data, and bioethics.

3. The in vivo repair model of focal articular cartilage defects in humans is based on ACI. This procedure involves obtaining, by arthroscopy, articular cartilage explants from low-weight-bearing areas. Chondrocytes are then isolated and grown in vitro to obtain a sufficient number of cells for implantation. In a second surgical intervention, the cultured chondrocytes are injected into the defect cavity, which is then closed with periosteal membrane from the same patient (5). ACI has several shortcomings; among them, the two most important are that it requires two surgical interventions: the first one to perform the arthroscopy and the second one to implant the chondrocytes. In the in vivo repair model, it is necessary to perform a third surgical intervention to obtain in vivo repair tissue biopsies for histological evaluation. In this way, both in vitro and in vivo models show similarities.

## Acknowledgments

This work was supported by grants: Servizo Galego de Saúde, Xunta de Galicia (PS07/84), Cátedra Bioiberica de la Universidade da Coruña, and Instituto de Salud Carlos III CIBER BBN CB06-01-0040. Silvia Diaz-Prado is beneficiary of an Isidro Parga Pondal contract from Xunta de Galicia, A Coruna, Spain.

### References

1. Reinholz GG, Lu L, Saris DB, Yaszemski MJ, O'Driscoll SW (2004) Animal models for cartilage reconstruction. Biomaterials 25: 1511–1521
2. Rahfoth B, Weisser J, Sternkopf F, Aigner T, von der Mark K, Brauer R (1998) Transplantation of allograft chondrocytes embedded in agarose gel into cartilage defects of rabbits. Osteoarthritis Cartilage 6:50–65
3. Narmoneva DA, Cheung HS, Wang JY, Howell DS, Setton LA (2002) Altered swelling behavior of femoral cartilage following joint immobilization in a canine model. J Orthop Res 20:83–91
4. Fuentes-Boquete I, López-Armada MJ, Maneiro E et al (2004) Pig chondrocyte xenoimplants for human chondral defect repair: an in vitro model. Wound Repair Regen 12:444–452
5. Fuentes-Boquete IM, Arufe Gonda MC, Díaz Prado SM, Hermida Gómez T, de Toro Santos FJ, Blanco FJ (2008) Cell and tissue transplant strategies for joint lesions. Open Transplant J 2:21–28
6. Schreiber RE, Ilten-Kirby BM, Dunkelman NS et al (1999) Repair of osteochondral defects with allogeneic tissue engineered cartilage implants. Clin Orthop Relat Res 367S:382–395
7. Wakitani S, Kimura T, Hirooka A et al (1989) Repair of rabbit articular surfaces with allograft chondrocytes embedded in collagen gel. J Bone Joint Surg Br 71:74–80
8. Ramallal M, Maneiro E, López E et al (2004) Xeno-implantation of pig chondrocytes into rabbit to treat localized articular cartilage defects: an animal model. Wound Repair Regen 12:337–345
9. Liebman J, Goldberg RL (2001) Chondrocyte culture and assay. Curr Protoc Pharmacol. Chapter 12:Unit 12.2. PMID: 21959754
10. Koga H, Shimaya M, Muneta T et al (2008) Local adherent technique for transplanting

mesenchymal stem cells as a potential treatment of cartilage defect. Arthritis Res Ther 10:R84

11. Instrumento de Ratificación del Convenio Europeo sobre protección de los animales vertebrados utilizados con fines experimentales y otros fines científicos, hecho en Estrasburgo el 18 de marzo de 1986 (1990) Disposiciones generales. Jefatura del Estado. Acuerdos Internacionales. Madrid. Boletín Oficial del Estado 256:31348–31362

# Chapter 17

# Potential Zoonotic Infection of Porcine Endogenous Retrovirus in Xenotransplantation

Giada Mattiuzzo, Yasuhiro Takeuchi, and Linda Scobie

## Abstract

Porcine endogenous retrovirus (PERV) is considered the major biosafety issue in xenotransplantation. Several techniques have been employed for the analysis of the PERV status in the animal donor and for the assessment of PERV transmission/infection in the xenograft recipient. In this chapter, methods to assess the expression of PERV and the potential for PERV transmission from a donor animal are described in addition to the identification of relevant loci within the porcine genome.

PERV detection can be carried out using several techniques of which quantitative polymerase chain reaction (PCR) and RT-PCR are the most sensitive. However, other procedures can be employed such as detection of reverse transcriptase activity (i.e. viral replication) in the sample or immunostaining of the infected cells using an anti-PERV antibody. The PERV transmission assay has been described to identify the transmission phenotype of the pig donor, and subsequent risk from a donor. This assay can, therefore, direct the selection of the most suitable animal. Finally, it is important to determine the presence of critical PERV loci involved in transmission in the pig genome and compare between different animals. One of the methods for the analysis of these PERV integration sites is described.

**Key words:** Porcine endogenous retrovirus, Infection assay, SYBR green-based quantitative PCR, Reverse transcriptase assay, In situ immunostaining, Splinkerette

## 1. Introduction

Xenotransplantation guidelines provided by the US Food and Drug Administration (FDA) and the World Health Organization indicate swine as the ideal animal donor (1). Aside from several economical and ethical reasons, the risk for transmission of infectious disease from pigs to humans is considered lower than that of more closely related non-human primates (1–4). Although most known pathogens can be eliminated by breeding donor animals in specific pathogen-free facilities, the presence of the porcine endogenous retrovirus (PERV) constitutes a major safety issue in

Cristina Costa and Rafael Máñez (eds.), *Xenotransplantation: Methods and Protocols,* Methods in Molecular Biology, vol. 885, DOI 10.1007/978-1-61779-845-0_17, © Springer Science+Business Media, LLC 2012

xenotransplantation (5, 6). Upon infection, retroviruses integrate into the host cell genome as a provirus. If this occurs in germ line cells, the provirus can be transmitted vertically to the offspring via Mendelian inheritance (7). Although through evolution most of the retroviral sequences have become defective, some are replication competent and can still generate infectious PERV which can be potentially transmitted from pigs. All infectious PERVs identified so far belong to a group of gammaretroviruses, PERV γ1. Three subgroups of PERV-γ1 with distinctive *env* genes have been identified in pig genomes (8, 9). PERV-A and PERV-B can infect several species, including human cells, while PERV-C tropism is limited to pig cells (10, 11). However, it has been shown that recombination between PERV-A and PERV-C occurs frequently, producing a high-titre, human tropic PERVA/C (11–13). These recombinant PERVs use the same receptor for cell entry as PERV-A, and they are almost exclusively the form of isolates derived by co-cultivation of porcine primary cells and human cells (12–14). This form is, therefore, considered to be most problematic (5, 6, 15).

PERV diagnostics are extremely important before xenotransplantation (to choose the safest animal donor) and after the procedure (to detect in the patient any sign of PERV transmission). Although it is extremely difficult (if not impossible) to eliminate PERV from the genome of the donor pig, it is possible to select an animal donor exhibiting a defined phenotype previously indicated as non-transmitter or null animal (12, 13, 16–19). These pigs do not produce PERV able to infect human cells (non-transmitter) or human and pig cells (null) *in vitro* (see Note 1). To determine the phenotype of these pigs, PERV transmission methods were employed based on the co-cultivation of activated PBMC derived from the candidate donor pig with human or porcine cells (12, 13, 16–19). These protocols are considered the "gold standard" for analysing the potential for PERV transmission and, therefore, could be employed to test future transgenic pigs engineered to have a reduced level of PERV production; however, this process is laborious and time consuming and alternatives are being sought.

In addition to careful selection of the donor, it is imperative that any patient receiving a xenotransplant has an adequate follow-up protocol as has been recommended (1). This would involve monitoring of any potential PERV infection in the xenotransplant recipient and also for the presence of circulating porcine cells which can complicate the results obtained if accurate diagnostics are not used.

The most sensitive methods for PERV detection are mainly molecular diagnostics, such as polymerase chain reaction (PCR); however, again, the use of accurate primer sets to identify subtypes and also the ability to distinguish true infection from contamination with porcine cellular material are of the utmost importance. Furthermore, the potential PERV infection in the patient can also be tested by looking for the presence of anti-PERV antibodies using Western blot (20) and an ELISA system (21–24). However, as this has been described previously and there has been some indication of cross reactivity, it is unclear if these methods are an accurate indication of a response to an active infection; therefore, these techniques will not be considered within this text (23).

## 2. Materials

### 2.1. Cells Lines and Culture Conditions

1. Human embryonic epithelial 293T cells (25).
2. Pig testis ST-IOWA cells (26).
3. Media for 293T cells: Dulbecco's modified Eagle's medium (DMEM) supplemented with 15% foetal bovine serum (FBS), and 100 U/mL penicillin and 100 μg/mL streptomycin.
4. Media for ST-IOWA cells: DMEM supplemented with 10% FBS, 100 U/mL penicillin and 100 μg/mL streptomycin, and 0.1 mM nonessential MEM amino acids.
5. Media to stimulate PBMC: RPMI-1640 supplemented with 10% FBS, 2.5 μg/mL phytohaemagglutin (PHA, Sigma, St Louis, MO, USA), and 100 U/mL penicillin and 100 μg/mL streptomycin. Add to the media 12-myristate-13-acetate (PMA, Sigma) at 1 ng/mL final concentration to attain full activation.

### 2.2. Quantitative PCR and RT-PCR

1. Mastercycler such as Eppendorf RealPlex 4 and related plastics: 96-well PCR plate and optically clear PCR film.
2. Plasmids to create standards: pCRII-PERVA14/220 (a construct based on vector pCRII-Blunt (Invitrogen, Carlsbad, CA, USA) containing the full-length genome of the high-titre, human tropic PERV-A/C recombinant (27) and the human housekeeping gene 18S ribosomal RNA (rRNA) nucleotide 102–702 (based on sequence GenBank acc. no. M10098.1) cloned into pCRII-Blunt (28).
3. Oligonucleotides (see Note 2): If provided in a lyophilised form, reconstitute in Tris–EDTA buffer (10 mM Tris–HCl, pH 8, 1 mM EDTA) at 0.1 mM concentration. To avoid freezing/thawing, prepare working concentration aliquots of 10 μM. Store at -20°C.
4. Mastermix reagent such as QuantiTect SYBR green PCR kit (Qiagen, Venlo, the Netherlands). Store at –20°C.

### 2.3. PERV Reverse Transcriptase Assay

1. C-type Reverse Transcriptase (RT) activity assay (Cavidi-tech, Uppsala, Sweden). The kit reagents are stored at –20°C, but once reconstituted, lyophilised component must be stored at 4°C and used within a week.

2. Triton X-100.
3. 50-mL Glass beaker.
4. Plastics: 10- and 1-L containers.
5. Multi-channel pipette (volume range 10–200 μL).
6. Incubator cabinet at 33°C.
7. Shaker suitable for 96-well plate, such as an orbital shaker.
8. ELISA plate reader (set up at 405 nm).

### 2.4. Immunostaining with Anti-PERV CA Antibody

1. Polybrene (hexadimethrine bromide, Sigma): It is a cationic polymer used to enhance enveloped virus infection by neutralizing the negative charges on the viral and cellular membrane. Prepare a stock solution at 8 mg/mL in water and sterilize by filtration (0.2-μm filter). A working aliquot can be stored at 4°C for a few weeks. For long storage, –20°C is recommended.
2. Methanol and acetone mixed 1:1 in a glass bottle and kept at –20°C.
3. Rabbit polyclonal anti-PERV capsid (CA) antibody: It is prepared by immunization of a rabbit with PERV CA as described (29). This antibody can recognise all three PERV subgroups and cross-reacts with murine leukaemia virus (MLV) CA. Store in aliquots (to avoid freeze/thaw cycle) at –20°C.
4. Secondary antibody alkaline phosphatase (AP)-conjugated anti-rabbit IgG. It should be stored at 4°C.
5. Substrate NBT/BCIP (Roche, Basel, Switzerland): It is a mixture of Nitro-blue tetrazolium chloride (NBT) and 5-Bromo-4-chloro-3-indolyl phosphate, toluidine salt (BCIP). Dilute one ready-to-use tablet in 10 mL of milliQ water and mix by inversion of the tube or vortex until all the powder is dissolved. Protect from light and use within 15 min. Store tablets at 4°C.

### 2.5. Splinkerette Protocol

1. Oligonucleotides: The following primers have been designed for the analysis of the flanking sequences of PERV-C (Scobie et al., unpublished data). For integration analysis of any other sub-types, it is only necessary to change the LTR primers to accommodate this.

   For the 5′-LTR:

   HMSpAa: 5′-CGA AGA GTA ACC GTT GCT AGG AGA GAC CGT GGC TGA ATG AGA CTG GTG TCG ACA CTA GTG G-3′.

   HMSpBb: 5′-GAT CCC ACT AGT GTC GAC ACC AGT CTC TAA TTT TTT TTT TCA AAA AAA-3′.

   HMSp1: 5′-CGA AGA GTA ACC GTT GCT AGG AGA GAC C-3′.

M5 (LTR): 5′-CATTTCAACATCTTTATGGCGCAA CCAG-3′.

HMSp2: 5′-GTGGCTGAATGAGACTGGTGTCGA-3′.

IP3-2 (LTR): 5′-TTC ATG CCT AGA GAC ATG TAC TCA G-3′.

M6 (LTR): 5′-CGGAAGTAAAATAGGCCCTGAGTAC-3′.

Her1 (LTR): 5′-GTGAACCCCATAAAAGCTGTC-3′.

2. Annealing buffer: 10 mM Tris, 1 mM EDTA, 150 mM NaCl.
3. *Pfu Turbo* polymerase (Agilent Technologies, Santa Clara, CA, USA) and Taq Core kit (Qiagen).
4. QIAquick PCR kit (Qiagen, see Note 3) and QIAquick Gel extraction kit (Qiagen).
5. pGEM-T Easy kit (Promega, Madison, WI, USA).
6. X2-Blue ultracompetent cells (Agilent Technologies).
7. BigDye Terminator v3.1 Cycle Sequencing Kit and the 3730xl DNA Analyzer (Applied Biosystems, Warrington, UK) or equivalent if sequencing is conducted in house.

## 3. Methods

Porcine retroviruses are labile and sensitive to inactivation even at –70°C storage (30); careless collection and storage of samples can lead to inaccurate results. Whole blood, sera, and tissues from the donor animal are required as would samples from the recipient pre- and post-xenotransplant. These samples should be collected within the guidelines indicated at the time of the procedure as defined by the FDA (1). When possible, preparation of PBMC from porcine whole blood should be carried out immediately and the cells frozen (to maintain viability) and stored under liquid nitrogen for assessment of their phenotype via co-culture. Serum should be stored at –80°C and in aliquots to avoid any freeze–thaw which can reduce the titre of the virus significantly.

### 3.1. PERV Transmission Assay by Co-culture with Porcine Cells (See Note 4)

1. Isolate pig PBMC from blood (see Note 5) and stimulate by adding to the media described for this purpose.
2. Seed $1 \times 10^7$ PBMC and allow to proliferate in 24-well plates for 5 days at 37°C, 5% $CO_2$.
3. Transfer then the PBMC to sub-confluent 25-cm$^2$ flask with target cells (either human 293T cells or porcine ST-IOWA cells) and keep in contact for 5 days.
4. Remove the PBMC and maintain the target cells in culture for up to 65 days (see Note 6).

### 3.2. Detection by Quantitative PCR/RT-PCR

Genomic DNA and viral or total RNA can be extracted from tissue/blood/cells by using commercially available kits, such as DNeasy Blood and Tissue kit, Viral RNA mini kit, or RNeasy mini kit (Qiagen) (see Note 7). The principle of SYBR green-based PCR has been previously described (31).

1. Quantify the concentration of DNA or RNA using a spectrophotometer. The ratio between absorbance at wavelength of 260 and 280 nm should be 1.8–2 for genomic DNA and >2 for RNA.
2. For RNA only: Produce cDNA by using commercially available kit (see Note 8).
3. Prepare the standards for the target genes. For PERV *gag*, the plasmid containing full-length genome of PERVA14/220 can be used (27) (see Note 9). Include standards for a housekeeping gene (e.g. 18S rRNA (28)) to normalise the amount of nucleic acid input among samples.
4. Prepare master mix for each target gene. Calculate the number of samples (run at least in duplicate, better in triplicate) including standards and negative control, and add an extra 20% to the final count. Use at least five points of standards, range between 10 and $10^5$ copies (see Note 10). For each sample, prepare the following reaction: 12.5 μL master mix Quantitect SYBR green PCR kit, 0.75 μL forward primer (10 μM), 0.75 μL reverse primer (10 μM) (see Note 2), and 8.5 μL molecular grade water. Then, aliquot 22.5 μL per well in a 96-well plate.
5. Add the nucleic acid template. For genomic DNA, use 50–200 ng, and for cDNA use the equivalent of 125 ng of total RNA (i.e. one-eighth of the reverse transcription reaction if the normal Quantitect RT kit protocol has been followed). Do not add more than 2.5 μL (≤10% of the final volume of the reaction) of template to the mix.
6. Cover the plate with optically clear adhesive film and briefly spin down.
7. Place the plate in the Eppendorf RealPlex4 mastercycler or equivalent.
8. Input the plate layout using SYBR green as dye and, if using a master mix which contains it, ROX as reference dye.

   Cycling conditions are (for Qiagen Quantitect SYBR green PCR kit) 2 min at 50°C, 10 min at 95°C, and repeat for 40 cycles: 15 s at 95°C, 30 s at 55°C, and 30 s at 72°C, where the fluorescence will be acquired (see Note 11).
9. Check that the standard curves obtained are reliable (i.e. $R^2 > 0.9$, one or two odd points can be deleted if necessary)

and then use them to calculate the copy number of your target gene in each sample. Normalise the results using the copy number of the housekeeping gene.

### *3.3. Detection by Screening for Reverse Transcriptase Activity in the Cell Supernatant*

Productive PERV infection of the target cells can be assessed by measuring PERV RT activity in the cell supernatant using the C-type RT activity assay. This kit can be used quantitatively (the amount of RT activity in each sample) or qualitatively (to screen for RT activity). To detect PERV in the infected cell supernatant, the screening for RT activity protocol is used. The following protocol is designed to screen up to 80 samples in duplicate (80 samples per plate, and there are 2 plates in a kit). The protocol requires 2–3 days to be completed.

1. On day 1, equilibrate an incubator at 33°C and thaw the following solutions: C-type sample reconstitution buffer (C2), C-type RT reaction components (C1, lyophilised), sample dilution buffer (B1), and C-type sample dilution components (B2, lyophilised).
2. Prepare the reaction mixture by adding 12 mL of C2 (Reconstitution Buffer) to each of the two vials of C1 (RT Reaction Components). Mix well by vortexing.
3. Transfer the contents of both C1 vials to a 50-mL glass beaker and add 12 mL of distilled water and mix thoroughly.
4. Prepare the dilution buffer by adding 5 mL of B1 (sample dilution buffer) to each of the two vials of B2 (sample dilution components). Transfer the B2 contents back to the B1 bottle. Repeat once to ensure that all material has been transferred.
5. The reaction mixture will be completed by adding 12 mL of dilution buffer to the C1/water solution and mix well by vortexing.
6. Take out the PolyA plates and add 200 μL of reaction mixture to each well.
7. Seal the plates and incubate at 33°C for 1 h.
8. In the meantime, prepare the standard controls by adding 420 μL of dilution buffer to vial D (MMuLV rRT standard). Mix well by vortexing.
9. Set up 12 tubes for serial dilutions. Leave the first tube empty, and add 250 μL of dilution buffer to the rest. Add 350 μL of MMuLV rRT standard to the first tube. Transfer 200 μL from tube 1 to tube 2. Then, transfer 200 μL from tube 2 to tube 3; keep going until the 12th test tube. Make sure to change the pipette tip after each dilution.
10. Collect 50 μL of each sample in a 96-well plate.
11. Take polyA plates from the incubator and add 10 μL of each sample to both plates until 80 samples are loaded. Reserve

wells A11–D12 for the standards (10 μL of each). Fill wells E12–H12 with 10 μL of dilution buffer.

12. Seal the plates and incubate with gentle agitation at 33°C overnight.
13. On day 2, thaw the following solutions: RT product tracers (O), AP substrate tablets (P1), AP substrate buffer (P2), and C-type plate wash buffer concentrated (E).
14. Prepare the wash buffer by pouring slowly 75 mL of Triton X-100 into 1 L of distilled water in continuous agitation for about 10 min.
15. Add 12 mL of the water/Triton solution to the RT product tracers (O) and vortex. Pool contents in a beaker and stand for at least 30 min.
16. Add 25 mL of solution E (concentrated wash buffer) to a 10-L container and add distilled water up to 10 L. Mix well.
17. Wash the plates. If you have an automatic plate washer, set it up to pass 3 mL per well, per cycle. Between washes, tap plate upside down. Wash twice and leave the plate to dry upside down for 5 min. If you wash manually the plate, fill each well with 200 μL of wash buffer, pour the liquid down into a sink, and tap plate upside down on paper until dry. Leave the plate to dry upside down for 5 min.
18. Add 100 μL of pooled RT product tracer to each well, seal plates, and incubate with agitation at 33°C for 90 min.
19. Prepare AP substrate by adding P1 (AP substrate tablets) to P2 (AP substrate buffer), shake occasionally, and allow 20 min to dissolve. Store at room temperature, protected from light.
20. Wash plates as described in step 17.
21. Add 125 μL AP substrate to each well without touching the bottom of the well. Cover plate and incubate with agitation at room temperature under dark cover.
22. Read absorbance at 405 nm with an ELISA plate reader after 30 min, 2 h, and overnight. It is advisable to take a reading after 15 min in case some produce colour quickly.
23. Absorbance (OD) of the standard controls should be plotted against RT activity and a linear regression should be calculated. By using the formula of the regression line, calculate the value of RT activity for each sample.

### 3.4. Detection by In Situ Immunostaining of PERV-Infected Cells

Production of infectious PERV from primary cells or cell lines can also be tested by the following infection assay.

1. The day prior to infection, seed $3 \times 10^4$ 293T cells per well in 0.5 mL of DMEM 10% FBS in a 48-well plate (see Note 12).

Change the media of the PERV-producing cells. If adherent cells are to be used, they should be at >90% of confluence.

2. Harvest the supernatant from PERV-producing cells and pass through a 0.45-μm filter to avoid cell contamination.
3. Prepare serial dilutions of the supernatant in DMEM supplemented with 10% FBS. PERV titres are normally low; therefore, a two- to threefold dilution is recommended. However, if using high-titre recombinant PERVs, five- to tenfold dilutions could be used. Dilution final volume is 0.5 mL and polybrene at the final concentration 8 μg/mL is to be added.
4. Incubate the cells at 37°C, 5% $CO_2$, for the next 2 days.
5. Remove the media and let the cells dry for 15 min at room temperature.
6. Fix the infected cells using 0.5 mL of cold methanol/acetone. This must be added very slowly to the cells to avoid disruption of the monolayer. Incubate for 10 min.
7. Wash once with 0.5 mL of PBS, added slowly.
8. Block the cells with 0.5 mL of PBS/10% FBS for 5 min.
9. Remove the blocking solution and add 200 μL per well of the anti-PERV CA antibody diluted 1:250 in PBS/2% FBS.
10. Incubate for 1 h.
11. Wash twice. Washes consist in adding 0.5 mL of PBS/2% FBS and then removing it.
12. Add 200 μL of the secondary antibody (AP-conjugated anti-rabbit IgG) diluted 1:250 in PBS/2% FBS and incubate for 1 h.
13. Wash twice with PBS/2% FBS and then twice with PBS.
14. Prepare the substrate NBT/BCIP (see Subheading 2.5) and add 300–500 μL per well.
15. Incubate at least for 5 min, protected from light. Check the cells under the visible light microscope for the appropriate duration of incubation. PERV-positive cells will appear brown. Do not allow the reaction to proceed for more than 10–15 min; otherwise, the dark background will make it difficult to count the positive colonies.
16. Wash once with PBS and add 0.5 mL of PBS.
17. Count positive colonies at the inverted optical microscope. Positive cells will appear as groups of 2–8 cells at the light microscope. This will constitute one colony. Titre will be determined using the following formula:

$$\text{titer (i.u./mL)} = \frac{\text{no. of colonies} \times \text{dilution factor}}{\text{volume of infection (mL)}}$$

### 3.5. PERV Integration Site Identification

For identification of insertion locations, a number of PCR methodologies have been employed including splinkerette, which are a couple of annealed oligonucleotides forming a hairpin structure on one strand (18, 32) (see Note 13 and Fig. 1).

1. Select restriction enzymes (see Note 14) and design primers according to the target sequence. For our studies, we used PERV-C (Genbank AM229312) as a reference sequence and primers were designed for use in both 5′- and 3′-LTR analysis of PERV-C loci and their accompanying flanking sequences.
2. Isolation of genomic DNA from pig cells can be performed using the DNeasy Blood and Tissue kit (Qiagen).
3. *For 5′-LTR analysis*, digest 3–5 μg of genomic DNA with one of the following enzymes: *Bst*YI, *Sau*3AI, or *Bam*HI. Incubate for 2 h with 5–10 U of enzyme at the appropriate temperature. Heat inactivate the enzyme after digestion and check 200 ng of your sample on an agarose gel to indicate efficient digestion.
4. Prepare the splinkerette adaptor by incubating 1.5 μL of 100 pmol of each of the oligonucleotides HMSpAa and HMSpBb in the annealing buffer, final volume of 100 μL for 10 min at 100°C. Allow the oligonucleotides to cool down slowly until they reach room temperature (~2–3 h) and anneal.
5. Ligate the annealed primers to 300 ng of the digested genomic DNA at an approximate molar ratio of 10:1 using 4 U of T4 ligase in ligation buffer for overnight incubation at 16°C in a final volume of 40 μL. Before setting up the ligation, it is optional to heat the DNA to 60°C to release any sticky ends.
6. After heat inactivation conducted at 70°C for 10 min, digest the whole reaction with restriction enzymes specific to remove any internal fragments that could be amplified. Use 10 U of enzyme for 5–6 h (for PERV C, *Nhe*I or *Kpn*I can be used).
7. Purify the reaction using QIAquick PCR kit (Qiagen, see Note 3).
8. Amplify regions flanking the PERV-C LTR sequences with the following nested PCR protocol. Add the following reagents to a mix: 100–200 ng digested DNA, 1 μL of HMSp1 primer (10 μM), 1 μL of M5 primer (10 μM), 1.6 μL of *PfuTurbo* DNA polymerase (2.5 U), and up to 50 μL with molecular grade water. Cycling conditions: two cycles of 94°C for 30 s, 68°C for 30 s, and 72°C for 2 min, and then 32 cycles of 94°C for 15 s, 65°C for 30 s, and 72°C for 2 min.
9. Second-round PCR is performed as follows using a mix of reagents: 2 μL of the previous PCR product, 1 μL of HMSp2 primer 10 μM, 1 μL of IP3-2 primer 10 μM, 0.5 μL of Taq Core kit, and up to 50 μL with molecular grade water. Cycling conditions: 30 cycles of 94°C for 15 s, 60°C for 30 s, and 72°C for 2 min.

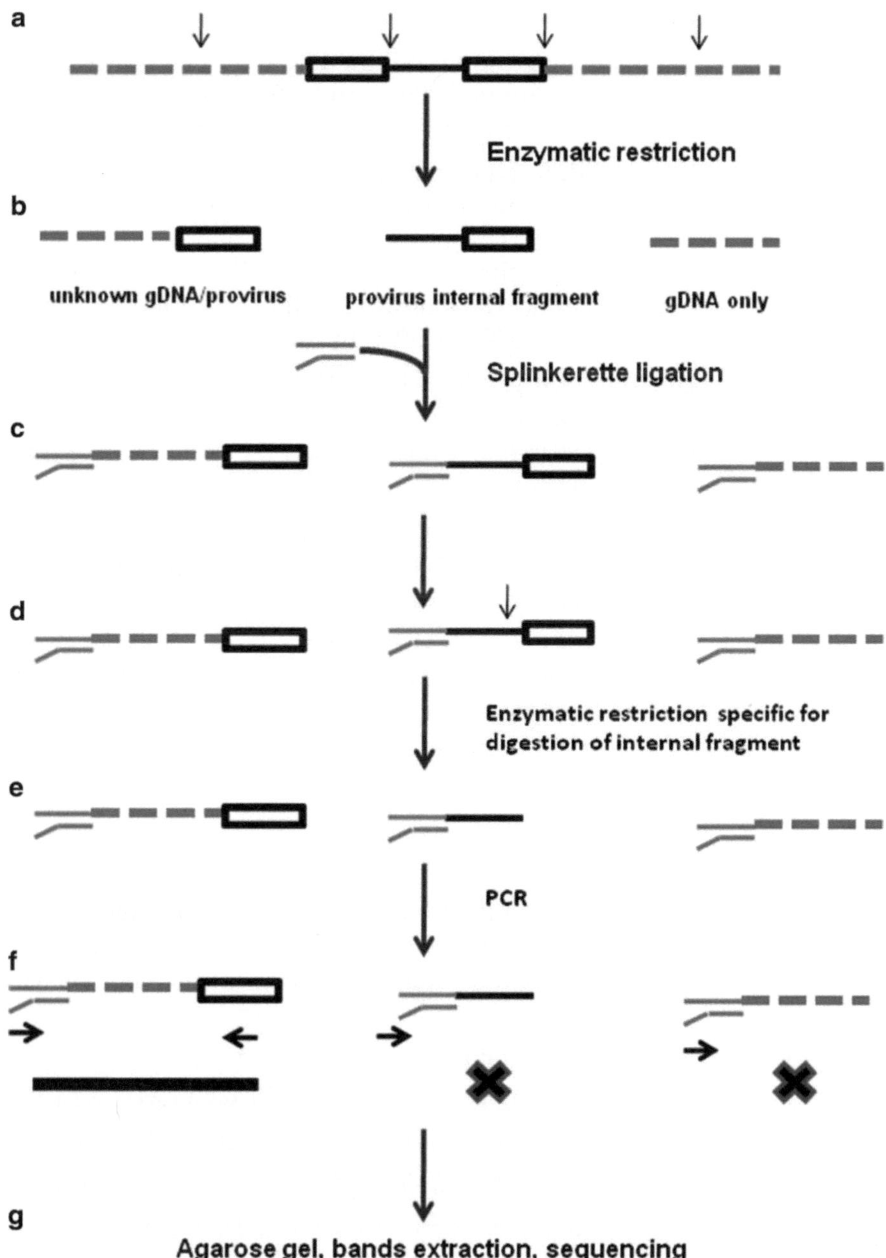


Fig. 1. Schematic diagram of the splinkerette protocol. Porcine genomic DNA (gDNA, *dotted gray*) containing porcine endogenous retrovirus provirus (*bold black*) is digested (**a**) resulting in the generation of different fragments (**b**). Annealed splinkerettes are ligated to the digested DNA (**c**). To prevent amplification of the proviral internal fragment, the DNA-splinkerette is digested with specific enzymes (**d**). The fragments are amplified by PCR and nested PCR using primers annealing to the splinkerette and to the proviral LTR (**e**). Only the fragments containing the junction of gDNA and provirus are amplified (**f**). PCR products are run on an agarose gel, and bands excised and cloned into a cloning vector before sequencing.

10. *For 3′-LTR analysis*, follow the protocol for 5′-LTR analysis with these modifications: Use enzymes *Eco*RI or *Hin*dIII to digest genomic DNA, and after ligation perform a second digestion with enzymes *Nhe*I or *Kpn*I.
11. As with 5′-LTR analysis, perform a nested PCR using the same conditions used to amplify regions flanking the PERV-C LTR sequences. Use primers HMSp1 and M6 for the first-round PCR. For the second-round PCR, use primers HMSp2 and Her1 under the following cycling conditions: 30 cycles of 94°C for 15 s, 60°C for 30 s, and 72°C for 2 min.
12. Run PCR products for both 5′-LTR and 3′-LTR analysis on a 0.8% agarose gel.
13. Excise all the bands and extract them using QIAquick Gel extraction kit.
14. Ligate the purified DNA into the vector pGEM-T Easy and transform into X2-Blue Ultracompetent cells according to the manufacturer's instructions.
15. Sequence the fragments on selected colonies using BigDye Terminator v3.1 Cycle Sequencing Kit and the 3730xl DNA Analyzer (see Note 15).

## 4. Notes

1. It has been observed that the non-transmitter phenotype is not stable and can revert into a transmitter one (13). Therefore, pigs with a null phenotype represent the safer choice as animal donor.
2. Primers sequences:

   PERV gag F: 5′-AGC CTA CTT GGG ATG ATT GTC AA-3′

   PERV gag R: 5′-GGC CCC AGG AAC ATT TTT TC-3′

   18S rRNA F: 5′-TCG AGG CCC TGT AAT TGG AA-3′

   18S rRNA R: 5′-CTT GCC CTC CAA TGG ATC CT-3′

   To test for microchimerism (i.e. detection of PERV sequences due to pig cell contamination) in non-pig samples, the following primers specific for porcine cytochrome c oxidase subunit II can be used (33):

   F: 5′-CGTATTCTTAATCAGCTCTTTAGTG-3′

   R: 5′-GCGGGTAGGATTGTTCAAATTGT-3′
3. QIAquick Gel Extraction kit can be used instead of the QIAquick PCR kit. Add three volumes of buffer GQ to the ligation and one volume of isoproponal. Mix well and load

the solution in the column. Proceed following the instruction in the kit manual.

4. An alternative protocol used in the literature employs irradiated pig PBMC in the co-culture (16, 17, 34). To this end, isolate pig PBMC from blood (see Note 5) and stimulate by adding to the media 2.5 μg/mL PHA and 1 ng/mL PMA. After at least 3 days, lethally irradiate $5 \times 10^6$ PBMC in 1 mL of serum-free media using a Pantac X-ray machine with an output of 240 KeV at a dose rate of 0.7 Gy/min. Incubate $4 \times 10^5$ irradiated PBMC with $4 \times 10^4$ target cells (PERV-susceptible human 293T cells or porcine ST-IOWA) in a 6-well plate using a mixture 1:1 of the RPMI and DMEM media. Culture by passing the cells for about 2 weeks to allow the irradiated PERV-producing cells to disappear.

5. The following protocol can be used to isolate PBMC: Dilute the pig blood 1:1 with PBS and transfer into 50-mL tubes. Add half of the original volume of LymphoPrep™ (Axis-shield). Centrifuge at $600 \times g$ for 20 min at 20°C with the brake off. Using a Pasteur pipette or p1000 pipette, carefully remove the interphase containing the PBMC between the plasma (top layer) and the LymphoPrep. Wash the cells twice by adding 50 mL of PBS and spin down for 5 min at $800 \times g$. Count the cells and resuspend them in RPMI-1640 supplemented with 10% FBS at the concentration of $2\text{–}3 \times 10^6$ cells/mL. Transfer into a 6-well plate, 2 mL per well, and incubate for 2 h at 37°C, 5% $CO_2$. Harvest cells in suspension and keep in culture in media supplemented with 2.5 μg/mL of PHA.

6. It is preferable to use a positive control in this assay to confirm PERV infection, such as PBMC from a pig with a known transmitter phenotype. PERV transmission in co-culture assay has been observed previously (12–14) from minipigs and there is no evidence to date of PERV transmission *in vitro* using PBMC isolated from other breeds (16, 17). More analysis is required in other animals to determine if the generation of recombinant PERV A/C is a phenomenon observed in co-culture of PBMC isolated from miniswine.

7. RNA from difficult samples, such as faeces, can be extracted using TRIZOL LS reagent (Invitrogen) as follows: Dilute the sample by adding an equal volume of PBS and vortex about every 15 min (for a minute each time) for 4 h. Keep the sample on ice. Spin down at $16{,}000 \times g$ for 15 min at 4°C. Collect supernatant, transfer into a new tube, and add one volume of PBS. Spin down at $16{,}000 \times g$ for 15 min at 4°C and pass the supernatant through a 0.45-μm filter. Add 750 μL of TRIZOL LS to 250 μL of the faeces supernatant and incubate for 5 min at room temperature. Add 200 μL of chloroform. Mix very well for 15 s and incubate for 10 min at room temperature.

Collect aqueous phase (top one) and mix with 2.5 μL of glycogen (20 mg/mL). Add 500 μL of isopropanol and incubate for 10 min at room temperature. Centrifuge at 12,000 × *g* for 10 min at 4°C. Wash pellet once by adding 75% ethanol (1 mL approximately), vortex, and spin down at 10,000 × *g* for 5 min. Air-dry for 15–20 min, and then resuspend in 20 μL of RNAse-free water.

8. QuantiTect Reverse Transcriptase kit (Qiagen) is specifically designed for the production of cDNA to be employed in real-time PCR. The primer mix allows the amplification of messenger RNA and other RNA species (e.g. ribosomal RNA). The kit includes also a genomic DNA removal step. For SYBR green-based quantitative PCR where the amplicons can be longer than probe-based quantitative PCR, prolong the incubation time with the RT up to 45 min.
9. This is an example of how to produce standards based on the plasmid pCR-PERVA14/220. Molecular weight of the plasmid is 7,427,788 Da, 1 mole = $6.022 \times 10^{23}$ molecules = 7,427,788 g, 1 molecule weights ($7.43 \times 10^{6}/6.022 \times 10^{23}$) $1.23 \times 10^{-17}$g. To prepare the stock of $10^{10}$ molecules/μL, add 6.15 μg of plasmid to 50 μL of TE. Prepare tenfold serial dilutions of the plasmid in salmon sperm DNA (100 ng/mL in water).
10. If performing a quantitative RT-PCR, the range recommended for 18S rRNA standards is between $10^5$ and $10^9$. 18S rRNA is a well-conserved gene among different species. The primers used (see Note 2) allow the amplification of 18S rRNA from many mammalian species, including primates and pigs.
11. SYBR green is a molecule which emits fluorescence once bound to double-strand DNA; therefore, primer dimers as well as any other unspecific products can produce a signal. To confirm the specificity of the amplicons, a melting curve can be run at the end of the 40th cycle. The melting temperature of the amplicon in the samples must be the same as that of the one obtained with the standards.
12. The amount of 293T cells indicated will produce a monolayer of cells after 72 h. If using different target cells, the number of cells to be seeded could vary. The assay can be shorted to 48 h. In this case, $4 \times 10^4$ 293T cells should be seeded. However, the absolute titres obtained after 72 h may be slightly higher than those after 48 h.
13. Other PCR techniques used to identify PERV integration site are LAM-PCR (35), classic inverse PCR (36), and LM-PCR (37). More recently, a new method called non-restrictive (nr) PCR has been developed and may circumvent the issues associated with omission of loci due to restriction sites located in the flanking genomic DNA (38).

14. It is important to carry out a restriction digestion of the target DNA with a number of enzymes to determine which would be the most appropriate for the assay as different breeds of animals can exhibit quite different restriction maps that can cause the assay to fail.
15. Sequencing service is offered by several companies, such as GATC Biotech (http://www.gatc-biotech.com/en/index.html). The results are provided within 24 h upon sample receipt and up to 1,000 bp can be read. These companies also provide universal primers (for pGEM-T Easy vector, M13 Forward and M13 Reverse can be used).

## Acknowledgments

The authors' PERV research has been funded by the European Sixth Framework Programme (Life Science, Genomics and Biotechnology for Health) funded project LSHB-CT-2006-037377.

### References

1. FDA (2001) PHS guideline on infectious disease issues in xenotransplantation: http://www.fda.gov/BiologicsBloodVaccines/GuidanceComplianceRegulatoryInformation/Guidances/Xenotransplantation/ucm074727.htm
2. Fishman JA, Patience C (2004) Xenotransplantation: infectious risk revisited. Am J Transplant 4:1383–1390
3. Magre S, Takeuchi Y, Bartosch B (2003) Xenotransplantation and pig endogenous retroviruses. Rev Med Virol 13:311–329
4. Mattiuzzo G, Scobie L, Takeuchi Y (2008) Strategies to enhance the safety profile of xenotransplantation: minimizing the risk of viral zoonoses. Curr Opin Organ Transplant 13: 184–188
5. Scobie L, Takeuchi Y (2009) Porcine endogenous retrovirus and other viruses in xenotransplantation. Curr Opin Organ Transplant 14: 175–179
6. Wilson CA (2008) Porcine endogenous retroviruses and xenotransplantation. Cell Mol Life Sci 65:3399–3412
7. Boeke J, Stoye JP (1997) Retrotransposons, endogenous retroviruses, and the evolution of retroelement. In: Coffin JM, Hughes SH, Varmus HE (eds) Retroviruses. Cold Spring Harbor Laboratory Press, Cold Spring Harbor, pp 345–362
8. Akiyoshi DE, Denaro M, Zhu H, Greenstein JL, Banerjee P, Fishman JA (1998) Identification of a full-length cDNA for an endogenous retrovirus of miniature swine. J Virol 72: 4503–4507
9. Le Tissier P, Stoye JP, Takeuchi Y, Patience C, Weiss RA (1997) Two sets of human-tropic pig retrovirus. Nature 389:681–682
10. Takeuchi Y, Patience C, Magre S, Weiss RA, Banerjee PT, Le Tissier P, Stoye JP (1998) Host range and interference studies of three classes of pig endogenous retrovirus. J Virol 72: 9986–9991
11. Wilson CA, Wong S, VanBrocklin M, Federspiel MJ (2000) Extended analysis of the in vitro tropism of porcine endogenous retrovirus. J Virol 74:49–56
12. Oldmixon BA, Wood JC, Ericsson TA, Wilson CA, White-Scharf ME, Andersson G, Greenstein JL, Schuurman HJ, Patience C (2002) Porcine endogenous retrovirus transmission characteristics of an inbred herd of miniature swine. J Virol 76:3045–3048
13. Wood JC, Quinn G, Suling KM, Oldmixon BA, Van Tine BA, Cina R, Arn S, Huang CA, Scobie L, Onions DE, Sachs DH, Schuurman HJ, Fishman JA, Patience C (2004) Identification of exogenous forms of human-tropic porcine endogenous retrovirus in miniature Swine. J Virol 78:2494–2501

14. Wilson CA, Wong S, Muller J, Davidson CE, Rose TM, Burd P (1998) Type C retrovirus released from porcine primary peripheral blood mononuclear cells infects human cells. J Virol 72:3082–3087
15. Denner J (2008) Recombinant porcine endogenous retroviruses (PERV-A/C): a new risk for xenotransplantation? Arch Virol 153:1421–1426
16. Garkavenko O, Wynyard S, Nathu D, Muzina M, Muzina Z, Scobie L, Hector RD, Croxson MC, Tan P, Elliott BR (2008) Porcine endogenous retrovirus transmission characteristics from a designated pathogen-free herd. Transplant Proc 40:590–593
17. Garkavenko O, Wynyard S, Nathu D, Simond D, Muzina M, Muzina Z, Scobie L, Hector RD, Croxson MC, Tan P, Elliott BR (2008) Porcine endogenous retrovirus (PERV) and its transmission characteristics: a study of the New Zealand designated pathogen-free herd. Cell Transplant 17:1381–1388
18. Hector RD, Meikle S, Grant L, Wilkinson RA, Fishman JA, Scobie L (2007) Pre-screening of miniature swine may reduce the risk of transmitting human tropic recombinant porcine endogenous retroviruses. Xenotransplantation 14:222–226
19. Scobie L, Taylor S, Wood JC, Suling KM, Quinn G, Meikle S, Patience C, Schuurman HJ, Onions DE (2004) Absence of replication-competent human-tropic porcine endogenous retroviruses in the germ line DNA of inbred miniature Swine. J Virol 78:2502–2509
20. Matthews AL, Brown J, Switzer W, Folks TM, Heneine W, Sandstrom PA (1999) Development and validation of a Western immunoblot assay for detection of antibodies to porcine endogenous retrovirus. Transplantation 67:939–943
21. Heneine W, Tibell A, Switzer WM, Sandstrom P, Rosales GV, Mathews A, Korsgren O, Chapman LE, Folks TM, Groth CG (1998) No evidence of infection with porcine endogenous retrovirus in recipients of porcine islet-cell xenografts. Lancet 352:695–699
22. Paradis K, Langford G, Long Z, Heneine W, Sandstrom P, Switzer WM, Chapman LE, Lockey C, Onions D, Otto E (1999) Search for cross-species transmission of porcine endogenous retrovirus in patients treated with living pig tissue. The XEN 111 Study Group. Science 285:1236–1241
23. Patience C, Patton GS, Takeuchi Y, Weiss RA, McClure MO, Rydberg L, Breimer ME (1998) No evidence of pig DNA or retroviral infection in patients with short-term extracorporeal connection to pig kidneys. Lancet 352: 699–701
24. Tacke SJ, Bodusch K, Berg A, Denner J (2001) Sensitive and specific immunological detection methods for porcine endogenous retroviruses applicable to experimental and clinical xenotransplantation. Xenotransplantation 8: 125–135
25. DuBridge RB, Tang P, Hsia HC, Leong PM, Miller JH, Calos MP (1987) Analysis of mutation in human cells by using an Epstein–Barr virus shuttle system. Mol Cell Biol 7:379–387
26. Quinn G, Wood JC, Ryan DJ, Suling KM, Moran KM, Kolber-Simonds DL, Greenstein JL, Schuurman HJ, Hawley RJ, Patience C (2004) Porcine endogenous retrovirus transmission characteristics of galactose alpha1–3 galactose-deficient pig cells. J Virol 78:5805–5811
27. Bartosch B, Stefanidis D, Myers R, Weiss R, Patience C, Takeuchi Y (2004) Evidence and consequence of porcine endogenous retrovirus recombination. J Virol 78:13880–13890
28. Mattiuzzo G, Matouskova M, Takeuchi Y (2007) Differential resistance to cell entry by porcine endogenous retrovirus subgroup A in rodent species. Retrovirology 4:93
29. Bartosch B, Weiss RA, Takeuchi Y (2002) PCR-based cloning and immunocytological titration of infectious porcine endogenous retrovirus subgroup A and B. J Gen Virol 83:2231–2240
30. Wilson CA (2006) Porcine retrovirus. In: Straw BE, Zimmerman JJ, D'Allaire S, Taylor DJ (eds) Diseases of swine 9th edn. Blackwell Publishing, Ames, pp 545–550
31. Morrison TB, Weis JJ, Wittwer CT (1998) Quantification of low-copy transcripts by continuous SYBR Green I monitoring during amplification. Biotechniques 24:954–958, 960, 962
32. Devon RS, Porteous DJ, Brookes AJ (1995) Splinkerettes-improved vectorettes for greater efficiency in PCR walking. Nucleic Acids Res 23:1644–1645
33. Issa NC, Wilkinson RA, Griesemer A, Cooper DK, Yamada K, Sachs DH, Fishman JA (2008) Absence of replication of porcine endogenous retrovirus and porcine lymphotropic herpesvirus type 1 with prolonged pig cell microchimerism after pig-to-baboon xenotransplantation. J Virol 82:12441–12448
34. Patience C, Takeuchi Y, Weiss RA (1997) Infection of human cells by an endogenous retrovirus of pigs. Nat Med 3:282–286
35. Schmidt M, Schwarzwaelder K, Bartholomae C, Zaoui K, Ball C, Pilz I, Braun S, Glimm H, von Kalle C (2007) High-resolution insertion-site analysis by linear amplification-mediated PCR (LAM-PCR). Nat Methods 4:1051–1057

36. Carteau S, Hoffmann C, Bushman F (1998) Chromosome structure and human immunodeficiency virus type 1 cDNA integration: centromeric alphoid repeats are a disfavored target. J Virol 72:4005–4014
37. Wu X, Li Y, Crise B, Burgess SM (2003) Transcription start regions in the human genome are favored targets for MLV integration. Science 300:1749–1751
38. Gabriel R, Eckenberg R, Paruzynski A, Bartholomae CC, Nowrouzi A, Arens A, Howe SJ, Recchia A, Cattoglio C, Wang W, Faber K, Schwarzwaelder K, Kirsten R, Deichmann A, Ball CR, Balaggan KS, Yanez-Munoz RJ, Ali RR, Gaspar HB, Biasco L, Aiuti A, Cesana D, Montini E, Naldini L, Cohen-Haguenauer O, Mavilio F, Thrasher AJ, Glimm H, von Kalle C, Saurin W, Schmidt M (2009) Comprehensive genomic access to vector integration in clinical gene therapy. Nat Med 15:1431–1436

# Chapter 18

# Ethical and Regulatory Issues for Clinical Trials in Xenotransplantation

**Jorge Guerra González**

## Abstract

Clinical trials in xenotransplantation (XTx) that have just started to fulfil a long delayed promise should certainly be performed under the same guarantees for the subjects involved as any other experimentation in human medicine. The most important is the absolute respect for their fundamental rights and freedoms, especially for their autonomy, which is expressed through their informed consent as essential requirement for the carrying out of any clinical trial. This chapter focuses on the legal and ethical adaption of the clinical trial's general rules to the particular conditions of xenografting. They are mainly related to the possibility that transmissible xenogeneic agents come into being and become a risk for third parties, even for the whole society. This aspect makes XTx different from any other therapy in (bio)medicine. According to most literature and norm proposals, such xenogeneic infection risk would justify important changes in clinical trial regulation: last but not least, it could mean fundamental right limitations for the xenografted subjects. However, an analysis of the present ethical and legal background at national and international levels shows that such special treatment would be awkwardly acceptable. Information and recommendations on XTx and on its chances and risks when consenting to the trial would be more advisable than right constraining approaches.

**Key words:** Biomedicine, Xenotransplantation, Xenografting, Clinical trial, Xenozoonosis, Xenosis, Xenogeneic infection risk, Informed consent, Fundamental rights limitation

## Abbreviations

| | |
|---|---|
| AIDS | Acquired immunodeficiency syndrome |
| BVerfG(E) | (Entscheidung vom) Bundesverfassungsgericht—(Decision of the) Constitutional Court (D) |
| CDC | Center for Disease Control and Prevention (USA) |
| CFR | Code of Federal Regulations |
| CHRB | Convention for the Protection of Human Rights and Dignity of the Human Being with regard to the Application of Biology and Medicine: Convention on Human Rights and Biomedicine, Oviedo, 4 Apr 1997 |

Cristina Costa and Rafael Máñez (eds.), *Xenotransplantation: Methods and Protocols,* Methods in Molecular Biology, vol. 885, DOI 10.1007/978-1-61779-845-0_18, 

| | |
|---|---|
| CHRB-APBR | Convention on Human Rights and Biomedicine—Additional Protocol Concerning Biomedical Research |
| CoE | Council of Europe |
| Comp. | Compare |
| D | Directive (EU) |
| e.g. | exempli gratia—for example (lat) |
| ECHR | European Convention on Human Rights (CoE) |
| ECHRB | European Convention on Human Rights and Biomedicine (CoE) |
| EMEA | European Agency for the Evaluation of Medicinal Products—Evaluation of Medicines for Human Use (EU) |
| EU | European Union |
| FDA | Food and Drug Administration (USA) |
| ff | and the following (pages/articles/sections) |
| HIV | Human immunodeficiency virus |
| i.e. | id and est—that is (lat) |
| IND | Investigational new drug |
| ISchG, | Infektionsschutzgesetz: Gesetz zur Verhütung und Bekämpfung von Infektionskrankheiten beim Menschen 20 Jul 2000—Protection Against Infections Act (D) |
| LOMESP | Ley Orgánica 13/1986, 14 Apr, de Medidas Especiales en Materia de Salud Pública—Special Measures in the Field of Public Health Organic Act (E) |
| NZ | New Zealand |
| OECD | Organisation for Economic Cooperation and Development |
| ONT | Organización Nacional de Trasplantes—National Transplantation Organisation (E) |
| PERV | Porcine endogenous retrovirus |
| PHCDA | Public Health (Control of Disease) Act 1984 (England) |
| PHS | Public Health Service (USA) |
| Rec | Recommendation |
| s | Section |
| s. | See |
| SACX | Secretary's Advisory Committee on Xenotransplantation (USA) |
| STC | Sentencia(s) del Tribunal Constitucional—decision(s) of the Constitutional Court (E) |
| UK | United Kingdom |
| UKXIRA | United Kingdom Xenotransplantation Interim Regulatory Authority |
| UN(O) | United Nations (Organisation) |
| US(A) | United States (of America) |
| v. | versus—against |
| WHO | World Health Organisation |
| WMA | World Medical Association |
| XIR | Xenogeneic infection risk |
| XTx | Xenotransplantation, xenografting |

## 1. Introduction

Clinical xenotransplantation (XTx) has its own character if compared to other therapies, biomedical or not, at least for two reasons: the use of animals as a grafting source and the possibility that new transmissible diseases (xenozoonoses) arise. Both characteristics ground the XTx challenge for ethics and law. Questions concerning animal welfare and appropriate breeding need, therefore, to be properly addressed, together with issues regarding the xenogeneic infection risk (XIR) (see Note 1) for public health, as this will affect xenograft recipients and eventually their close contacts or perhaps the whole society as soon as XTx is available.

The time for XTx has just arrived; this therapy has recently reached the clinical phase. Hence, a normative framework has to be ready for its first clinical experiments, including the reduction of the XIR from this moment on.

This chapter treats the general regulatory and ethical issues of clinical experimentation; it focuses on the ways and possibilities of XIR reduction during this phase. The reason is that there is enough legal literature and norms on clinical trials, their goal being the protection of the subjects involved. However, there is a limited amount of information pertaining to issues, their goal being additionally the protection of third parties from xenogeneic infections. In this context, plenty of analysis, especially from the ethical perspective, have opened a wide field of questions that must be addressed when considering clinical XTx. However, they have contributed to confusion on some important aspects as well, such as consent withdrawal or post-transplant preventive obligations for the xenograft recipients. It is necessary for everyone to consider them from the legal point of view, either as a medical professional or a health authority. Patients themselves also need to be orientated about the way those specific concerns are regulated.

Finally, the present chapter draws conclusions of the results obtained and analyses their probable consequences.

## 2. Xenotransplantation Clinical Trials

### 2.1. General Issues

#### 2.1.1. Xenotransplantation Is Fulfilling Its Promise

XTx is a promising therapy; when it is ready for clinical use, its health-improving potential would be enormous and such would be its market too. We cannot say yet though when this will happen, but we know that the door is opening little by little. First clinical experiments concerning pancreatic islet XTx have already started to become a reality (1–3) (see Note 2). The future is thus closer, although it will not come for all XTx therapies and activities at once.

That is why it is important to have a solid ethical and mainly legal background when this time comes—perhaps the only advantage of the long time lapse from the first forecasts on XTx to its real clinical use.

XTx is a unique challenge in many senses. If clinical trials normally affect only patients (and eventually their offspring), the medical staff, and the usual medical products or procedures, XTx confronts medicine, law, and ethics due to its latent XIR and additionally with issues that also involve society in general.

#### *2.1.2. First Steps: Clinical Trials*

Clinical trials can be defined as *any investigation in human subjects intended to discover or verify the clinical, pharmacological, and/or other pharmacodynamic effects of one or more investigational medicinal product(s), to identify any adverse reactions to one or more investigational medicinal product(s), and/or to study absorption, distribution, metabolism, and excretion of one or more investigational medicinal product(s) with the object of ascertaining its (their) safety and/or efficacy* (see Note 3). Clinical trials have to follow strict norms nowadays. It is unavoidable to start an analysis of these norms by referring to their origin and foundation: to guarantee the maximum respect for the patients concerned and for their fundamental rights. Hence, all these norms are conceived for their protection, especially their autonomy on decisions concerning their person, so as to assure that those individuals freely agree to participate in such trials. This approach is relatively recent. The Nuremberg Code 1946 is widely considered as its beginning. Its goal was to ban forever the unhuman medical experiments during the Nazi domination in Germany.

Two aspects of clinical trial regulations are to be underlined. The first one is the relative expansion of autonomy even beyond other values that could have been more important so far, as the right to life, or the interests of the State, partners, relatives, etc. (4) (see Note 4). The second one is its western origin, which may come into conflict with other cultures, where the individuals may need to be complemented by their family or community (5) (see Note 5).

The central concept in clinical trial regulations, the first point in the named Code, is therefore *voluntary consent* or, today more usually, *informed consent* (see Note 6). Concerning clinical trials, it must be ensured to actually face the free will of the concerned patients. For instance, they have to know previously about the implications of the clinical trial—chances and risks, possible alternatives, complications, etc.—in order to take a free decision on their participation. They are allowed to withdraw this consent at any time, as they might discontinue any clinical treatment whenever they deem it necessary.

Regarding the essentially important legal provisions on clinical trials in Europe, I will briefly refer to the regulatory framework of the two most relevant Institutions—the EU and the Council of

Europe (CoE)—which are certainly analogue to most applicable international regulations. In this ground, root also the legal proposals on XTx clinical trials, which are explained subsequently.

Clinical trial regulation in the EU is chiefly based on its directives D 2001/20/EC (see Note 7) and D 2005/28/EC (see Note 8). They are considered the minimum safeguard level accepted for the patients concerned, as every Member State can provide them with a more comprehensive protection. Paramount, also for the purpose of this chapter, is art 2.1 D 2005/28: *The interests and welfare of the human being participating in research shall prevail over the sole interest of society or science* (see Note 9). Clinical trials can only be carried on if the expected therapeutic and public health benefits justify their risks. This has to be checked by the competent authority and/or the relevant Ethics Committee—art. 3.2(a) D 2001/20/EC. The trial subjects or, when applicable, their legal representative must have the opportunity, in a prior interview with the investigator or a member of the investigating team, to understand the objectives, risks, and inconveniences of the trial and the conditions under which it is to be conducted. They have to be also informed of their right to withdraw from the trial at any time—art. 3.2(b) and (e). In this situation, the rights of subjects to physical and mental integrity, to privacy, and to the protection of the data concerning them have to be especially safeguarded—art. 3.2(c). In case it comes to any, mainly, personal damage, it must be foreseen that liability of the responsible persons must be covered by insurance or indemnity—art. 3.2(f). Clinical trials *shall be conducted in accordance with the Declaration of Helsinki on Ethical Principles for Medical Research Involving Human Subjects, adopted by the General Assembly of the World Medical Association (1996)* (see Note 10)—art. 3 D 2005/28/EC. This provision makes of such Declaration and its versions immediate part of the EU Law.

Concerning the CoE (see Note 11), the Convention on Human Rights and Biomedicine (CHRB) as well as its Additional Protocol concerning Biomedical Research (CHRB-APBR) are directly applicable law among its members. Art. 5 CHRB-APBR: *Research on human beings may only be undertaken if there is no alternative of comparable effectiveness* (see Note 12). Also as guarantee for the patient indicates art. 4 CHRB (Professional standards) *that any intervention in the health field, including research, must be carried out in accordance with relevant professional obligations and standards.*

Clinical trial or investigation in the USA is ruled by a complex body of regulations and recommendations. It is important to make the difference whether research is privately or federally funded. In this last case, the 1991 Common Rule (The Federal Policy for the Protection of Human Subjects) will be applicable. Concerning investigational new drugs (INDs), the regulations and recommendations of the Food and Drug Administration (FDA) would have to be followed. Most IND research is privately funded. Were instead IND research federally funded, so both regulations would have to be complied with by investigators (see Note 13). As XTx is within the scope of IND, I will focus on the FDA approach.

Under the term "good clinical practice", the FDA includes all relevant issues that are to be observed in clinical experimentation. Its basis is the Federal Food, Drug, and Cosmetic Act 1938. A respectful treatment of study subjects is a relevant part of it, mainly through the informed consent, that has to be written and documented (see Note 14): *For studies that are subject to the requirements of the FDA regulations, the informed consent documents should meet the requirements of 21 CFR 50.20 and contain the information required by each of the eight basic elements of 21 CFR 50.25(a), and each of the six elements of 21 CFR 50.25(b) that is appropriate to the study* (see Note 15). Two elements can be highlighted in this regard: The possibility of compensation if injury to the participating subject occurs is foreseen in 21 CFR 50.25(a) (6), a fact they have to be made aware of in the consent procedure, as well as 21 CFR 50.25(a) (8), most important and relevant for the analysis of this chapter, that underlines that the study subject has to be informed *that participation is voluntary, that refusal to participate will involve no penalty or loss of benefits to which the subject is otherwise entitled, and that the subject may discontinue participation at any time without penalty or loss of benefits to which the subject is otherwise entitled* (see Note 16).

### 2.2. Specific Issues

Clinical trial regulations concern also XTx on its way to medical treatment. Certainly, they start to be only applicable when XTx begins to be considered as medical therapy according to strong preclinical evidence on its feasibility, that is to say, when the xenograft has adapted and performs properly in recipients (physiological aspects), and when their body's immune reaction has been brought acceptably under control (immunological aspects) (6, 7).

XTx implies though the possibility of development of xenogeneic diseases (xenozoonoses). This question is imperative, although it does not affect its clinical feasibility. However, it makes XTx a relevant issue for public health safety, given that it could have negative effects not only in xenograft recipients but also in society as a whole. These microbiological aspects challenge medicine and society in an unknown manner so far. According to most authors, this challenge is so significant that the regulation of XTx clinical trials should be correspondingly adapted in order to assure the maximum benefit for recipients at the minimum risk for public health.

In this regard, the following has to be pointed out:

- Xeno-therapies are quite different from each other concerning the XIR—e.g. perfusion, islet-cell, or solid-organ XTx. Hence, it would be bold to make statements that were applicable to all of them. General affirmations would be possible though, assuming that any measures that were not appropriate to the potentially most risky XTx-therapies (i.e. whole-organ XTx) (see Note 17) would neither be appropriate to the other, thus less risky, xeno-therapies or activities (see Note 18).

- There will always be a certain XIR when carrying out XTx, given that, as any other risk to any legal good, it cannot be completely ruled out (see Note 19).
- According to the preclinical and clinical experience so far, scientists do not assess the XIR as being very high (8–11) (see Note 20). Taking this into account, the conclusion would be that the XIR should be considered acceptable if it were outbalanced by the benefits for society of a clinically feasible XTx. That is, we should compare the real benefit for many patients (e.g. without medical alternative, on organ waiting lists) with such hypothetical risk. Consequently, XTx could be carried out as soon as it is deemed clinically applicable, as the XIR would not seem to outweigh its advantages.

It does not mean at all that the XIR should not be taken care of. On the contrary, it should be reduced as much as possible. One strategy is *scientific research*. The more we know about potential xenogeneic pathogens, as much as on their detection, deactivation, elimination, etc., the better for society, as it would be more prepared to prevent or to react against future negative consequences (9).

Another strategy, previous to xenografting and based on that scientific research, concerns the use of *animal sources* (see Note 21). If it is feasible, they must be completely free of potentially harmful pathogens. This goal can be achieved through animal selection (12) (see Note 22), gene engineering (see Note 23), and/or by carrying out animal breeding in special hygienic conditions (13). In this regard, public health safety must be attained with the highest possible level of animal welfare.

The third strategy can only be accomplished post transplant and concerns *xenograft recipients* (and eventually other persons, e.g. their close/intimate contacts). They would have to observe prophylactic measures after XTx in order to prevent the spreading of possible xenogeneic pathogens to third parties. According to most norms or recommendations on XTx, this observation should be explicitly considered in the recipients' informed consent procedure, thus directly affecting the regulation of XTx clinical trials. This is also the main issue that this chapter is dealing with.

#### *2.2.1. Xenotransplantation Clinical Trials: Post-transplant XIR Reduction. The Role of the Informed Consent*

The recommendations and norms specifically made for XTx clinical trials follow and expand the guarantees exposed in the section above (5, 14–18). The protection of the human subjects involved, of their fundamental rights and freedoms, and thus the respect for their freedom to choose are the main goals of these norms and recommendations too. The informed consent document should facilitate its understanding (it should avoid unfamiliar medical jargon, use short and plainly worded sentences, etc.); different teams should obtain consent and carry out the clinical trial in order to avoid undue influence or instrumentalisation; trials must be conducted by experts in the field with the appropriate scientific training and qualifications; etc.

Public health safety due to the XIR, the unique challenge of XTx analysed here, opens different, partially opposite concerns. For example, it is expected that XTx recipients contribute to the effort of reducing the XIR as much as possible. The best instrument to implement this goal would be the consent form that the clinical trial should begin with.

As it is mandatory in any clinical research, patients should be previously informed of all general or particular aspects of XTx, on its chances, risks, alternatives, and so on, in order for them to agree in freedom if they wish to take part in the trial. Consequent with the respect for their autonomy, they can withdraw this consent at any time before xenografting.

Nevertheless, with regard to the XIR and public health protection and according to the proposal of most recommendations and many eminent authors, XTx patients should understand that they have some duties before third persons and before society in general. For this reason, they should oblige themselves to observe lifelong third-party-protecting preventive measures just after xenografting. For public safety reasons, they should not be allowed to withdraw their consent to follow those measures once this xenogeneic transplantation has been carried out.

As a result, patients would agree in the same informed consent form to participate in a xenografting trial as well as that they would adhere to comply with the post-transplant third-party-protecting measures that were regarded necessary.

Concerning the post-protocol responsibilities that should be expected and required of xenograft recipients, we will consider the last and perhaps most meaningful of them, as they represent the opinion of the International Xenotransplantation Association (IXA) (5): "*(i) regular post-clinical research checkups; (ii) informing researchers of future changes of address/contact numbers; (iii) timely reporting of all unexplained illnesses; (iv) following present and updated behavioral guidelines with respect to exchanges of body fluids with intimate contacts; (v) no future donations of blood, sperm, or other body fluids or tissues; (vi) autopsy at time of death; (vii) education of family members and intimate contacts about their need to take precautions associated with infectious disease risks—that includes offered educational assistance from the research team; (viii) disclosure to future healthcare providers that subjects have received an XTx product; (ix) willingness to accept possible isolation and possible quarantine if necessary for public health; and (x) arrangements for assistance in meeting future responsibilities should the subject lose decision-making capacity"*. *Moreover, "subjects should be informed that they may withdraw from the medical interventions of the protocol, but not from their post-protocol responsibilities"* (14, 19, 20) (see Note 24).

With regard to the XIR and its reduction, some specifications have been proposed that are not directly related to recipients: the archiving of patient plasma, tissue specimens, or health records for at least 50 years (15); that XTx should only be accomplished in places where it has been thoroughly regulated (see Note 25); that a coordinated system for vigilance and surveillance at local, national, and global levels is to be set so as to react on time to possible xenogeneic infections; etc. (4, 20–25) (see Note 26).

The following subsection analyses the legal feasibility of those proposals, particularly of those that affect the fundamental rights of xenograft patients.

*2.2.2. Xenotransplantation Clinical Trials: Xenogeneic Infection Risk Reduction and Informed Consent. Legal Review*

Most of these proposals come from an ethical context or precautionary legal approaches aiming at giving preventively public health protection an unusual weight before individual rights and freedoms. Such proposals would challenge law as it is nationally and internationally established at present, though. It is, therefore, due to examine them from a legal perspective in order to find out if they could be still legally acceptable or whether law should be changed if it were necessary. Such analyses are hard to find in XTx literature, although they would be indispensable before carrying out first xenografting clinical trials. For this reason, this subsection undertakes this task and is highlighted as the focus of this chapter.

1. *Consideration of third-party-protecting agreements previous to XTx clinical trials in private law.*
   The goal of the informed consent institution in clinical trials is to protect and respect the participating human subjects. Only if they autonomously, freely, and explicitly agree may be a clinical trial carried on.

   The nature of informed consent is similar to that of permission or acquiescence. It means in short: "I allow this intervention in order to recover my health". If the clinical trial—as any medical treatment—is performed without the patient's consent, so potentially beneficial for the patient it may be, those who undertake the intervention—who invade their private sphere, the integrity of their body, etc.—will face criminal and civil responsibility. As it was mentioned above, the right to withdraw at any time from the clinical trial is consequent to that respect for the patient's autonomy.

   The named third-party-protecting duties to be observed in xenografting follow a different logic. Their goal is not patient but public health protection, and their nature, a "sui-generis" contractual one. What is more, those obligations constrain directly the fundamental rights of the individuals involved—if being consequent, without any possibility to withdraw.

If the nature of these third-party-protecting responsibilities is contractual, then it would regard the area of private law, i.e. more detailed of contract law. In *contract law*, subjects interact at the same horizontal level and are free to act within a given regulatory framework (contractual/party autonomy, freedom of contract). Once an agreement takes place though, it becomes obligatory for the concerned parties (*lex contractus*) and its clauses must be fulfilled accordingly (*pacta sunt servanda*). Paradoxically, this is the part that makes contract law interesting in the context of infection safety and XTx: only the aspect that allows individual freedom or autonomy to constrain themselves—here even, lifelong.

The problem is that contract law is foreseen to pursue economical interests, not public goals (i.e. society health protection)—at least not directly. The obligations that are derived from its agreements cannot affect the (most personal) fundamental rights of the contracting subjects, or else they would be unlawful (e.g. immoral)—and/or just void. As a result, the responsibilities that a third-party-protecting agreement should contain cannot be legally covered by contract law (10).

We would arrive to similar conclusions if we look for agreements at the border or beyond contract law, like the "Ulysses-Contracts" proposed in this context (19, 23, 26–28) (see Note 27). The aim of these agreements, anyway polemic (4, 29) (see Note 28), is to avoid that the (mental) health treatment of momentarily incapacitated persons be discontinued in case they should change their mind. Their consent to renounce for a while to some fundamental rights—communication, privacy, etc.—is given in a lucid phase, and can be revoked by the next capable time, or just by itself, when they recover their (mental) health completely.

Many factors make the Ulysses proposal inapplicable for XTxs from the legal point of view: its goal—a direct benefit for the consenting patient instead of a benefit for society without any personal gain (see Note 29); the possibility to withdraw—that cannot be given in XTx, as it is not clear when the XIR would reach an acceptable minimum along the life of the xenograft recipient; and the fact that Ulysses-agreements are foreseen for the time when patients are mentally incapacitated, whereas xenograft recipients would be—in principle—in the same healthy mental condition as when consenting XTx.

With clinical trial and public health safety having a different nature and objectives, it could be advisable to consider not one informed consent document, but two: one as the usual permission to the XTx clinical trial, and the other as agreement to observe third-party-protecting measures. This separation would also make sense in case the legal viability of that last agreement had to be challenged—as it seems to be the case—that should not affect the validity of the first one (see Note 30).

A final comment should close this argumentation. The rationale of third-party-protecting agreements would argue against any legal admissibility of XTx: if the infection risk posed to society were that high—contrary to the opinion of most scientists (s. above)—that such safeguards were needed, this should definitely compromise clinical XTx.

2. *Consideration of third-party-protecting agreements previous to XTx clinical trials in public law.*
   If the goal of third-party-protecting responsibilities is to protect public health, this general aim concerns only public authorities. Thus, their legal field would not be private but public law, which rules at a vertical level the relations between the state and its citizens. Public authorities are to act—or not—only in accordance with constitutional parliamentary acts or with norms directly derived from them (see Note 31). No private agreement may substitute the allowance for public authorities to act or to omit action. For this reason, private agreements would be redundant if they are conforming to those public laws allowing norms or void otherwise. In any case, they may be directly ignored by public authorities, to which the named third-party-protecting agreements would be in fact addressed.
3. *Post-XTx third-party responsibilities and fundamental rights and freedoms. General view.*

   Most of the named responsibilities for patients after XTx affect mediately and immediately their fundamental rights and freedoms (see Note 32). Fundamental rights and freedoms are not absolute and can be constrained by public authorities—mainly when their protection is in conflict with general interests, e.g. public health risks (see Note 33). However, taking into account that democracy and the rule of law are built on those individual rights and freedoms, that they are to be sheltered from the arbitrarity of public authorities (i.e. when they act *ultra vires*), and that they are potentially similar to understand and protect all over the world (see Note 34), only in very particular cases these rights and freedoms may be constrained.

   It is understandable that society and also public authorities aim at reducing risks for their citizens as much as possible. Moreover, it is their duty and they must fill it to the best of their ability. But fundamental right or freedom protection must be the rule and, therefore, only exceptional circumstances of immediate general danger may justify their restraining.

In case it is necessary to constrain fundamental rights, public authorities do not have boundless means to fulfil such duty. Any restriction of fundamental rights must be reasonable, i.e. conform to the proportionality principle (30, 31) (see Note 35). Finally and due to the highest esteem for human dignity, the core and essence of those rights (indeed the dignity of the individuals concerned, considered as an absolute value) may never be affected by any public action (10).

The result of these argumentations is clear: mere damage hypothesis, or residual risks, cannot outweigh the importance of fundamental rights and justify their limitation. Thus, the previously preventive time, from xenografting until the first signs of dangerous transmissible xenozoonosis arise, would allow no fundamental right restriction, as such restrictions would require an imminent risk for public health (19) (see Note 36). From this moment on and only in proportion to the given risk (or rather, danger), fundamental right limitations would start to be lawful (19) (see Note 37).

If otherwise just residual or hypothetical risks would suffice to constrain fundamental rights and freedoms, their central legal position in democratic states would be challenged and the rule of law would be easy to undermine and become exceptional itself: if any suspicion could be enough to limit fundamental rights, public authorities could always have arguments to justify their suspension—given that zero-risk actions or omissions are utopian, as no action is completely neutral to any legal good, even if the damage possibility may be marginal. Therefore, no precautionary approach can be legally acceptable, at least concerning the constraining of personal fundamental rights.

In this sense, actions and omissions of public authorities (e.g. in Criminal Law, Police Law, etc.) already follow the respectful approach exposed: Just to have travelled to high-infectious areas is no legal support for fundamental right restraining unless clear evidence of infection danger for third parties is detected. Also treatments of pathogen carriers (s. HIV, mycobacterium tuberculosis) prove how deep the respect is for those fundamental rights (see Note 38).

For these reasons, no particular argument would sustain that XTx should receive a different legal treatment. Certainly, if the XIR becomes a real danger for society, health authorities will be allowed (and also obliged) to act within the respective legal framework so as to preserve public health (see Note 39). This means that the IXA post-xenografting duties to prevent the XIR cannot be freely implemented by health authorities but probably very restrictively, depending on the level of risk and the fundamental rights concerned. Some of them can be imposed under no circumstances though, as they directly concern the core of the recipients' personal fundamental rights (20, 29) (see Note 40).

4. *Post-XTx third-party responsibilities and fundamental rights and freedoms. Considering some examples.*
With regard to the responsibility for recipients to inform their close contacts and health authorities about their condition of being xenografted, it is doubtful that this can become a duty during the previously preventive time, even if it can be expected from them for obvious reasons (their own health improvement, or the protection of their closest ones). This statement would apply as well to a possible obligation to identify any close or intimate contacts or to inform of future changes of address or their communication numbers. At posterior risk levels, these responsibilities could start to become legal duties, always in concordance with the proportionality principle and the respect for human dignity. Such restrictions could finally be applicable concerning the confidentiality duty that health professionals owe to their (xenografted) patients too, which could exceptionally be allowed (or even mandatory) to be breached.
Concerning the prohibition for xenograft recipients to donate tissues or bodily fluids after XTx, such interdiction may affect directly fundamental rights (e.g. free development, freedom to dispose over the own body). However, there is no obligation for anyone to accept donations nor there is a subjective right to become (e.g. organ or blood) donor. As a result, it would not be problematic to accept the graft donation prohibition for XTx recipients and to inform them about such legal impossibility during the consent procedure. Nevertheless, it can be doubted if this interdiction can be absolute, i.e. if an (hypothetical) infection risk would always be relevant enough to impede donations in any case. Some exceptional circumstances (e.g. concerning ex vivo donation to family members) or especially accurate screening methods, etc. could justify a different approach. For this reason, weaker formulations (e.g. prohibited "in principle") could be more suitable in the consent form.
Finally, the only objection concerning the archiving of patient plasma, tissue specimens, or health records would be the consequent preservation of the patient confidentiality and the prevention of any kind of stigmatisation.

5. *Third-party-protecting responsibilities in XTx: hard alternatives.*
For all these reasons, if xenograft recipients decide to disregard the third-party-protecting measures named above just after XTx

and before any xenogeneic transmission danger for society springs up, there can be no legal way to oblige them to act in one way or the other. There is no problem if they voluntarily assume such responsibilities—during such time, the term "responsibility" would probably not be appropriate—precisely as an exercise of their fundamental rights and freedoms, what can be expected though, as it was seen. But if they do not, no previous xeno agreement can become a basis for legal coercion.

Even if we could speculate with the possibility of "blackmailing" clinical trial subjects ("either you agree to their observation or you will not obtain a xenograft"), this dubious (see Note 41) precondition would be of no use in the post-transplant time. If they are well informed that this agreement will be irrelevant, they could acquiesce simply to obtain the xenograft. If they do not, they will realise sooner or later that they can interrupt the observance of such "responsibilities" without facing any consequence.

Other blackmailing alternatives (e.g. that medical authorities discontinue clinical support if patients do no comply in the aftermath) (19) (see Note 42) would be legally even more questionable (s. above). Still worse: The benefit for society of this option would also be uncertain given that public health protection is in fact the aim to pursue, and those patients could not be under surveillance or control anymore.

6. *Third-party-protecting responsibilities in XTx: soft alternatives.* As a result, the informed consent form for clinical trial XTxs should just *advise* or *recommend* what to do in order for patients to prevent negative infectious episodes to themselves and their spreading to other persons, and how to act in these cases. Ergo, it may not impose any obligation or condition to the participation in the trial. Certainly, recipients can be informed on their responsibilities and of those of public health authorities, provided that they become a dangerous infection focus. In this case, they should be particularly concerned with the duty *neminem laedere* (do not harm others)—as should any other capable subject: they could be found liable and/or criminally responsible if it were proven that after that time they were causally responsible (through willingfullness or negligence) for xenogeneic damages to third persons owing to their actions or omissions. In this regard, however, it is important to point out that the act for the patient of not having observed third-party-protecting behavioural guidelines even in case of xenogeneic infection danger could not be sufficient for them to assume responsibility for many reasons: break of the causality chain, absence of evidence, legal assumption of civil (strict) liability by other actors (animal breeder, importer), epidemiological inopportunity, etc. (4, 32).

7. *XTx clinical trials and minors or incapacitated persons.* It has been discussed if minors or incapacitated persons may enrol in XTx clinical trials. Doubtlessly, they are due especial protection as they belong to the most vulnerable patient groups (see Note 43). Some authors and institutions argue for this reason, but also for the especial lifelong commitments concerning patients after xenografting, that minors or incapacitated persons should not be first considered in XTx clinical trials. Nonetheless, we can observe that the main differences between XTx and other therapies in the (bio-)medical research area are those related to the XIR; and that the deep concerns prohibiting the participation of children and incapacitated adults mostly refer to public health safety rather than to their additional protection. These premises together with arguments above bring to the conclusion that no firm justification would sustain a ban or a more restrictive treatment for these citizen groups to XTx clinical trials. Their protection should not prevent them to take advantage of a clinically mature XTx, provided certainly that a prima facie benefit for them can be objectively expected. In this sense, the general provisions for clinical trials regarding minors and incapacitated adults are a sufficient guarantee and may be applicable as well to XTx (see Note 44).

## 3. Conclusions, Consequences

From the analyses above, the following conclusions and consequences can be derived:

- Even if XTx means a challenge in medical experimentation, especially with regard to the XIR, and despite the amount of proposals in this regard, no solid reason would support that the regulation of XTx clinical trials should substantially differ from that of other potential medical therapies. In other words, it would not be necessary to change law in order to face clinical XTx. Some minor adaptations would suffice.
- It is in the interest of xenograft recipients and their closest human fellows to know in advance about the chances and risks of XTx; therefore, they have to be informed on any aspect concerning the XIR. Mainly, they have to be aware of the possible transmission processes of xenogeneic pathogens and the ways to prevent them. The recommendations that are deemed necessary in this matter should be made. Finally, it is also important to clarify the responsibility they carry in case they harbour dangerous transmissible xenogeneic agents and do not act correspondingly to protect third parties, as much as that law can in

some exceptional situations constrain their fundamental rights (within a firm guarantee framework) for this reason in order to protect public health.

- Consent to clinical trials as a kind of hidden agreement is not the appropriate way to ensure preventive behaviours that fall into the area of personal decisions. It would be not only effectless and irrelevant for public authorities. Such insisting in constraining rather than in promoting can undermine the relationship of physician–patient as well as be epidemiologically contraproductive afterwards.
- The fact that previously preventive duties contained in such agreement were compulsory would contradict the logic of any authorisation of XTx: either XTx is not a safe therapy, so it should not reach the clinical phase, or it is safe, and those mandatory behaviours would be disproportional.
- The benefits of a clinically admissible XTx would have to be unquestionable for public health so as to be allowed. The XIR is to be taken into account in this assessment.
- Even if XTx is considered by science an epidemiologically safe therapy, the XIR has always to be reduced as much as possible.
- Negative views of human nature should not be the leading force behind the XTx debate. It can actually be expected that xenograft recipients look for health recovery—otherwise, they would not opt for this therapy—and that they do not pretend to suffer from xenozoonoses, nor to infect their close ones—so they would have an interest in complying with any necessary preventive measure in this sense. Hence, patients support rather than their constraining should be the main approach (also in Law) to follow.
- The Changsha Communiqué 2008, as a result of the First WHO Global Consultation on Regulatory Requirements for Xenotransplantation Clinical Trials, constitutes a broad but reasonable base to build upon. It does not stress patient obligations and underlines the role of support, cooperation, and coordination at all action levels (31, 32) (see Note 45).

## 4. Notes

1. The infection risk that concerns new pathogens originated in the body of patients as a consequence of XTx.
2. The Minister of Health in New Zealand, David Cunliffe, conditionally approved a clinical trial for eight diabetes sufferers by the company Living Cell Technologies Ltd. on 23 Oct 2008. After fulfilling the conditions (Medicines Act 1981, Part 7A s

96 E), first trials were carried out in Oct 2009. The New Zealand Data Safety and Monitoring Board made a positive assessment of the first four patients to receive the Diabecell implant, which consists of encapsulated insulin-producing pig cells (1–3).

3. Art. 2 (a) D 2001/20/EC: "Clinical investigation" *means any experiment that involves a test article and one or more human subjects and that either is subject to requirements for prior submission to the Food and Drug Administration under section 505(i), 507(d), or 520(g) of the act, or is not subject to requirements for prior submission to the Food and Drug Administration under these sections of the act, but the results of which are intended to be submitted later to, or held for inspection by, the Food and Drug Administration as part of an application for a research or marketing permit. The term does not include experiments that are subject to the provisions of Part 58 of this chapter, regarding nonclinical laboratory studies* (FDA 21CFR50.3. Protection of Human Subjects. Informed Consent).
4. This evolution of ethical values can be perceived, for instance, in fields like abortion or the autonomy to decide on one's health or even death under certain conditions (4).
5. It seems although that the worldwide expansion of individual concerns and especially self-determination has become a solid trend that hat transcended borders-as the successive (universal) fundamental right declarations show, which focus on the protection individual subjects from other (familiar, societal, public) interests (5).
6. For the scope of this paper, it may be defined with art. 2 (j) D 2001/20/EC as a *decision, which must be written, dated and signed, to take part in a clinical trial, taken freely after being duly informed of its nature, significance, implications and risks and appropriately documented, by any person capable of giving consent or, where the person is not capable of giving consent, by his or her legal representative.*

   FDA 21CFR50.20. Protection of Human Subjects. Informed Consent: Subpart B—Informed Consent of Human Subjects: FD&C § 50.20 General requirements for informed consent: *Except as provided in § 50.23, no investigator may involve a human being as a subject in research covered by these regulations unless the investigator has obtained the legally effective informed consent of the subject or the subject's legally authorized representative. An investigator shall seek such consent only under circumstances that provide the prospective subject or the representative sufficient opportunity to consider whether or not to participate and that minimize the possibility of coercion or undue influence. The information that is given to the subject or the representative shall be in language understandable to the subject or the representative. No informed consent, whether oral or written, may include any exculpatory language through which the subject or the representative is made to waive or appear to waive any of the subject's legal rights, or releases or appears to release the investigator, the sponsor, the institution, or its agents from liability for negligence.*

7. Directive 2001/20/EC of the European Parliament and the Council of 6 November 2001 on the Community code relating to medicinal products for human use.
8. Commission Directive 2005/28/EC of 8 Apr 2005 laying down principles and detailed guidelines for good clinical practice as regards investigational medicinal products for human use, as well as the requirements for authorisation of the manufacturing or importation of such products.
9. Or art. 3 CHRB-APBR, Primacy of the human being; art. 2 CHRB.
10. Adopted by the 18th WMA General Assembly, Helsinki, Finland, June 1964. Current version 2008: http://www.wma.net/en/30publications/10policies/b3/. Accessed 1 July 2011.
11. It counts at present 47 members, including all 27 EU States.
12. Other provisions underline and partially explain values that have already been mentioned (s. especially arts. 5–17 CHRB; arts. 6–17 CHRB-APRB).
13. Green, Allan M; Steinmetz, Niel FDA Regulations for Clinical Investigators. Summary. http://www.fdaregs.com/index_files/Page1061.htm. Accessed 1 July 2011.
14. Sec. 50.27 Documentation of informed consent. *(a) Except as provided in 56.109(c), informed consent shall be documented by the use of a written consent form approved by the IRB and signed and dated by the subject or the subject's legally authorized representative at the time of consent. A copy shall be given to the person signing the form.*
15. FDA, A guide to informed consent—information sheet. Guidance for Institutional Review Boards and Clinical Investigators, http://www.fda.gov/RegulatoryInformation/Guidances/ucm126431.htm. Accessed 1 July 2011.
16. Sec. 50.25 Elements of informed consent:

    (a) *Basic elements of informed consent. In seeking informed consent, the following information shall be provided to each subject: (1) A statement that the study involves research, an explanation of the purposes of the research and the expected duration of the subject's participation, a description of the procedures to be followed, and identification of any procedures which are experimental. (2) A description of any reasonably foreseeable risks or discomforts to the subject. (3) A description of any benefits to the subject or to others which may reasonably be expected from the research. (4) A disclosure of appropriate alternative procedures or courses of treatment, if any, that might be advantageous to the subject. (5) A statement describing the extent, if any, to which confidentiality of records identifying the subject will be maintained and that notes the possibility that the Food and Drug Administration may inspect the records. (6) For research involving more than minimal risk, an explanation as to whether any compensation and an explanation as to whether any medical treatments are available if injury occurs and, if so, what they consist of, or where further information may be obtained. (7) An explanation of whom to contact for answers to pertinent questions about the research and research subjects' rights, and whom to contact in the event of a research-related injury to the subject. (8) A statement that participation is voluntary, that refusal to participate will involve no penalty or loss of benefits to which the subject is otherwise entitled, and that the subject may discontinue participation at any time without penalty or loss of benefits to which the subject is otherwise entitled.*

(b) *Additional elements of informed consent. When appropriate, one or more of the following elements of information shall also be provided to each subject: (1)A statement that the particular treatment or procedure may involve risks to the subject (or to the embryo or fetus, if the subject is or may become pregnant) which are currently unforeseeable. (2) Anticipated circumstances under which the subject's participation may be terminated by the investigator without regard to the subject's consent. (3) Any additional costs to the subject that may result from participation in the research. (4) The consequences of a subject's decision to withdraw from the research and procedures for orderly termination of participation by the subject. (5) A statement that significant new findings developed during the course of the research which may relate to the subject's willingness to continue participation will be provided to the subject. (6) The approximate number of subjects involved in the study.*

17. Because whole organs are vascularised, their epidemiological cleanliness is more difficult to check (e.g. concerning undiscovered agents) and the possibility of interaction of own with xenogeneic agents would be higher.
18. For this reason and unless it is indicated differently, this chapter means whole-organ XTx when referring to XTx.

19. It seems, hence, unrealistic and far too precautious the provision of art. 5 Rec (2003) 10 CoE of the Committee of Ministers on XTx that xenografting *should not be authorised other than in clinical research unless, on the basis of clinical data: i. there is adequate evidence, in accordance with internationally accepted scientific standards, that no risks, in particular of infection, to the general population exist (...).*
20. No transmission to animals or human beings has been evidenced so far (1, 2, 4, 8–10). Therefore, this risk would be called residual or hypothetical in legal categories. The XIR would radically diminish the more important the damage considered—S. Coffin in ref. 11: possible adverse events following transplantation with porcine endogenous retrovirus (PERV)-containing cells (The Stoye Scale): *1. Expression of infectious virus—very likely; 2. localized infection of host cells—likely; 3. spreading infection in the host—unlikely; 4. persistent viremia—very unlikely; 5. disease (e.g. lymphoma, "AIDS")—very, very unlikely; 6. transmission to close contacts—very, very, very unlikely; 7. spreading, epidemic transmission—very, very, very, very unlikely.*

    In any case, the highest possible level of scientific objectivity should be the foundation on any decision to take on XTx (s. e.g. art. 8 CHRB-APRB).
21. S. Annex I Analytical, Pharmacotoxicological and Clinical Standards and Protocols in Respect of the Testing of Medicinal Products, D 2001/83/EC. It was through D 2003/63/EC and 2009/120/EC that it became directly applicable to XTx: *For xenogeneic cell-based products, information on the source of animals (such as geographical origin, animal husbandry, age), specific acceptance criteria, measures to prevent and monitor infections in the source/donor animals, testing of the animals for infectious agents, including vertically transmitted micro-organisms and viruses, and evidence of the suitability of the animal facilities shall be provided* (Part IV, 3.3.2.1. (d)).
22. "null" pigs or pigs that do not have a transmittable PERV.
23. An example would be the knocking out of PERVs from the swine genome, which is however not possible for science to date.
24. Recipients—and close contacts in part—have to comply with the following: *regular check-ups throughout their lives (...). Lifelong compliance with the post-operative regime for any xeno-transplant: to remain in the surveillance programme whether or not the xenotransplantation was successful; to have samples taken at specified intervals; to use barrier contraception consistently and for life; to refrain from pregnancy/fathering a child; to undertake to be contactable; to allow the relevant Health Authorities to be notified when moving abroad; to be willing to identify possible contacts beyond known close contacts*

*in case of adverse events; to storage of data and storage of samples indefinitely; to divulging of anonymised information to others for research/analysis; that any doctor they register with will be told of their status as a xeno-recipient; that they should declare their status to any doctor of whom they might seek emergency advice. Patients must refrain from blood donation/organ donation. Patients must agree to post mortem (and to tell their family of this agreement) and to storage of samples post mortem. Additionally, they must agree to their contacts being informed about the patient's condition, the xeno-procedure, subsequent treatment, possible side effects in the patient and symptoms in themselves* (14, 19, 20).

25. Changsa Communiqué, principle 4.
26. Changsa Communiqué, principles 8 and 9. Art. 14–15 Regulation (EU) No 1394/2007 of 17.11. on advanced therapy medicinal products and amending D 2001/83/EC and Regulation (EC) No 726/2004; Title IX (Pharmacovigilance) D 2001/83/EC of 06.11. on the Community code relating to medicinal products for human use (4, 20–25).
27. Their name is taken from Homer's hero of the Odyssey. Ulysses wanted to enjoy the singing of the sirens without suffering their enchantment. Hence, he commanded his crew to tie him at the mast, to plug their ears, and to ignore any order he could later give (19, 23, 26–28).
28. Ulysses-agreements are not generally accepted (4, 29).
29. It cannot be seen as benefit for patients to be allowed to participate in the trial, as such or similar conditions are not required for their participation in other clinical trials.
30. Analogue to the fact that non-essential void clauses cannot vitiate contracts, and can just be ignored.
31. Rule of Law. There are plenty of definitions of it. I prefer the ones that lean on law as the foundation of the state so that it is allowed to act only within the given legal framework (Gabler Wirtschaftslexikon 2010). This means that any action of public authorities (executive power) has to be previously authorised by the legislative power, and that such actions are to be controlled by the judiciary power in order to prevent misuse.
32. For instance, with regard to the ten IXA responsibilities above: (i), (ii), (iv), (vi), (vii), (viii), and (ix).
33. S. e.g. European Convention on Human Rights: Article 5—Right to liberty and security: *1. Everyone has the right to liberty and security of person. No one shall be deprived of his liberty save in the following cases and in accordance with a procedure prescribed by law: (...) e. the lawful detention of persons for the prevention of the spreading of infectious diseases, of persons of unsound mind, alcoholics or drug addicts or vagrants.*

34. S. among many others the American Declaration of the Rights and Duties of Man (April 1948); the Universal Declaration of Human Rights of the United Nations General Assembly (10.12.1948); the European Convention on Human Rights—The Convention for the Protection of Human Rights and Fundamental Freedoms (04.11.1950 in force 03.09.1953); or the Charter of Fundamental Rights of the EU (07.12.2000, in force with the Treaty of Lisbon 01.12.2009). The precedents of the time of Enlightment and liberal individualism can be mentioned too: the English Bill of Rights—An Act Declaring the Rights and Liberties of the Subject and Settling the Succession of the Crown (December 1689); the Virginia Declaration of Rights (12.06.1776); and the Déclaration des droits de l'homme et du citoyen (26.08.1789).
35. The applicability conditions of the proportionality principle are the following: (1) *Appropriateness*: The action of public authorities can prevent the potential damage. (2) *Necessariness*: Public authorities must choose among all possibilities the one action that constrains the least fundamental rights or freedoms. (3) *Reasonableness* or *proportionality* strictu sensu: The goal of public action and its consequences must be in reasonable proportion (e.g. BVerfGE 65 1/54; 67, 157/173; STC 22/1981; 158/1993; etc.). Comp. Acmanne and Others v. Belgium (1983) 40 DR 251, 10435/83, 10.12.1984 (30, 31).
36. United States Supreme Court O'Connor v. Donaldson, 422 U.S. 563 (1975), concerning the acceptability of individual right suspensions for reasons of public health: *(1) the risks posed are subject to rigorous scientific assessment; (2) the restrictions are targeted to avoid undue burdens; (...) (4) procedural due process is protected; and (5) the least restrictive possible means of achieving the desired public health outcomes is used* (19).
37. *If a XTx recipient acquires an infectious disease that poses a serious and immediate threat to others, public health laws could necessitate isolation or quarantine.* It is certainly irrespective if a patient's previous willingness to accept such restrictions—e.g. that isolation or quarantine—was given or not (19).
38. They are let free to act, and only respond before law if they abuse the confidence they are granted. S. Guerra 2008, 230 ff. Comp. STC 22/1984, 17.2; 254/1988, 21.12; BVerfGE 7, 198/204; 50, 290/336 ff; R v. Metropolitan Police Comr, ex p Blackburn (1968) 2 QB 118; Tarasoff v. Regents of the University of California 529 P 2d 55 (Cal, 1974); on appeal 551 P 2d 334 (Cal, 1976).
39. S. §§ 16 ff ISchG; arts. 1–3 LOMESP; ss 45B, 45C ff PHCDA, etc. or art. 5 Convention of Fundamental Rights.

40. *Following present and updated behavioural guidelines with respect to exchanges of body fluids with intimate contacts*—as it cannot be controlled without affecting the core of personality fundamental rights. If any, just responsibility in the aftermath (of a damage, most probably) can be made effective, but the preventive measure cannot be checked or, even less, imposed. The same applies to some partially old proposals (to refrain from pregnancy/fathering a child or even to use barrier contraception consistently and for life) (20, 29).
41. It would be critical according to Smith/Grady v. UK (27.09.1999) to hinder a patient access to a clinically acceptable and feasible XTx. Comp. Art. 15–16 Rec (2003) 10.
42. Concerning the specific agreement to follow (adhere) to responsibilities: *Your dropping out (withdrawing) from this study may result in the discontinuation of financial support for lifetime checkups and other responsibilities (...)*(19).
43. Comp. art. 4 (minors), art. 5 (incapacitated adults) EU D 2001/20/EC relating to the implementation of good clinical practice in the conduct of clinical trials on medicinal products for human use. Or art. 17 (in relation to 6, 7, 16) CHRB Chapter V CHRB-APBR (25.1.2005). S. 21 CFR 50.23, 50.51–56.
44. Comp. art. 19 Rec (2003) 10. We can lean for instance on the Chapter V CHRB-APBR (Protection of persons not able to consent to research), Article 15—Protection of persons not able to consent to research.
    1. *Research on a person without the capacity to consent to research may be undertaken only if all the following specific conditions are met: i. the results of the research have the potential to produce real and direct benefit to his or her health; ii. research of comparable effectiveness cannot be carried out on individuals capable of giving consent; iii. the person undergoing research has been informed of his or her rights and the safeguards prescribed by law for his or her protection, unless this person is not in a state to receive the information; iv. the necessary authorisation has been given specifically and in writing by the legal representative or an authority, person, or body provided for by law, and after having received the information required by Article 16, taking into account the person's previously expressed wishes or objections. An adult not able to consent shall as far as possible take part in the authorisation procedure. The opinion of a minor shall be taken into consideration as an increasingly determining factor in proportion to age and degree of maturity; v. the person concerned does not object.*
    2. *Exceptionally and under the protective conditions prescribed by law, where the research has not the potential to produce results of direct benefit to the health of the person concerned,*

*such research may be authorised subject to the conditions laid down in paragraph 1, sub-paragraphs ii, iii, iv, and v above, and to the following additional conditions: i. the research has the aim of contributing, through significant improvement in the scientific understanding of the individual's condition, disease or disorder, to the ultimate attainment of results capable of conferring benefit to the person concerned or to other persons in the same age category or afflicted with the same disease or disorder or having the same condition; ii. the research entails only minimal risk and minimal burden for the individual concerned; and any consideration of additional potential benefits of the research shall not be used to justify an increased level of risk or burden.*

3. S. also art. 16 (Information prior to authorisation) and art. 17 (Research with minimal risk and minimal burden).

45. Changsa Communiqué, principle 6: (...) *Patients and close contacts should be effectively educated about their treatment to encourage compliance, and to minimize risks for themselves and for society* (31, 32).

## References


1. Garkavenko O, Wynyard S, Nathu D, Quane T, Durbin K, Denner J, Elliot RB (2011) The first clinical xenotransplantation trial in New Zealand: efficacy and safety. 14. Minisymposium xenotransplantation, RKI, Berlin, 14 Jun 2011
2. Wynyard S, Garkavenko O, Elliot RB (2011) Multiplex high resolution melting assay for estimation of porcine endogenous retrovirus (PERV) relative gene dosage in pigs and detection of PERV infection in xenograft recipients. J Virol Methods 175:95–100
3. Pierson RN III, Dorling A, Ayares D, Rees MA, Seebach JD, Fishman JA, Hering BJ, Cooper DKC (2009) Current status of xenotransplantation and prospects for clinical application. Xenotransplantation 16:263–280
4. Guerra González J (2008) Xenotransplantation: prävention des xenogenen Infektionsrisikos und die individuellen Grundrechte, insbesondere das Selbstbestimmungsrecht der Person, im spanischen und im deutschen Recht. Peter-Lang, Hamburg
5. Vanderpool HY (2009) The International Xenotransplantation Association consensus statement on conditions for undertaking clinical trials of porcine islet products in type 1 diabetes—Chapter 7: informed consent and xenotransplantation clinical trials. Xenotransplantation 16:255–262
6. Cooper DKC, Casu A (2009) The International Xenotransplantation Association consensus statement on conditions for undertaking clinical trials of porcine islet products in type 1 diabetes—Chapter 4: pre-clinical efficacy and complication data required to justify a clinical trial. Xenotransplantation 16:229–238
7. Máñez Mendiluce R (2002) Xenotrasplante: los retos científicos. In: Romeo Casabona CM (ed) Los xenotrasplantes. Aspectos científicos, éticos y jurídicos, Comares, Granada
8. Denner J (2011) Infectious risk in xenotransplantation – what post-transplant screening for the human recipient? Xenotransplantation 18:151–157
9. Denner J, Schuurman HJ, Patience C (2009) The International Xenotransplantation Association consensus statement on conditions for undertaking clinical trials of porcine islet products in type 1 diabetes—Chapter 5: strategies to prevent transmission of porcine endogenous retroviruses. Xenotransplantation 16:239–248
10. Guerra González J (2010) Infection risk through xenotransplantations and its prevention concerning the xenograft recipients in European Law – main focus: Spanish and German Law, special consideration: English Law. Dr. Kovač, Hamburg
11. SACX (2001) Meeting of the Secretary's Advisory Committee on xenotransplantation.

U.S. Department of Health and Human Services, 20 Feb 2001

12. Garkavenko O, Durbin K, Tan P, Elliot RB (2010) Islets transplantation: New Zealand experience. 13. Minisymposium xenotransplantation, RKI, Berlin 4 Jun 2010
13. Schuurman HJ (2009) The International Xenotransplantation Association consensus statement on conditions for undertaking clinical trials of porcine islet products in type 1 diabetes—Chapter 2: source pigs. Xenotransplantation 16:215–222
14. FDA (2003) Guidance for industry: source animal, product, preclinical, and clinical issues concerning the use of xenotransplantation products in humans, final guidance. US Department of Health and Human Services, FDA, Center for Biologics Evaluation and Research, Rockville, USA
15. PHS (2001) Guideline on infectious disease issues in xenotransplantation. PHS, 19th Jan 2001
16. ONT (1999) Xenotrasplante: Informe de la Subcomisión de Xenotrasplante de la Comisión Permanente de Trasplantes del Consejo Interterritorial del Sistema Nacional de la Salud. MSC, Madrid
17. Romeo Casabona CM, Urruela Mora A (2002) Aspectos jurídicos del xenotrasplante. La configuración de un marco jurídico para el xenotrasplante en el Derecho Español. In: Romeo Casabona CM (ed) Los xenotrasplantes. Aspectos científicos, éticos y jurídicos, Comares, Granada
18. Beckmann JP, Brem G, Eigler FW, Günzburg W, Hammer C, Müller-Rüchholtz W, Neumann-Held EM, Schreiber H-L (2000) Xenotransplantation von Zellen, Geweben und Organen. Wissenschaftliche Entwicklungen und ethische Implikationen. Springer Verlag, Berlin
19. SACX (2004) Informed consent in clinical research concerning xenotransplantation. US Department on Health and Human Services, SACX, Bethesda
20. UKXIRA (1999) Draft report of the infection surveillance. Steering Group of the UKXIRA
21. UKXIRA (2006) Xenotransplantation guidance. Gateway ref. number: 7345, 12th Dec 2006
22. WHO (2004) Human organ and tissue transplantation. Resolution of the World Health Assembly, WHA 57.18, Geneva, Switzerland, 22 May 2004
23. EMEA (2003) Points to consider on xenogenic cell therapy medicinal products. Committee for Proprietary Medicinal Products, CPMP/1199/02, 17 Dec 2003
24. WHO (2001) WHO guidance on xenogeneic infection/disease surveillance and response: a strategy for international cooperation and coordination. World Health Organization, Department of Communicable Disease Surveillance and Response, WHO/CDS/CSR/EPH/ 2001.2
25. OECD/WHO (2001) Consultation on xenotransplantation surveillance: summary. DSTI/STP/BIO(2001)11/FINAL OECD-Directorate for Science, Technology and Industry, Committee for Scientific and Technological Policy, Working Party on Biotechnology
26. Cozzi E et al (2009) The International Xenotransplantation Association consensus statement on conditions for undertaking clinical trials of porcine islet products in type 1 diabetes—Chapter 1: key ethical requirements and progress toward the definition of an international regulatory framework. Xenotransplantation 16:203–214
27. Daar AS (1999) Xenotransplantation: informed consent/contract and patient surveillance. Biomed Ethics 4:3
28. Spillman MA, Sade RM (2007) Clinical trials of xenotransplantation: waiver of the right to withdraw from a clinical trial should be required. J Law Med Ethics 35:265–272
29. Spellecy R (2003) Reviving Ulysses contracts. Kennedy Inst Ethics J 13:373–392
30. Caulfield TA, Robertson GB (2001) Xenotransplantation: consent, public health and charter issues. Med Law Intl 5:81–99
31. Mclean SAM, Williamson L (2005) Xenotransplantation. Law and ethics. Ashgate, Aldershot
32. Guerra González J (2002) Xenotrasplante y responsabilidad civil en el caso de xenozoonosis según los derechos alemán y español. In: Casabona R (ed) Los xenotrasplantes. Aspectos científicos, éticos y jurídicos, Comares, Granada

# Chapter 19

# Some Ethical, Social, and Legal Considerations of Xenotransplantation

María Jorqui Azofra and Carlos María Romeo Casabona

## Abstract

Xenotransplantation has changed its focus from solid organs to cells and tissues, and it is now mainly conceived of and regulated as a pharmaceutical product. Animal cell therapies are showing promising results and may involve fewer risks than organs. However, countries should be cautious about allowing xenotransplantation clinical trials to develop. Regulatory frameworks should contain specific conditions about the safety of the source animals, of the xenotransplantation product, and of the manufacturing process. In turn, these frameworks should ensure that preclinical studies indicate safety and efficacy of the procedure and that risk-management protocols are in place to identify, contain, and combat any outbreak of infection in a timely manner. The fragile balance between individual and collective rights and the tensions of globalization make necessary a coordinated international action to harmonize global practices in this field. Xenotransplantation clinical trials should be carried out in a context in which specific safety and ethical issues are addressed, and in an environment in which specific practices that facilitate public engagement as a form of shared responsibility for regulatory decision making are promoted as well.

**Key words:** Xenotransplantation clinical trials, Individual rights, Public health, Infectious disease, Globalization, Public participation, Risk–benefit evaluation

## 1. Introduction

Xenotransplantation has changed its focus from solid organs to cells and tissues, and it is now mainly conceived of and regulated as a pharmaceutical product (see Note 1). Indeed, researchers predict that, in the short-to-medium term, animal cell therapies (such as pancreatic islet cells) or animal external therapies (such as devices using animal liver cells, or skin grafts) are more likely to be successful. Hopes have been placed on these xenogeneic cell therapies (see Note 2) because cell transplants and procedures may cause less immune

Cristina Costa and Rafael Máñez (eds.), *Xenotransplantation: Methods and Protocols,* Methods in Molecular Biology, vol. 885,
DOI 10.1007/978-1-61779-845-0_19, 

rejection than organ transplants, can be encapsulated, and present fewer structural and functional problems (see Note 3). Certainly, cell therapies are showing promising results and have been recently tested in some clinical trials. Pig islets represent one of the most studied models of cell xenotransplantation in the primate (1). It is hoped that the enormous progress achieved in this field may ultimately allow islet xenotransplantation to be performed without chronic immunosuppression, and hence extend its applicability to the broader populations of diabetic patients (2, 3). Certainly, xenotransplantation research holds considerable promise for people with type 1 diabetes, in this area of islet transplantation. For instance, in 2009, the New Zealand Government approved a phase I/IIa clinical trial using porcine islet cells. In October 2009, the first New Zealand patient was transplanted with DIABECELL—which consists of encapsulated insulin-producing pig cells—. To date, the trial is meeting important objectives, such as safety issues— related to the absence of evidence of pig viruses' transmission to the recipients (4). Therefore, some success has been achieved in these studies. However, further preclinical studies are now being carried out to discover not only how to promote the function of the cells as well as their survival, but also to confirm safety of the administered islets. These safety parameters have not been confirmed by other investigators. To this end, investigators and/or sponsors planning to conduct clinical trials using these or other xenotransplantation products should make use of the various regulations and guidance documents published by international and regulatory authorities, as well as request interaction with them for additional advice on a suitable preclinical program.

Therefore, the great promise of xenotransplantation is that it may result in important clinical benefit to diseased recipients. However, uncertainties still remain, for instance, about the risks of transmission of animal viruses to recipients, their close contacts, and the wider community. Therefore, before the initiation of xenotransplantation clinical trials, it is crucial that several ethical and regulatory conditions are satisfied, including public involvement procedures, since the potential risks associated with clinical xenotransplantation procedures go beyond the treated subject. Xenotransplantation raises a number of issues ranging from animal ethics—that will not be analyzed here, since it would exceed the limits of this chapter—to human rights and informed consent to public health risks and democratic acceptance of this emerging biomedical technology. In this chapter, we briefly review the ethical, legal, and social conditions that should be met to enable the initiation of xenotransplantation clinical trials. We provide an overview of the regulatory situation in the USA and Europe. Next, we identify difficulties in the current regulations, and discuss regulatory approaches that have been undertaken to address some of these difficulties, particularly at international level. We analyze the merits and drawbacks of such regulatory approaches. Finally, we conclude with some recommendations for ensuring the initiation of xenotransplantation clinical trials within a coherent and appropriate ethical and regulatory framework.

## 2. Xenotransplantation Clinical Trials: Some Ethical, Social, and Legal Considerations

### *2.1. Criteria for Evaluating Safety Concerns*

Xenotransplantation products present important challenges and opportunities as medicinal products, since they might bring benefits to patients in terms of increases in quality and length of life. However, the human use of those products is associated with difficult obstacles, such as those related with the potential risk of introducing new infectious diseases into the general population. Therefore, it is crucial to ensure that xenotransplantation procedures are carried out in a context in which specific safety and ethical issues are addressed, such as those related with the need of minimizing the potential risks of infections while respecting fundamental rights of recipients and close contacts, in addition to protecting general population and source animals. Next, we provide an overview of the regulatory framework ensuring that a comprehensive evaluation of the xenotransplantation products is performed and any potential safety concerns addressed before a clinical trial is allowed to proceed. To this end, we take into account the regulatory landscape in the USA and Europe by exploring the steps taken by the Food and Drug Administration (FDA) and the European Medicines Agency (EMA, former EMEA) with the publication of some key guidance documents.

#### *2.1.1. The European Regulatory Approach*

In the European Union (EU), Directive 2001/20/EC relating to the implementation of good clinical practice in the conduct of clinical trials on medicinal products for human use is relevant to xenotransplantation because it contains specific rules concerning the authorization process before commencing clinical trials involving medicinal products for xenogeneic cell therapy (see Note 4). This Directive (updated by Commission Directive 2005/28/EC) is of importance as it explicitly addresses the use of xenogeneic cell therapy and contains several key conditions for the conduct of clinical trials in humans (see Note 5). However, as we will see later, potential risks to public health have implications for many aspects of clinical research in xenotransplantation, including the informed consent process, the right to privacy, and the right of the research subject to remove himself or herself from a research study in which long-term infectious monitoring is mandated as a protective measure for other members of the society (5).

Also in 2001, Directive 2001/83/EC on the Community code relating to medicinal products for human use was established.

It becomes also relevant to xenotransplantation after it was amended by Commission Directive 2003/63/EC to take account of scientific and technical progress and new requirements resulting from recent legislation. Specially, the added part IV deals with advanced therapy medicinal products (ATMPs) and concern-specific requirements for gene therapy medicinal products using xenogeneic system and cell therapy medicinal products both of human and animal origin and xenogeneic transplantation medicinal products. The Directive contains a "Specific Statement on Xeno-transplantation medicinal products" (Part IV.4) (see Note 6).

In 2008, a new regulation (Regulation 1394/2007) on ATMPs came into force (see Note 7). ATMPs are medicinal products for human use based on genes (gene therapy), cells (cell therapy), or tissues (tissue engineering) and include products from autologous (patient's own cells), allogeneic (donor cells), or xenogeneic (animals) origin (see Note 8). They are at the forefront of innovation, offering potential treatment opportunities for diseases that currently have limited or no effective therapeutic options. ATMPs have, therefore, been subject to considerable interest and debate. Regulations for cell-based therapies and tissue engineering products have been developed separately from xenotransplants. However, by passing Regulation 1394/2007, the EU has one single regulatory provision that covers all ATMPs (human and animal). A centralized marketing authorization, once granted by the European commission, is valid in all EU and European Economic Area and the European Free Trade Association States. This centralized authorization procedure was installed through the Committee for Advanced Therapies (CAT) within the EMA. Indeed, central to this new legislation is the establishment of the CAT. It is a multidisciplinary scientific committee of experts representing all Member States of the EU and countries from the European Economic Area and the European Free Trade Association (Iceland and Norway are currently represented in the CAT), as well as representative from patient and medical associations. The CAT gathers dedicated European experts to review the quality, safety, and efficacy of ATMPs according to standards established by regulatory authorities, and to debate scientific developments in this field (see Note 9).

Guidelines for implementing ATMPs' regulation in the EU have been also developed. The key guidance documents for xenogeneic cell-based therapies are the following ones: "Points to Consider Xenogeneic Cell Therapy Medicinal Products" (2003); "Concept Paper on the Revision of the Points to Consider Xenogeneic Cell Therapy Medicinal Products" (2007); and "Guideline on Xenogeneic Cell-based Medicinal Products" (2009). This guideline is an annex to the guideline on human cell-based medicinal products (see Note 10), and deals specifically with requirements unique to xenogeneic specificities. It is intended to provide general principles to be taken into account for the development and assessment of xenogeneic cell-based products without prejudice to medical practice or national legislation, which may be applicable (see Note 11).

#### *2.1.2. The US Regulatory Approach*

In September 1996, the Department of Health and Human Services (DHHS) published for public comment the *Draft PHS Guideline on Infectious disease Issues in Xenotransplantation* to address the infectious disease concerns raised by xenotransplantation. The Draft guideline was a product of the DHHS that formed several interagency working groups to address potential public health risks and to establish safety standards for xenotransplantation (see Note 12).

The DHHS also established the Secretary's Advisory Committee on Xenotransplantation (SACX) that is administered through the National Institutes of Health (NIH). The purpose of this committee of independent advisors was to consider the full range of complex scientific, medical, social, and ethical issues, public health concerns raised by xenotransplantation, address patient-specific issues, and make recommendation to the Secretary on policy and procedures. Diverse open public discussions took place as part of SACX and the FDA Biological Response Modifiers Advisory Committee (BRMAC) whose main purpose was to address issues related to xenotransplantation products used in clinical trials. These discussions were crucial for DHHS and FDA to publish and finalize different guidance documents that address the manufacturing, preclinical, and clinical safety issues raised by xenotransplantation (see Note 13).

The FDA is the sole agency responsible for regulation of clinical xenotransplantation in the USA, and as such has published diverse guidance documents with recommendations for sponsors of xenotransplantation clinical trials. These documents are consistent with the above-mentioned PHS Guideline (see Note 14).

#### *2.1.3. Scientific and Technical Key Points*

In the light of the steps taken by the EMA and FDA to facilitate the clinical evaluation of xenotransplantation products, we can point out several criteria that have paved the way toward addressing the special needs of xenotransplantation, especially with regard to minimizing the potential risks of infections. These risks may be minimized—but not completely ruled out—by careful choice of donor animals, reproducible manufacturing process, accurate preclinical and clinical testing and monitoring, as well as risk management program with regard to infections agents.

##### Biosafety Standards for Source Animals

Source animals should be specific pathogen free (SPF) and raised in SPF conditions. Effective animal qualification includes herd management (e.g., closed herds housed in a controlled environment) and programs for prevention and screening for infectious diseases. Program testing protocols should be updated periodically to reflect advances in the knowledge of infectious diseases. In addition to adequate and validated methodologies for surveillance of infectious agents from the source animals, monitoring of herd health and tracking of veterinary care must be exercised (6). Appropriate quality assurance standards for the facilities—designed to minimize the potential for the introduction of infectious agents into the final xenotransplantation product—and adequate qualifications of personnel must be also considered. On the other hand, recommendations for producers and developers of genetic engineering animals (see Note 15), harvest of xenotransplantation products, transportation of source animals, sample archiving, herd records, and establishment of an identification system that allows traceability from source animal to recipient must be provided (see Note 16). These objectives should be achieved while strictly abiding by animal welfare rules (see Note 17).

#### Quality and Manufacturing Aspects

Procedures must be in place that will prevent cross-contamination of harvested and/or processed xenotransplantation products. To accomplish this, records of xenotransplantation product processing, storage, and shipping must be retained. There should be a restricted access and nominated person/persons who is/are responsible for the archives. All xenotransplantation products should be labeled so that the corresponding samples in the archives can be traced. Requirements for characterization and quality control of xenotransplantation products are considered necessary (see Note 18).

#### Clinical Development of Xenotransplantation Products

As recommended by international organizations and most regulatory authorities, there must be adequate preclinical data to support clinical development of a xenotransplantation product (see Note 19). Preclinical information will be vital in the planning of the first phase 1 trials (see Note 20). In addition, the result of initial clinical trials should provide a much better understanding of potential risks and benefits and will be crucial to the design of subsequent studies. Xenotransplantation clinical trials must be designed to ensure that subjects are not exposed to unreasonable risk, while maximizing the information derived from such studies. As stated by the EMA and the FDA, the clinical development of xenotransplantation products should involve initially patients with serious life-threatening disease for whom adequately safe and effective alternative therapies are not available or where there is a potential for a clinically relevant benefit (7, 8). Several factors unique to xenotransplant trials, such as the requirement for lifelong monitoring, should be also considered. The evaluation of risks and benefits of xenotransplantation should address both recipient and public health concerns.

The criteria that must be addressed regarding xenotransplantation product recipients, their close contacts, and public health implications can be summarized as follows (see Note 21): counseling regarding behavior modification and other issues associated with the risk of infection should be provided to the patient and made available to the patient's family and other close contacts prior to and at the time of consent, and such counseling should continue to be available thereafter; informed consent process for recipients and the education and counseling process for recipients and their close contacts, including associated health care professionals, should be carried out; detailed plan for screening, treating, and control of unexpected infectious diseases should be in place; procedures to ensure the appropriate care and follow-up of the recipients, and the necessary safety measures for all close contacts during the initial period after treatment, must be undertaken; a system to follow up close contacts of the recipients and health care professionals may be required when relevant; and archiving of patient plasma, tissue specimens, and medical records, which should contain all relevant information for both safety and maintenance of efficacy profiling, should be ensured.

#### *2.1.4. Ethical, Legal, and Social Issues*

As outlined above, there is a need for thorough monitoring and surveillance of xenotransplantation product recipients, with regular testing for signs and symptoms of disease. This monitoring and surveillance plan should be followed and updated during the clinical development and the post-marketing phases (see Note 22). The effectiveness and efficiency of public health action are directly related to the quality—and for many purposes the quantity—of relevant surveillance data so as to provide a foundation for a rapid response to emerging infectious disease (9). The best hope for tracking, studying the development of, implementing treatments for, and creating vaccines to combat infectious agents lies in carrying out routine physical evaluations and collecting and archiving tissue and/or body fluid specimens obtained from each recipient. To this end, active compliance with long-term surveillance is crucial. This requires recipients to actively participate in the monitoring program. Indeed, the prospective xenotransplantation product recipients should be informed about the importance of complying with this long-term surveillance plan and the corresponding responsibilities, all of which should be covered by their consent. Therefore, the risk of transmission of infection in the course of treatment with a xenogeneic medicinal product makes necessary that possible limitations or conditioning factors in the recipient's subsequent life—in addition to those safety measures that may intrude on their family and intimate contacts—be explained during the informed consent process (see Note 23). Among the monitoring and surveillance responsibilities expected and required of subjects include, for instance, those related with the need of regular post-clinical checkups, informing about possible changes of address/contact numbers, timely reporting of all unexplained illnesses, following specific behavioral guidelines with respect to exchanges of body fluids with intimate contacts, education of family members and intimate contacts about their need to take precautions associated with infectious disease risks, possible isolation and quarantine, etc. (10). Therefore, the subjection to rigorous post-operational monitoring by the medical staff may compromise recipient's right to liberty, bodily integrity, privacy, etc.

However, those monitoring and surveillance responsibilities are extraordinarily difficult to ensure due to both problems with predicting and controlling subjects' behavior and the potential for violating their fundamental rights. Additional factors, such as demographics, have proven ineffective in prediction of patient compliance, making it nearly impossible to forecast which subjects will remain faithful to long-term monitoring. Furthermore, the impact of xenotransplantation goes beyond national boundaries. To date, no mechanism is in place to monitor (or prevent) a recipient from traveling freely to other countries that may lack the resources to execute a specific surveillance plan or the infrastructure to deal with any possible pandemic from xenotransplantation. An adequate legal framework for well-monitored global application of xenotransplantation practices would be also necessary due to xenotransplantation expansion into the medical tourism industry. The risk of "xenotourism" which is the travel of prospective xenotransplant candidates to countries, where xenotransplantation procedures are less stringently regulated, represents an important concern that should be addressed (2, 11).

The harmonization between individual and collective rights in matters of health has proved difficult to achieve. The rights of individuals to liberty, bodily integrity, privacy, etc. must be respected to the fullest extent possible, consistent with maintaining and preserving the public's health and security. Long-term surveillance might compromise recipient's fundamental rights and freedoms (e.g., freedom of movement within and across jurisdictions, privacy, and freedom to withdraw from the trial). However, allowing a recipient to drop out the monitoring program would handicap the scientific community's ability to track and treat the disease, thus subjecting the public to risk of infection. For this reason, as is the opinion of some authors, a clear statement that subjects may not withdraw from the monitoring and surveillance responsibilities should be included in the topics about which recipients should be well informed in the consent process and in consent documents (10). In xenotransplantation clinical trials, the need of complying with a specific surveillance plan arguably outweighs the individual's right to withdraw.

Certainly, there are important reasons that militate against characterizing the recipient's consent as an irrevocable agreement, including the importance of protecting recipient autonomy and bodily integrity. However, xenotransplantation clinical trials raise specific questions that hardly ever emerge in other areas of medical research. Among these issues is, on the one hand, the possibility that xenotransplantation product recipients might lead to a new infectious disease. This accounts for the above-mentioned surveillance responsibilities. On the other hand, waiving the right to withdraw from those responsibilities seems to abrogate one of the standard principles of research ethics based on freedom to discontinue involvement in a clinical trial at any time without reprisal (see Note 24). The proposed lifelong surveillance of xenotransplantation product recipients is often cited as being contrary to this fundamental cornerstone of the Declaration of Helsinki. However, recipients might risk the health of the entire community by refusing to observe monitoring protocols. In this respect, it may be acceptable to withdraw respect from self-autonomy in the face of circumstances, where there is a wide agreement that respecting it potentially undermines the health and safety of others (12). In these clinical trials, the possibility of imposing the monitoring to the subjects might be justifiable to ensure public safety (13).

In addition to the interference in the subjects' fundamental rights, another argument contrary to a coactive monitoring is based on the fact that the risk of introducing new infectious diseases does not constitute a certain and effective risk (see Note 25). However, the risk of transferring viruses from animals to humans via xenotransplantation cannot be eliminated completely and it is one of the major concerns in the developing area of xenotransplantation (see Note 26). For this reason, all risks must be managed appropriately (see Note 29). In this respect, the guideline on xenogeneic cell-based medicinal products includes several measures that should be taken into account in the planning of long-term follow-up of the patients (7). Furthermore, this guidance document states that a risk-based approach should also consider the special characteristics of xenogeneic medicinal products, since the different levels of risks associated with each individual product and the proposed therapeutic use can vary a lot (see Note 27). It is important to note that follow-up of efficacy and adverse reactions is a crucial aspect of the regulation of ATMPs. The requirements of a Risk Management Plan have to be considered in the light of relevant national and EU legislation (7).

All the mentioned measures are part of a precautionary approach that proceeds when risks are uncertain. When there are reasonable signs, it is necessary to act with precaution or caution. The precautionary principle justifies the public intervention so as to avoid or palliate a risk, although that intervention may be based on signs.

Triple judgment that application of proportionality principle involves in health authorities' intervention should be considered: that is, measures should be adequate to the end (adequacy); the least restrictive of the human rights among all the adequate ones that could be applied (necessity); and proportional stricto sensu, that is to say, to keep the balance between the costs and the benefits it causes (proportionality stricto sensu). Concerning the first one (adequacy), measures based on carrying out the monitoring and surveillance plan might be invasive and might compromise recipients' fundamental rights, but, as outlined before, a detailed plan for screening, treating, and control of unexpected infectious diseases should be in place, given the possible danger for the health of the population derived from a new infectious disease. With regard to the second one (necessity), it is argued that the "least restrictive" means should be used to achieve health care objectives. It is certainly preferable, before proceeding with the possibility of imposing the monitoring to the recipient against his/her wish, the voluntary collaboration with the health authorities. Of course, as the risks of transmitting new infectious diseases into the population are reassessed and minimized, the post-monitoring responsibilities should be revised accordingly (e.g., time frame for checkups). On the other hand, coercive isolation and quarantine should only be used as a "last resort." These measures would require the highest level of evidentiary justification. Coercive social distancing is only justified if it results in net benefits to society as a whole (14). This being said, measures adopted should be linked, as far as possible, with the respect of fundamental rights. For instance, those measures related with the need of collecting and archiving tissue and/or body fluid specimens obtained from each recipient should be respectful with the individuals' rights included in the data protection normative. Once it has been stated that the measures have complied with the adequacy and necessity requirements, it should be determined whether they are reasonable stricto sensu or not. As has been indicated, the restriction of a right is proportional stricto sensu if it is balanced because more benefits for the general interest are derived from it than damages against other goods or values in conflict. In this case, recipients' rights—such as the right to withdraw from the xenotransplantation clinical trial at any time by revoking his/her informed consent—have to be confronted with the serious risk for public health derived from an uncontrolled transmission of known and unknown pathogens or the risk of introducing new infectious diseases into the general population, which can be considered of greater priority (15).

Another important issue that should be considered makes reference to the possibility of monitoring close contacts and health care providers. In this respect, it is stated that all efforts should be made to adequately inform close contacts if transmission of infectious agents cannot be excluded. It may be, in rare circumstances, necessary to collect blood samples of close contacts (e.g., family) prior to the procedure and store for retrospective testing. This raises some questions, such as the difficulty of tracking "intimate contacts" due to the fact that a person's contacts and relationships change over time and how to assure that they are as fully informed and educated as possible. On the other hand, as is the opinion of some authors, researchers should assess the possibility of excluding subjects from the clinical development of xenogeneic cell-based products or from a xenotransplant clinical trial when they are unlikely to abide by critically important public health responsibilities (10). One wonders if the prospective recipient selection criteria should include the willingness or unwillingness of close contacts to accept surveillance and monitoring responsibilities. If the refusal of close contacts to participate in a surveillance program was a compelling reason to reject a prospective patient, third parties would be given an authority that had always lacked in cases where a competent person freely decides on matters of his/her health. The involvement of relatives and close contacts takes on a new significance in this context.

On the other hand, there is an ethical argument that, because xenotransplantation clinical trials have the potential to impact a wider population beyond the individual participants, a form of collective consent should be also required. Indeed, the potential risks that xenotransplantation pose to society as a whole, the fragile balance between individual and collective rights in matters of health, and other ethical issues discussed suggest the need to include an active participation and engagement of citizens in decision-making and regulatory procedures concerning this emerging biomedical technology. To this end, important initiatives of public consultation have taken place. In the early 2000s, Canada carried out the first important nationwide public consultation, "Animal-to-human transplantation: Should Canada proceed?" (16). The Canadian Government stressed that citizens should be given an adequate role in scientific matters affecting society as a whole. In 2001, the Australian National Health and Medical Research Council (NHMRC) established the Xenotransplantation Working Party (XWP) to provide advice on the scientific and normative aspects of xenotransplantation, produce guidelines for clinical trials, and consult with the community. In July 2002, the XWP released a Discussion Paper to promote an informed community discussion that took place through two rounds of consultation (17). In recent years, important initiatives of public consultation have taken place. For instance, in 2008, the National Health Committee (NHC) invited written public submission on an application by Living Cell Technologies (LCT) —an international biotechnology company—to conduct a clinical trial of pig cell transplantation in New Zealand (see Note 30) (18). Between 2007 and 2009, the European Commission launched through the Web several advertised public consultations, asking for public comments on the technical aspects of guidelines for the management of risks, good clinical practices, etc., concerning advanced therapies (19). These current experiences are different from the Canadian and

Australian ones. These latter aimed at framing a normative context, where citizens could become responsible partners in policy making. Public engagement is seen as collective learning process for all involved parties. However, the New Zealand case and the European regulatory procedures were mainly prepared to seek approval or implement regulatory measures concerning specific applications for clinical trials involving xenotransplantation or ATMPs.

Although some countries may decide not to proceed with xenotransplantation clinical trials, the potential risk of infectious diseases would not be restricted to the countries in which those trials are performed. Globalization clearly presents important difficulties for crafting consensus on regulatory policy in xenotransplantation. It becomes increasingly apparent that both national and international regulatory responses are required. Regardless of a state's position on xenotransplantation, certain facilities and provisions are key to protecting public health from the risks of novel infections (see Note 26). In what follows, we examine what kind of measures should be addressed to protect individual and public interests at the state level while acknowledging the challenges presented by globalization.

### 2.2. Toward an International Regulatory Framework

States should seek to establish their own policies and regulations on xenotransplantation. In turn, this decision needs to be influenced and guided by the international community to harmonize global practices. In the absence of internationally agreed regulations and monitoring procedures, the most assiduous safety measures of any nation, or group of nations, are likely to be ineffective or unsuccessful. Therefore, clinical xenotransplantation trials should be conducted only in countries where regulations and adequate resources empower a national health authority to effectively regulate these trials, according to such internationally agreed regulations.

Firstly, in proposed clinical trials of xenotransplantation, there should not be an "unacceptable risk" to the health or safety of the public (see Note 32). We have seen that some regulatory agencies, such as FDA and EMA, have been proactive in addressing issues that may contribute to rational development of xenotransplantation products. Those issues make reference to important areas, such as safety, efficacy, public health, and ethical considerations. Proposers of trials should provide all the information required by the regulatory authorities to assess the risks and determine how the risks can be minimized. As has been outlined, certain measures—such as qualification and control of the source animals, reproducible manufacturing process, appropriate preclinical studies that provide support regarding the safety and effectiveness of xenotransplantation products, risk management program, etc.—may be the basis to demonstrate a favorable risk–benefit ratio or to convince, for example, that the risk of introducing new infectious diseases into the general population is extremely low. This implies that assessment of xenotransplantation clinical trials or treatments with xenogeneic medicinal products is a multidisciplinary exercise, requiring expertise for characterization, manufacturing and control, and nonclinical, clinical, and ethical aspects. All these domains should be represented in the corresponding committees involved in the evaluation process of the proposed trials.

On the other hand, any possible risk posed by the conduct of the proposed trial should be appropriately managed (20). Lifelong monitoring is crucial for managing the risk to the public of a potential infectious disease. Therefore, effective monitoring and surveillance will require cooperation at local, national, and international levels—e.g., monitoring of patients and trial participants will require that patients and clinical teams are aware of the importance of reporting any problems to local authorities. Certainly, compliance with a requirement to submit to lifelong monitoring may be unlikely to be enforceable in some states. However, as noted earlier, when an individual fails to voluntarily take required precautions against the possibility of infecting others, legal sanctions and/or the use of force may be justified based on proportionality principle requirements (see Note 33). States have the power to limit personal freedoms and rights when public health or security is at stake (see Note 28). At national level, health authorities will be responsible for implementing a surveillance system as part of their own regulatory framework, possibly including a patient register. Indeed, a national register and biological sample archive are also among the key provisions to protect public health from the risks of novel infections. They provide the basis for traceability in the event that a novel infection is detected. Therefore, there should be rigorous analysis of trial outcomes and all recipients should be registered in an appropriate database while their privacy should be protected. In addition, there should be a designed response plan for xenotransplantation-related infectious disease.

Other important criteria that should be considered and acknowledged make reference to the ethical and social issues involved in the proposed trials. Informed consent and selection of patients remain controversial due to the special characteristics of these trials. As noted earlier, the need for lifelong monitoring and restrictions on participants and possibly their close contacts, at least in the initial trials, are important issues to consider. Not only would participants be expected to submit to unusual limitations on their individual rights and liberties, but also others would have a direct impact on the availability of the treatment, and would also be required to make significant modifications to their lifestyles. As outlined before, patient selection should be on the basis of informed consent from motivated patients willing to accept the special conditions that will be required by the trial. In this regard, patients and close contacts should be effectively educated about their treatment to encourage compliance and minimize risks for themselves and for society. Certainly, the need for adherence to lifelong monitoring directly conflicts with regulations that require the subject to be free to withdraw from the study at any point. However, the mentioned consent-based questions go beyond the individual and certainly test the accepted legal understanding of consent. The question is how to overcome the tensions of informed consent and human rights at a national and international level.

In addition, there should be government commitments to support and develop different forms of collaboration within the civil society, and between civil society organizations and the international scientific community, about this emerging biomedical technology. This way of framing public involvement would require not only simple forms of citizen consultations, but also more complex practices which involve learning processes, interaction with experts, deliberations, etc. These latter would require an intellectual environment in which citizens would be encouraged to bring their knowledge to bear on the resolution of issues surrounding xenotransplantation. A lot of effort should be made to create a more knowledgeable and more responsible citizenship, to favor an open dialogue between scientists and citizens, and to make policy making more transparent and accessible. All these efforts may apply to public health protection, as it has been demonstrated that forms of participatory management of infectious diseases may work better than practices based merely on control and risk assessment. As it has been suggested, the next step in this regard should go through a more complex meaning of public participation. The shift that is taking place goes from mere passive approaches to develop consensus and risk acceptance to a shared accountability for risk control and regulatory decision making (21).

It is also important to note that global threat of infectious diseases forces states to cooperate with each other in order to develop global governance, which involves certainly the interaction of states, international organizations, and non-state actors to shape values, policies, and rules. To this end, important steps have been taken toward the harmonization of xenotransplantation procedures and policies. For instance, in May 2004, 192 countries at the World Health Assembly (WHA) adopted Resolution 57.18. This resolution urges Member States to allow xenotransplantation "only when effective national regulatory control and surveillance overseen by national health authorities are in place" (22). In 2008, the WHO, in collaboration with the Chinese Ministry of Health, the University of Central-South China, and the International Xenotransplantation Association (IXA), conducted a global consultation on clinical trials in xenotransplantation. This consultation produced a communiqué—known as the Changsha Communiqué—which listed a set of principles relating to xenotransplantation. In addition, there are recommendations to WHO, Member States of the WHA, and investigators and supporters of xenotransplantation procedures and trials (see Note 34).

The role played by the WHO and other international agencies, although important, may not be sufficiently strong to make a difference in terms of compliance with desirable legal standards. Governmental regulatory bodies have the sovereign power to interpret those principles on conducting xenotransplantation clinical trials that do not necessarily translate into adequate protection for individual rights and public health. International cooperation on xenotransplantation that is effective will need to consider ways of combining legally binding measures with statements of principle at local, national, and international levels (23).

## 3. Conclusions

Because public health risks posed by xenotransplantation transcend national boundaries, regional and intergovernmental bodies have underscored the necessity for cooperation between countries to manage the risks of this emerging biomedical technology. In parallel with many other national efforts, FDA and EMA recognized the global concerns over the conduct of clinical trials of xenotransplantation in the absence of regulatory oversight. Therefore, these regulatory agencies have been active participants in open dialogues with international organizations—such as the WHO and the Organization for Economic Development Council of Europe—as well as with regulatory bodies of individual countries, regarding the regulatory requirements for xenotransplantation clinical trials.

In general, the consensus is that the risk posed by animal viruses is low and can be managed via the above-mentioned requirements, provided there is a regulatory mechanism to obligate compliance (see Note 31). Regulatory frameworks should contain specific conditions about the source animal and how that animal is housed, fed, and tested for infectious agents. In turn, these frameworks should ensure that preclinical studies indicate safety and efficacy of the procedure and that risk-management protocols are in place to identify, contain, and combat any outbreak of infection in a timely manner (see Note 36). Therefore, under such guidelines, only xenotransplantation clinical trial proposals that are deemed safe, provide a real possibility of therapeutic success, and have protocols that ensure the highest ethical and regulatory standards for both human and animal participants should be allowed to proceed. If these precautions are taken under the international supervision, the potential risks could be reduced to that which may be considered acceptable and outweighed by the benefits of the technology.

Despite the progress and related efforts, some of the questions outlined earlier would still remain open. Given these difficult issues, —such as those regarding an appropriate informed consent process, how to ensure that patients can legally be monitored on a lifelong basis for infectious diseases, or how to develop supportive national and multinational actions for preventing unregulated xenotourism and inappropriate unilateral responses with international consequences—, (see Note 35) a coordinated international action should be certainly necessary to harmonize global practices in accordance with the highest ethical and regulatory standards. However, specific practices and instruments to facilitate public engagement as a form of shared accountability for regulatory decision making should be promoted as well.

## 4. Notes

1. Xenotransplantation is defined in line with international terminology to include a range of procedures that involve the use of living animal products in human therapies. The term xenotransplantation refers to the transplantation, implantation, or infusion into a human recipient of live cells, tissues, or organs derived from nonhuman animals. The procedure includes the use of human body fluids, cells, tissues, or organs that have had ex vivo contact with live nonhuman animal cells, tissues, or organs (8, 17, 24–27).
2. Xenogeneic cell-based therapy is defined in EMA (2009) Guideline on xenogeneic cell-based medicinal products. EMEA/CHMP/CPWP/83508/2009:3 (7).
3. Animal organ transplants will be the hardest procedures to perfect due to the problems of rejection and other structural and functional problems (28). Cell-based xenotransplantation products imply a significantly smaller risk of virus transmission than xenotransplants of vascularized organs. Cells can also be best screened for a spectrum of infectious agents in advance. Moreover, xenogeneic cell transplant barriers to immunology, such as the encapsulation techniques, may control viral transmission as well (29).
4. Directive 2001/20/EC of the European Parliament and of the Council of 4 April 2001 on the approximation of the laws, regulations, and administrative provisions of the Member States relating to the implementation of good clinical practice on the conduct of clinical trials on medicinal products for human use: For each clinical trial, approval of the protocol by an Ethics Committee and competent regulatory authority is mandatory. This Ethics Committee has a limit time for giving its reasoned opinion to the applicant and the competent authority in the Member State (Article 6, paragraph 5). However, in the case of xenogeneic cell therapy, there shall be no limit to the authorization period (Article 6, paragraph 7).

On the other hand, it is required a written authorization before commencing clinical trials involving medicinal products for xenogeneic cell therapy (Article 9, paragraph 6).

5. These include, for instance: the initiation of a clinical trial only if the Ethics Committee and/or the competent authority comes to the conclusion that the anticipated therapeutic and public health benefits justify the risks (Article 3.2.a), the rights of the subject to physical and mental integrity, to privacy and to protection of the data concerning him in accordance with Directive 95/46/EC are safeguard (Article 3.2.c), and the subject may without any resulting detriment withdraw from the clinical trial at any time by revoking his informed consent (Article 3.2.e).
6. Commission Directive 2003/63/EC of 25 June 2003 amending Directive 2001/83/EC of the European Parliament and of the Council on the Community code relating to medicinal products for human use, in which Part IV.4—"Specific Statement on Xeno-transplantation medicinal products"—states that "Specific emphasis shall be paid to the starting materials" and "In this respect, detailed information related to the following items shall be provided according to specific guidelines: Sourcing of the animals; Animal husbandry and care; Genetically modified animals (…); Measures to prevent and monitor infections in the source /donor animals; Testing for infectious agents; Facilities; Control of starting and raw materials; Traceability."
7. Regulation (EC) No 1394/2007 of the European Parliament and of the Council of 13 November 2007 on ATMPs and amending Directive 2001/83/EC and Regulation (EC) No 726/2004.
8. For legal definitions, see Chapter 1 Article 2 of Regulation (EC) No 1394/2007.
9. One of the main responsibilities of the CAT is to prepare a draft opinion on each ATMP application submitted to the EMA, before the EMA's Committee for Medicinal Products for Human Use (CHMP) adopts a final opinion on the granting, variation, suspension, or revocation of a marketing authorization for the medicine concerned.
10. Under the umbrella of this "mother guideline" (30), the Cell-Based Products Working Party (CPWP), together with the Committee for Human Medicinal Products for Human Use (CHMP), the CAT, and the Biologics Working Party (BWP), issued further guidance, including, for instance, a multidisciplinary guidance on xenogeneic cell-based medicinal products.
11. The guideline addresses issues, such as sourcing and testing of animals, manufacture, quality control, nonclinical and clinical

development, surveillance, and animal health and welfare in the sourcing materials. It also contains requirement for the development and marketing authorization of xenogeneic cell-based medicinal products for human use. Particularly, this guideline is intended for products entering the marketing authorization procedure. However, the principles laid down in the guideline should be considered by applicants entering into clinical trials. These products contain viable animal cells or tissues, which might be sourced from (non)transgenic animals or are genetically modified cells. Where genetically modified cells are used, the available guidance on gene transfer medicinal products should be also taken into account (CHMP/GTWP/234523/09, CHPM/GTWP/405681/06).

12. This guideline was a product of the DHHS Interagency Working Group on Xenotransplantation, which was formed in 1996 to develop a unified departmental approach to xenotransplantation and to provide sound policy recommendations to the US Secretary of Health and Human Services and is composed of representatives from the FDA, NIH, the Centres for Disease Control and Prevention (CDC), the Health Resource Services Administration (HRSA), and staff from the Office of the Assistant Secretary for Planning and Evaluation (OASPE).

13. These include, for instance: refs. 8, 26; See http://www.fda.gov/BiologicsBloodVaccines/GuidanceCompliance RegulatoryInformation/Guidances/Xenotransplantation/default.htm (Accessed 21 April 2011). Others are still in draft form.

14. In this document are described some specific procedures for infectious disease control. It is aimed at minimizing the risks to the public of human disease due to known and new diseases arising from xenotransplantation. It suggests safety measures for the procurement, screening, and use of xenotransplantation products as well as clinical care requirements for recipients. It recommends maintaining systematic health records and storage of designated biological specimens from both the source animal and the patient in the event of a public health investigation. In addition to this guideline, the aforementioned guidance documents offer further recommendations regarding product issues, good manufacturing practices, etc.

15. FDA clarified its requirements and recommendation for producers and developers of genetic engineering animals and their products—included those used as source materials for xenotransplantation (31). In addition to this latter, there are other guidelines and laws that may apply to genetic engineering animals, such as FDA (2003) in reference (8). Center for Veterinary Medicine (CVM) and Center for Biologics Evaluation and Research (CBER) collaborate to ensure the safety and consistency of xenotransplantation

product. Concerning European legislation, animal cells from genetically modified animals used as active substance should comply with EMEA 2001 in reference (32). EMEA 2007, in ref. (33), contains guidance that can be applied to xenogeneic cell therapy as well.

16. EMA states that the archiving strategy should be adequate for the intended use of these types of products. Records should be kept for 30 years (7). FDA recommends that all tissues and animal records should be archived for a minimum of 50 years (8).
17. The new Directive 2010/63/EU on the protection of animals used for scientific purposes permits Member States to implement measures affording more extensive protection according to our knowledge of factors influencing animal welfare as well as the capacity of animals to sense and express pain, suffering, distress, and lasting harm. See Recital (7) of Directive 2010/63/EU. See also ref. (34).
18. As stated by the EMA, the requirements set for the manufacturing process of human cell-based products are also applicable for medicinal products containing xenogeneic cells or tissues (30). The FDA has stated that the same general principles of current good manufacturing practices (CGMPs) that apply to human pharmaceuticals also apply to xenotransplantation products (35).
19. Generalized information on the kind of preclinical studies that should be done can be found in ref. (8). See also ref. (36).
20. Nonclinical programs should be performed, wherever possible, in relevant animal models that provide genetic, metabolic, and physical profiles that most closely resemble the human situation. Both FDA and EMA count on expert advice regarding the use of available relevant animal models to evaluate the safety and effectiveness of xenotransplantation products for the treatment of specific diseases.
21. These criteria for evaluating clinical issues in xenotransplantation are included, for instance, in refs. (7, 8, 26).
22. "The plan may have to be modified according to new scientific information on the infectious agents and their epidemiology. It has to be defined prior to marketing authorization which tests should be performed on a regular basis. It may be acceptable that certain tests will only be performed when clinically indicated (e.g., in the case of a suspected transmission of an infectious agent)": see ref. (7).
23. The ethical challenges involved in the process of fully informed consent in xenotransplant clinical trials account for why the SACX committee recommended that the process of informed consent should consist of a "team" of knowledgeable

communicators and should involve several carefully timed and paced conversations (27).

24. Principle 24 of the Declaration of Helsinki—adopted by the World Medical Association (WMA) as a statement of ethical principles for medical research involving human subjects—includes that "the potential subject must be informed of the right to refuse to participate in the study or to withdraw consent to participate at any time without reprisal." WMA Declaration of Helsinki: Ethical Principles for Medical Research Involving Human Subjects (amended policy adopted October 2008).
25. Concerning the application that has been made by LCT to conduct clinical trials of pig cell transplantation in New Zealand, the NHC is convinced by the evidence that an infectious agent is very unlikely to pose any risk to the public, therefore voluntary compliance should be seen as sufficient. However, should new evidence emerge that the procedure is not safe, the Data Safety Monitoring Board would be able to stop the trial. See ref. (18).
26. Indeed, undetectable organisms constitute the greatest concern of all, particularly if they can remain in a latent state within the source animal and recipient for indefinite time (7, 20).
27. For instance, in relation to xenogeneic cell-based medicinal products, there are various adventitious agents (viral, bacterial, parasitical infections, and infestations) that need to be considered. Also malignancies and other potential long-term adverse effects, associated medical devices, and biomaterials have to be taken into account (7).
28. See Article 14 of the Regulation (EC) No 1394/2007 and refs. (7, 37).
29. For instance, in the context of xenogeneic cell-based therapies, the infections may be caused by human pathogens, pathogens originating from the xenogeneic cells, or pathogen that could emerge through recombination. There may be a significant delay of clinical manifestations of infection and the symptoms of an infection may be atypical (e.g., organ dysfunction, or hyperacute, foudroyant forms). Thus, when the etiology of a recipient's posttreatment illness or reasons for a failure of the xenogeneic cell therapy remain unclear, appropriate testing should be conducted.
30. This consultation related only to the application made by LCT for this particular trial; it is not a more general consultation about xenotransplantation.
31. As is the opinion of some authors, there needs to be methods of identifying, dealing with, and monitoring xenotourists and xenotransplantation product recipients (27). Others suggest

that xenotransplantation clinical trials should not proceed until "the detection, treatment and follow-up measures as proposed by xenotransplant regulation authorities are 'geoethically' implemented in all nations" (29).

32. An assessment of whether there is an unacceptable risk requires a comprehensive assessment of costs and benefits to be undertaken. In general terms, the term "unacceptable risk" may describe the likelihood of an event (e.g., the possibility of introducing new infectious diseases into the general population), whose probability of occurrence is high and whose consequences are so considerable (e.g., high morbidity and mortality rates) that individuals or groups in society are not willing to take, despite the possible benefits (perceived or real).

33. Romeo Casabona CM and Urruela Mora A underline the importance of adopting measures in order to prevent the propagation and the risk of pandemics. They emphasize that these measures could adopt a coactive expression. It would be necessary to dispose of an adequate legal framework in order to conciliate the rapidity and effectiveness of the response with respect and the minimal restriction to the rights of the affected persons (38).

34. For instance, the WHO should have in place a system for the identification of and response to any xenotransplantation infectious disease outbreak in a timely manner, and it should maintain a register of xenotransplantation trials. Likewise, Member States are encouraged to take immediate steps to identify any xenotransplantation practices in their territory and ban those that are unregulated. They should ensure that public health officials are aware of the infection risks of xenotransplantation, including those associated with patients travelling to receive xenotransplantation products outside their territories and have plans in place to timely identify, combat, and control any such infection. On the other hand, investigators proposing clinical xenotransplantation trials should use source animals with the highest biosafety profiles, generate adequate preclinical data on safety and efficacy, carefully select trial participants, and ensure long-term patient follow-up and sample archiving.

35. "Inadequacies or weaknesses in national regulations not only may produce undesired local effects but may also have international implications, both in terms of ethics and safety. In turn, lack of international implementation of rules or loose interpretation of standards may adversely affect already disadvantaged groups and populations, and give rise to potential worldwide risks": see ref. (23).

36. As noted earlier, among the requirements that the applicant should include in his or her application are the following ones: data on the safety of the source animal; safety of the xenotransplantation product and the manufacturing process;

protocols for the clinical study design; risk assessment protocols and evaluations; patient monitoring; informed consent and additional patient information processes; and risk-management protocols, including, for instance, register of recipients, specimen archival, post-transplant monitoring of recipients, public health surveillance strategy and response, etc.

## Acknowledgment

This work has been part-financed by the European Commission's Sixth Framework Programme, under the priority thematic area "Life Sciences, Genomics and Biotechnology for Health", contract no. LSHB-CT-2006-037377, Xenome.

### References

1. Rood PP, Buhler LH et al (2006) Pig-to-nonhuman primate islet xenotransplantation: a review of current problems. Cell Transplant 15:89–104
2. Cozzi E, Tallacchini M et al (2009) The International Xenotransplantation Association consensus statement on conditions for undertaking clinical trials of porcine islet products in type 1 diabetes—Chapter 1: key ethical requirements and progress toward the definition of an international regulatory framework. Xenotransplantation 16:203–214
3. Arcidiacono JA, Evdokimov E et al (2010) Regulation of xenogeneic porcine pancreatic islets. Xenotransplantation 17:336
4. Garvenko O, Durbin K, Tan P, Elliot R (2011) Islets transplantation: New Zealand experience. Xenotransplantation 18:60
5. Yang Y, Sykes M (2007) Xenotransplantation: current status and a perspective on the future. Nat Rev 7:519–531
6. Guerra-González J (2010) Infection risk and limitation of fundamental rights by animal-to-human transplantations. EU, Spanish and German Law with Special Consideration of English Law, Medizinrecht in Forschung und Praxis, Bd. 21 Hamburg, 134:23–87
7. EMEA (2009) Guideline on xenogeneic cell-based medicinal products. EMEA/CHMP/CPWP/83508/2009
8. FDA (2003) Guidance for Industry: Source Animal, Product, Preclinical, and Clinical Issues Concerning the Use of Xenotransplantation Products in Humans, Final Guidance
9. Florencio PS, Ramanathan ED (2004) Legal enforcement of xenotransplantation public health safeguards. J Law Med Ethics 32:118
10. Vanderpool HY (2009) The International Xenotransplantation Association consensus statement on conditions for undertaking clinical trials of porcine islet products type 1 diabetes —Chapter 7: informed consent and xenotransplantation clinical trials. Xenotransplantation 16:255–262
11. Cook PS (2007) Informed consent and human rights: some regulatory challenges of xenotransplantation. Soc Altern 26:29–34
12. Rothblatt M (2004) Your life or mine: how geoethics can resolve the conflict between public and private interests in xenotransplantation. Ashgate, Aldershot
13. Spillman MA, Sade RM (2007) Clinical trials of xenotransplantation: Waiver of the right to withdraw from a clinical trial should be required. J Law Med Ethics 35:265–272
14. Selgelid MJ (2009) Pandethics. Pub Health 123:258
15. Urruela-Mora A, Romeo-Casabona CM (2002) Los dilemas éticos del xenotrasplante. In: Romeo-Casabona CM (Coord.) Los Xenotrasplantes. Aspectos Científicos, Éticos y Jurídicos. Editorial Comares, Granada
16. Canadian Public Health Association (2001) Animal-to-human transplantation: should Canada proceed? A public consultation on xenotransplantation
17. XWP (2003) Animal-to-human transplantation research: how should Australia proceed? Response to the 2002 public consultation on

xenotransplantation on draft guidelines and discussion paper on xenotransplantation. Public consultation 2003/4 NHMRC, Australian Government, Commonwealth of Australia, Canberra

18. NHC (2008) Advice on living cell technologies application for xenotransplantation clinical trials in New Zealand
19. http://ec.europa.eu/health/human-use/advanced-therapies/developments/index_en.htm. Accessed 19 Apr 2011
20. NHMRC (2009) Xenotransplantation: a review of the parameters, risks and benefits. Discussion paper
21. Tallacchini M (2007) Community and public participation in the risk assessment of experimental clinical trials. Xenotransplantation 14:356–358
22. World Health Assembly (WHA) (2005) Eighth plenary meeting of the fifty-seventh WHA in Geneva: human organ and tissue transplantation. Transplantation 79:635
23. Tallacchini M (2008) Defining an appropriate ethical, social and regulatory framework for clinical xenotransplantation. Curr Opin Organ Transplant 13:163
24. WHO (2001) WHO guidance on xenogeneic infection /disease surveillance and response: a strategy for international cooperation and coordination. WHO/CDS/CSP/EPH/ 2001-2
25. Council of Europe (2003) Report on the state of the art in the field of Xenotransplantation Council of Europe: Strasbourg
26. FDA (2001) Department of Health and Human Services (DHHS) Public Health Services (PHS) Guideline on infectious disease issues in xenotransplantation
27. SACX (2004) Draft report on the state of the science in xenotransplantation
28. Romeo-Casabona CM (1999) New challenges for organ transplantation. Eur J Health Law 6:207
29. Ravelingien A (2007) Xenotransplantation and the harm principle: factoring out foreseen risk. J Evol Technol 16:127–149
30. EMEA (2008) Guideline on human cell-based medicinal products EMEA/CHMP/ 410869/ 2006
31. FDA (2009) Guidance for industry regulation of genetically engineered animals concerning heritable rDNA constructs
32. EMEA (2001) Note for guidance on the quality, preclinical and clinical aspects of gene transfer medicinal products CPM/ BWP/3088/99
33. EMEA (2007) Guideline on scientific requirements for the environmental risk assessment of gene therapy medicinal products
34. Jorqui-Azofra M, Romeo-Casabona CM (2010) Some ethical aspects of xenotransplantation in light of the proposed European directive on the protection of animals used for scientific purposes. Transplant Proc 42:2122–2125
35. FDA (2008) Guidance for industry CGMP for phase 1 investigational drugs: 4
36. Sykes M, D'Apice A, Sandrin M (2003) Position paper of the Ethics Committee of the International Xenotransplantation Association. Xenotransplantation 10:201
37. EMEA (2008) Guideline on safety and efficacy follow-up—risk management of advanced therapy medicinal products (draft for public consultation). EMEA149995/2008
38. Romeo-Casabona CM, Urruela-Mora A (2008) New legal developments in xenotransplantation: the Spanish approach. Law Hum Genome Rev 29:123

ERRATUM

# In Vitro Repair Model of Focal Articular Cartilage Defects in Humans

Díaz Prado SM, Fuentes-Boquete IM, and Blanco FJ

Cristina Costa and Rafael Máñez (eds.), *Xenotransplantation: Methods and Protocols,* Methods in Molecular Biology, vol. 885, DOI 10.1007/978-1-61779-845-0_16, 

---

**DOI 10.1007/978-1-61779-845-0_20**

On PubMed the authors of book chapter, *In Vitro Repair Model of Focal Articular Cartilage Defects in Humans*, are incorrectly displayed as:
**Prado SD, Fuentes-Boquete I, Blanco F.**

The correct appearance should be:
**Díaz Prado SM, Fuentes-Boquete IM, Blanco FJ.**

---

The online version of the original chapter can be found at http://dx.doi.org/10.1007/978-1-61779-845-0_16

# Index

## A



## B



## C




Cristina Costa and Rafael Máñez (eds.), *Xenotransplantation: Methods and Protocols,* Methods in Molecular Biology, vol. 885, DOI 10.1007/978-1-61779-845-0, © Springer Science+Business Media, LLC 2012

## Q



## R



## S



## T

## V



## W



## X



## Z

MIX
Papier aus verantwortungsvollen Quellen
Paper from responsible sources
FSC® C105338

If you have any concerns about our products, you can contact us on
**ProductSafety@springernature.com**

In case Publisher is established outside the EU, the EU authorized representative is:
**Springer Nature Customer Service Center GmbH**
**Europaplatz 3, 69115 Heidelberg, Germany**

Printed by Libri Plureos GmbH
in Hamburg, Germany